HERBIVOROUS FISHES

Culture and Use for Weed Management

by

Karol Opuszynski

and

Jerome V. Shireman

Department of Fisheries and Aquatic Sciences
Institute of Food and Agricultural Sciences
University of Florida

In Cooperation With
USFWS National Fisheries Research Center
Gainesville, Florida
James E. Weaver, Director

CRC Press

Boca Raton Ann Arbor London Tokyo

Library of Congress Cataloging-in-Publication Data

Opuszynski, Karol.
Herbivorous fishes : culture and use for weed management / by Karol Opuszynski and Jerome V. Shireman ; in cooperation with USFWS National Fisheries Research Center.
p. cm.
Includes bibliographical references (p.) and index.
ISBN 0-8493-4988-5
1. Carp. 2. Aquatic weeds—Biological control. 3. Fish-culture.
4. Herbivores. I. Shireman, J. V. II. National Fishery Research Center (U.S.)
III. Title. IV. Title: Herbivorous fish.
SH167.C3068 1994
639.3′—dc20 94-13708
CIP

International Standard Book Number 0-8493-4988-5
Library of Congress Card Number 94-13708
Printed in the United States of America 1 2 3 4 5 6 7 8 9 0
Printed on acid-free paper

PREFACE

Herbivorous fishes are important not only as food fish, but also as a means for biological vegetation control. They are difficult to categorize because of their varied food habits. It is often difficult to determine the proportion of animal food in their diets which may be ingested either purposely or accidentally along with the vegetation they consume. Realizing that they will consume different amounts of animal food, we placed fish into the herbivorous category if they consumed greater than 50% plant material. Fish herbivory is discussed in detail including feeding methods, digestive processes, and energy relationships.

Although some marine herbivorous fishes are important economically, we focused more attention on the freshwater herbivores, primarily the Chinese carps, which have been widely introduced. Culture, use for food, aquatic weed management, and the effects they might have on the environment are discussed.

This book is not designed as a culture manual for herbivorous fishes, but the reader is directed to other sources. This book should meet the needs that exist for a comprehensive publication on herbivorous fishes.

Karol Opuszynski
Jerome V. Shireman

THE AUTHORS

Dr. Karol Opuszynski received his Ph.D. degree in 1968 at the Institute of Applied Biology Agricultural Academy in Olsztyn, Poland. He served as Head of the Department of Pond Fish Culture and Head of the Fish Rearing and Breeding Laboratory at the Institute of Freshwater Fisheries under the Ministry of Agriculture. His research interests include warm-water pond culture management; hatchery operations; rearing larvae and fry of cultivated fishes; the influence of temperature on fish, and the utilization of heated effluents for fishery purposes. In 1970, his doctoral dissertation was awarded first prize by the Polish Hydrobiological Association. In 1976, he was given a special award by the State Agricultural and Forestry Publishing House for the best article published on fisheries in that year. He has traveled widely throughout the world giving lectures and short courses concerning fisheries. He is author of over 80 papers printed in Polish and English. He has published a textbook titled "Fundamentals of Fish Biology".

Dr. Jerome V. Shireman received his Ph.D. degree in 1967 from Iowa State University in Fisheries Science. He has been at the University of Florida since 1973 and has served as Chairman of the Fisheries and Aquatic Sciences Department since 1984. His research interests include the use of Phytophagous fishes for weed control, fish vegetation relationships, aquaculture effluent management, and fish reproduction. He is author of over 85 scientific articles and was editor of the Grass Carp Symposium held in Gainesville, Florida, in 1978.

ACKNOWLEDGMENT

This book was written in cooperation with USFWS National Fisheries Research Center, Gainesville, Florida, James E. Weaver, Director.

DEDICATION

The authors dedicate this book in memory of Roger W. Rottmann, whose inspiration, dedication, and positive attitude toward our goals made this book possible. Not only did Roger play a major role in this accomplishment, but he was truly a treasured colleague and friend.

TABLE OF CONTENTS

Chapter 1

Introduction

Herbivorous fishes are among the most abundant fishes even though they comprise a minor proportion of the total number of fish species. They are interesting and important because of their specific anatomical and physiological adaptations for plant feeding, as well as their role in aquatic ecosystems as direct consumers of primary production, their importance as food fishes, and their potential for biological management of aquatic macrophytes and algae.

Fishes, according to their food habits, are classified as herbivores, omnivores, and carnivores. The value of this classification can be limited, however, because of the complexity of fish food habits. Fishes feed on a striking variety of foods, including bacteria, algae, macrophytes, detritus (dead organic matter, mostly of plant origin), and animals from protozoans to vertebrates. Even humans are not excluded as 12 species of sharks and smaller fishes, such as the voracious piranha (*Serrasalmo* sp.) and barracuda (*Sphyraena* sp.) may attack people. Fish food habits change depending on size, developmental stage, season, and geographic distribution. Moreover, most fishes are opportunistic, changing their diet to whatever food items are most abundant. Despite the vast amount of information available pertaining to herbivorous fishes, it is difficult to define clearly and concisely which species should be included in the herbivorous fishes group. Coincident with macrophyte and phytoplankton feeding they ingest zooplankton and animals that live in detritus, on mats of algae, and on leaves and plant stems. These animals can constitute a substantial share of their food.

Another difficulty in determining and categorizing herbivorous fishes arises because the food habits of many fish species are either unknown or have not been analyzed quantitatively. Therefore, the proportions of animal and plant food in their diets are difficult to evaluate. In order to circumvent this difficulty we propose a general definition (of herbivores) that places fishes into the herbivorous category if they consume >50% plant material. Additional indications of herbivory come from morphological and physiological adaptations, especially those of the alimentary tract that are described in this book.

The authors focus greater attention on the freshwater herbivores, because more data are available pertaining to their biology, usage, and commercial importance as food fishes. Herbivores, particularly in the family Cyprinidae (carp), Cichlidae (tilapia), Mugilidae (mullet), and Chanidae (milkfish) are important world aquaculture species.

Marine herbivores are not as important as aquaculture species, but some of the pelagic clupeids, which feed on planktonic algae, are important economically. Menhaden, for example, are collected for their oil and fish meal. Generally, the biology of marine herbivores has been less intensively studied, except for that of herbivores living in tropical coral reefs.

Among the freshwater herbivores, Chinese carp (Asiatic carp) are the most important. This group includes grass carp (*Ctenopharyngodon idella*), silver carp (*Hypophthalmichthys molitrix*), and bighead carp (*H. nobilis*). The grass and silver carp are not only native to China, but also inhabit the middle and lower Amur River basin in the Commonwealth of Independent States (formerly U.S.S.R.). The Indian major carp species, which include catla *(Catla catla),* rohu *(Labeo rohita),* and mrigala *(Cirrhinus mrigala),* are also herbivorous, but have not been as widely introduced as the Chinese carp.

After 1960 Chinese carp were introduced worldwide for aquaculture and aquatic vegetation management purposes. Chinese carp are reported to spawn in some locations outside their natural range, but in most cases they must be sustained through artificial propagation. Different techniques to artificially propagate Chinese carp are described, including the most contemporary methods of sterile fish production. In the U.S., in locations where the threat of unwanted natural reproduction exists, only sterile Chinese carp are allowed for management of aquatic weeds.

In addition to reproduction, the culture of larval Chinese carp, production of fingerlings for open waters, and culture of food fishes are also discussed. Special attention has been given to the use of grass carp for aquatic weed management and to the side effects of their stocking on indigenous fishes and aquatic ecosystems. Due to the growing interest in using Chinese filter feeding carp to control algae blooms to counteract eutrophication, this subject is also discussed.

While writing this book, the authors have drawn upon their considerable experience with Chinese carp in temperate and subtropical conditions in different parts of the world, the abundant literature scattered among scientific journals, proceedings of symposiums, extension pamphlets and manuals, and other sources.

This book is not designed as a manual for the culture and use of herbivorous fishes, although these subjects are covered and discussed. Detailed information is not included when the reader can be directed to other sources. This book is written to meet the needs that exist for a comprehensive publication on herbivorous fishes. It should be valuable for students, ecologists, botanists, zoologists, fishery biologists, fish culturists, weed managers, and sport fishermen.

Chapter 2

Systematics, Distribution, Diversity, and Abundance of Herbivorous Fishes

I. TAXONOMY AND MORPHOLOGY

The three species of the Chinese carp are the grass carp, the silver carp, and the bighead carp. They belong to the family Cyprinidae. The grass carp is included in the subfamily Leuciscinae and is the sole member of the genus *Ctenopharyngodon.*[1,2] The other two species are included by Howes[3] in the subfamily of Hypophthalmichthyinae. However, other authors such as Kryzanovskij[4] placed *Hypophthalmichthys* in the subfamily Leuciscinae with the genus *Ctenopharyngodon.* Taxonomic ambiguity also exists concerning the generic classification of these species. Oshima[5] considered morphological differences between the silver carp and bighead carp as being significant enough to include the latter species in the separate genus *Aristichthys.* Also, Gosline[6] argued that clear indications of a cultrin deviation of *Hypophthalmichthys* and *Aristichthys* existed. Howes,[3] however, disagreed with these authors and stated that the taxa *H. nobilis* and *H. molitrix* "possess unique synamorphies, and consequently belong to *Hypophthalmichthys*". *H. nobilis* is the name accepted by the American Fisheries Society's Committee on Names of Fish,[7] and, therefore, it is used in this book. Nevertheless, a synonymous name, *Aristichthys nobilis,* can also be found, especially in the older literature.

A. GRASS CARP, *C. idella*[8]

It is a large fish, reaching a weight of over 45 kg and a length of over 1 m. The dorsal area is dark gray, and the sides of the body are lighter with a slightly gold shine. The fins are either greenish-gray or buff in color. The cycloid scales are moderate to large with a dark brown base. The body is oblong with a rounded belly and broad head (Figure 1). The depth in standard length (SL) is 3.8 to 4.8. The head in SL is 3.4 to 4.5. The forehead is very wide. The eye is located in or above the axis of the body. The mouth is subterminal to terminal and somewhat oblique, the jaws have simple lips. The upper jaw is slightly protractile. The lateral line is complete and slightly decurved, extending along the middle of the depth of the tail.

The dorsal and anal fins are short and without spines. The dorsal fin is placed opposite to the pelvic fins. The origin of the anal fin is well behind the posterior edge of the dorsal. The numerical description of the fin's rays and scales is expressed by the following formula:[9]

$$D\,\mathrm{III}/7, A\,\mathrm{III}/8, L.L.\ 42-45, \quad 6-7/5$$

Berry and Low[10] gave a more complete meristic description, but the description was based only on the examination of 20 young specimens 9.1 to 18.5 cm SL:

$$D\,\mathrm{I}-\mathrm{II}/8, P\,\mathrm{I}/14-16, V\,\mathrm{I}/7-8, A\,\mathrm{II}/8-9, C\,\mathrm{V}-\mathrm{VI}/17/\mathrm{IV}-\mathrm{VI},$$

$$L.L.\ 40-42, \quad 6-7/5$$

Figure 1 Photograph of a grass carp *(Ctenopharyngodon idella).* (Photograph by K. Opuszynski.)

For the dorsal *(D)*, pectoral *(P)*, ventral *(V)*, and anal *(A)* fins the first set of Roman figures represents the number of unbranched rays. The Arabic numbers represent the number of soft, branched, articulated rays. In the caudal fin *(C)* the number of branched rays is represented by the second set of numbers, while the number of unbranched rays above and below are represented by the first and third sets of numbers, respectively. The number of scales along the lateral line is indicated by *L.L.*, while the numerator and denominator show scale counts above and below the lateral line, respectively.

The gill rakers number about 12; they are unfused, short, and widely set. The number and arrangement of the pharyngeal teeth (see also Section IV) is 2.5 to 4.2, 2.4 to 4.2, 2.4 to 5.2, or 1.4 to 5.2.[11]

Ojima et al.[12] determined the diploid chromosome number of grass carp to be $n = 48$, which consisted of eight pairs of nearly metacentric, ten pairs of submetacentric and subtelocentric, and six pairs of acrocentric and telocentric chromosomes. The number of chromosome arms (AN) was 84. Similar results were reported by Manna and Khuda-Bukhsh[13] and Márián and Krasznai,[14] although small differences in chromosome types were found (Table 1). Beck et al.[15] determined 15 pairs of metacentric, 9 pairs of submetacentric, and 0 pairs of acrocentric chromosomes. They suggested that study methods may account for these differences.

Biochemical analyses of the five tissues (serum, eye, liver, heart, and muscle) of cultured grass carp in the U.S. revealed an estimated 49 loci.[16] The three polymorphic loci identified included the enzymes phosphoglucose isomerase, esterase, and the serum protein haptoglobin. The small amount of genetic variation (the average heterozygosity per locus was equal to 0.021) was probably due to inbreeding of the fish examined. It may be expected that the proportion of polymorphic loci be considerably higher in wild populations. In another study of a culture population 17 enzyme systems were encoded and 30 loci were determined.[17]

The chemical composition of grass carp is given in Table 2. Body fat content shows the greatest variation. The distribution of the various weight compositions was determined as follows: head, 19.9%; body, 62%; fillet, 55%; intestine, 10.2%; bones, 7%; and fins and scales, 7.9%.[18] The flesh is bony, but in regions where carp are consumed as food, the grass carp is considered a fine food fish.

B. SILVER CARP, *H. molitrix*[8]

The largest fish caught in the Yangtze River in China was a silver carp, weighing about 20 kg at the age of 12 to 15 years. The silver carp is a deep-bodied and distinctively

Table 1 **Comparison of important karyological features of Chinese carp**

Carp species	2n	Metacentric	Submetacentric	Telocentic	AR[a]	Total length (μm)
Grass	48	10	8	6	84	59.15
Bighead	48	10	8	6	84	73.91
Silver	48	11	7	6	84	71.7

[a]AR = arm number.

From Márián, T. and Krasznai, Z., *Aquaculture*, 18, 325, 1979. With permission.

Table 2 **Proximate chemical analysis in percentage of eviscerated grass and silver carp**

Water	Protein	Lipid	Ash	Ref.
		Grass Carp		
77.2	—	1.4	—	668
74.5	19.9	4.6	1.0	669
76.8	17.9	4.2	1.2	18
73.0–75.1	16.1–18.7	5.2–6.7	1.4–1.6	670
		Silver Carp		
76.4	18.5	3.4	1.3	669
79.0	14.8	5.1	1.1	436
75.7	16.1–18.3	4.5–23.5	1.2–2.1	670
57–79	15–19	2–28	1	436

laterally compressed fish (Figure 2). The scales are cycloid and very small. The scales are silver in younger fishes with a grayish-lead coloration in older ones. A darker band extends along the dorsal part of the body. The mouth is upturned, with a horizontal line extending from the end of the mouth through the middle of the eye. A sharp keel begins just after the head and extends to the base of the anal fin. The following formula describes fin and scale meristic characteristics:[9]

$$D\ \mathrm{III}/7, A\ \mathrm{II}-\mathrm{III}/12-14, L.L.\ 110-124 \quad 28-33/16-28$$

The pharyngeal tooth formula is 4–4. Gill rakers are modified and arranged into a complex filtering apparatus.

Karyological characteristics of silver carp are given in Table 1. Electrophoretically detectable genetic markers of silver carp protein systems were described by Slechtova et al.[19] They analyzed 12 protein systems representing 21 presumptive loci and divided these systems into three groups. The first group consisted of species-specific proteins useful in distinguishing silver carp, bighead carp, and their F_1 hybrids. The second group included polymorphic protein systems, with some alleles common for both species and the others

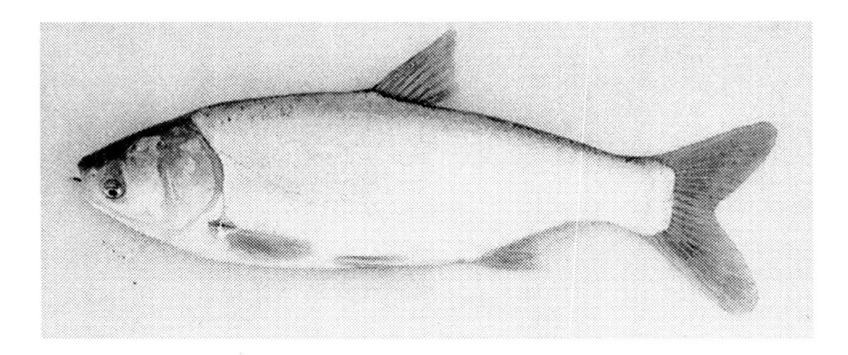

Figure 2 Photograph of a silver carp *(Hypophthalmichthys molitrix).* (Photograph by K. Opuszynski.)

being species-specific (Table 3). The third group of proteins included monomorphic loci common to both species.

The proximate chemical composition of silver carp from different culture populations is shown in Table 2. While fat content varies noticeably, the other parameter values are similar. Fat content tends to increase with size, but in fishes of equal size, those grown in ponds are usually fatter than wild fishes.[20] The weight composition of silver carp was determined as follows:[18] head, 24.2%; body, 59.7%; fillet, 53%; intestine, 9.55%; bones, 7%; and fins and scales, 6.3%. Although silver carp production is highest among the three Chinese carp in question, its palatability is the lowest. Nevertheless, Pasteur and Herzberg[20] claimed that this fish compares favorably in composition as well as in storage potential (it can be stored on ice for about 9 to 11 d) with fine food fishes such as cod, haddock, or halibut. The problem of boniness can be alleviated by increased marketing of larger sized fishes (4 to 5 kg and over).

C. BIGHEAD CARP, *H. nobilis*[21]

This fish attains a large size; however, individuals over 30 to 40 kg are not common. The body is deep and laterally compressed. As the name implies, the bighead carp has a disproportionately large head (Figure 3). The mouth is large and upturned, with the lower jaw longer than the upper. The eyes are located anteriorly on the head and are positioned ventrally. The scales are cycloid and small. The coloration of older fishes is dark gray above and off-white below, with dark gray to black and sometimes reddish-brown splotches that are irregularly shaped and positioned over the entire body. Besides the general body shape and color, position of the keel is distinctively different between the bighead and silver carp. In bighead the keel is shorter, extending from between the base of the ventral fins to the anal fin base.

Antalfi and Tölg[9] gave the following formula for the fin rays and scale arrangement:

$$D\,\text{III}/10, A\,\text{III}/15-17, L.L.\,114-120 28-32/16-28$$

The pharyngeal tooth formula is 4–4. Gill rakers are closely arranged in a comb-like structure that allows filtering of fine food particles. The size of the particles that can be retained is larger, however, than in silver carp.

The karyological characteristics of bighead carp are similar to those of the other two Chinese carp (Table 1). Biochemical markers characteristic of bighead carp are shown in

Table 3 **Species specific and other polymorphic protein systems in silver and bighead carp**

Protein	Alleles	
	Silver carp	Bighead carp
Superoxide dismutase (SOD)	*SOD*100*	*SOD*200*
Malate dehydrogenase (MDH)	*sMDH-2*100*	*sMDH-2*67,*86*
Nonspecific dehydrogenase (NDH)	*NDH-1*100*	*NDH-1*166*
Albumin (ALB)	*ALB*100*	*ALB*91*
Prealbumin (PA)	*PA*100*	*PA*108*
Muscle protein-myogene (MYO)	*MYO-II*100*	*MO-II*115*
	*MYO-III*100*	*MYO-III*118*
Creatine kinase (CK)	*CK-2*100*y	*CK-2*118*
Esterase (EST)	*EST-II*100*	*EST-II*94,*97*
Isocitrate dehydrogenase (IDHP)	*sIDHP-1*100,*122*[a]	*sIDHP-1*100*
	*sIDHP-2*100*	*sIDHP-2*72,*[a]**100*
Glucose phosphate isomerase (GPI)	*GPI-2*50,*[a]**100*	*GPI-2*100,*150*[a]
Lactate dehydrogenase (LDH)	*LDH-C*100,*500*[a]	*LDH-C*100*

Note: Transferrin (TF) not included in table.

[a] Auxiliary criterion in polymorphic protein system for respective species.

After Slechtova, V. et al., *J. Fish. Biol.*, 39(Suppl. A), 349, 1991.

Figure 3 Photograph of a bighead carp *(Hypophthalmichthys nobilis).* (Photograph by K. Opuszynski.)

Table 3. Slechtova et al.[19] conducted morphometric and biochemical studies of bighead and silver carp from Czech ponds. They found that silver carp could be unambiguously determined using only morphometric characteristics because all individuals determined in this manner as straight silver carp were confirmed by electrophoretic methods. In a group that was morphometrically determined as bigheads, the electrophoretic analysis was found to be consistent with morphometric determinations 27% of the time. The author concluded that most of the bighead carp were hybrids between bighead and silver carp and that those hybrids could not be distinguished using morphological analysis alone.

The bighead carp is important worldwide as an aquaculture species. Even in the U.S., where carp are not consumed to a great extent, a marketability study revealed that bighead may have commercial value in certain areas of the country. Production costs are low, and their palatability is comparable to channel catfish *(Ictalurus punctatus)* or bigmouth buffalo *(Ictiobus cyprinellus)*.[22] Most Americans do not eat them, however, because of the number of bones in the flesh.

II. DIVERSITY, DISTRIBUTION, AND ABUNDANCE

A. MARINE HERBIVORES

Marine herbivores are mostly dwellers of shallow water coastal habitats, where they feed largely on bottom plants restricted to inshore areas. Pelagic herbivores are planktivorous. They are rare because phytoplankton populations in the open ocean are sparse and do not provide enough food for filter feeding fishes. Therefore, most of the phytoplanktivorous fishes live in coastal regions, especially in areas of strong upwellings, where large colonial algae such as diatoms *Chaetoceros* and *Fragilaria* spp. are abundant. These diatoms often constitute the bulk of their food. The main phytoplanktivores in the sea are clupeids such as the Atlantic and Gulf menhaden, *Brevoortia tyrannus* and *B. patronus;* the South African sardine, *Sardinops ocellata;* the anchovy, *Engraulis capensis;* the Indian oil sardine, *Sardinella longiceps;* and the Peruvian anchoveta, *E. ringens*. These fishes form large schools and make up the major clupeid fisheries of the world.[23]

Herbivorous fishes constitute a small proportion of the total diversity of marine fishes. Excluding phytoplanktivores, only 19 families include herbivorous fishes[24] (Table 4). Of these families, 15 are in the order Perciformes. Because 409 families of teleostean fishes are recognized,[25] only 5% of the families include herbivorous species. It is difficult, however, to estimate the number of herbivorous species because the food habits of many species are not known or not known accurately enough to classify them even employing the very broad definition of herbivory used here.

It is widely recognized that many more herbivorous species exist in tropical waters as compared to temperate waters.[26-28] The occurrence of herbivory is inversely correlated with latitude in terms of both the absolute numbers ($r = -0.80$, $p < 0.01$) and proportions ($r = -0.92$, $p < 0.01$) of herbivores in the community.[24] The effects of low temperature on consumption rate and fish digestion processes may account for the rarity of herbivorous fish species in temperate, boreal, and arctic regions. With some exceptions, strictly herbivorous fishes appear to live between latitudes 40°N to about 40°S. The stichaeid *(Xiphister mucosus)*, is one of these exceptions.[24] It lives in central Californian waters and is known to occur as far north as latitude 55°N in southeastern Alaska. This stichaeid is a year-round herbivore in Californian waters, but it is not known if it maintains its herbivorous diet in Alaska. A few fishes living beyond the 40° latitudes consume algae. It is believed, however, that these fishes consume algae simultaneously with invertebrate prey, thus removing them from the herbivore classification.

Although the number of herbivores may be greater in tropical waters, there is no indication that they are more abundant in tropical seas than in temperate regions. Horn[24] compiled data on the density of herbivores from tropical and temperate regions and found great variation, ranging from 8 to 7640 fishes per 0.1 ha, with no apparent differences in abundance between the two regions. Horn concluded that temperate herbivorous fishes achieved densities as high as those of tropical species primarily during the summer months, but that the temperate species become more omnivorous in their food habits and move offshore during the winter.

Table 4 **Principal teleostean fish families containing marine herbivorous species; order, suborder, family distribution, and total species**

Order	Suborder	Family	Family distribution	Total no. sp.	Estimated no. herbivorous sp.[a]
Gonorynchiformes	Chanoidei	Chanidae	Tropical	1	1
Atheriniformes	Exocoetoidei	Hemiramphidae	Tropical	80[b]	?
Perciformes	Percoidei	Sparidae	Tropical temperate	100	?
		Girellidae	Temperate	20	20
		Kyphosidae	Tropical/ temperate	10	10
		Scorpididae	Tropical/ temperate	15	?
		Pomacanthidae	Tropical	74	?
		Pomacentridae	Tropical/ temperate	235	?
		Aplodactylidae	Temperate	5	5
	Mugiloidei	Mugilidae	Tropical/ temperate	70[b]	70
	Labroidei	Odacidae	Temperate	12	5
		Scaridae	Tropical	68	68
	Zoarcoidei	Stichaeidae	Temperate	60	≥2
	Blennioidie	Blenniidae	Tropical/ temperate	301[b]	?
	Gobioidei	Gobiidae	Tropical/ temperate	1500[b]	?
	Acanthuroidei	Acanthuridae	Tropical	76	?
		Siganidae	Tropical	25	25
Tetraodontiformes	Balistoidei	Balistidae	Tropical/ temperate	135	?
	Tetraodontoidei	Tetraodontidae	Tropical	118[b]	?

[a]Estimates are included where all species are expected to be herbivores with two exceptions. Total number of species in the Odacidae is from Gomon and Paxton[671] and the number of herbivorous odacids from Choat and Ayling[672] and M. J. Kingsford.[770] The minimum number of herbivorous stichaeids is based on Horn et al.[673]

[b]Totals include freshwater species.

B. FRESHWATER HERBIVORES

Freshwater fishes reported to ingest plant material are found in 24 families (in roughly 6% of the teleostean families), but only two families, the Cyprinidae and Cichlidae, are rich in plant-eating species (Table 5). Some of these species are of great commercial importance, but the ingestion of plant material does not necessarily mean that these fishes should be classified as herbivores. The contribution of plants to the diet of some European lake fishes is shown in Table 6. Of these fishes only the ide *(Leuciscus idus),* roach

(Rutilus rutilus), and rudd *(Scardinius erythrophthalmus)* can be classified as true herbivores.

The most important herbivorous cyprinids are the Chinese carp (grass, silver, and bighead) as well as some of the Indian carp belonging to the two genera *Labeo* and *Cirrhinus.* These fishes contribute greatly to the world catch of freshwater fishes and have been cultured for centuries in warm water ponds in polyculture with other fish species, increasing total yield due to utilization of primary production. The catch of silver carp was 1,359,724 t in 1989, which means this fish ranks first in the total world catch in inland waters. The herbivorous Chinese carp accounted for 26% (silver, 12%; grass, 8%; and bighead, 6%) of the total catch in inland waters in 1989.[29]

The major herbivorous cichlids belong to three genera, *Oreochromis, Sarotherodon,* and *Tilapia,* which are commonly referred to as tilapia. Tilapia are confined to tropical and subtropical waters because their lower lethal temperature threshold is about 10°C.[30] The natural distribution of the genus *Oreochromis* is in the rivers and lakes of East Africa and the Middle East, including the Jordan River Valley. The genus *Sarotherodon* occurs in West Africa, from Zaire to Senegal. The range of members of the *Tilapia* genus overlaps these two genera in the Zambezi River basin and southward. *Tilapia* and *Sarotherodon* are also each represented by a single species, redbelly tilapia *(T. zilli)* and the Galilee cichlid *(S. galilaeus),* in the Nile River drainage and in the Middle East.[31] Because of interest in tilapia for food and aquarium fishes, different tilapia have been introduced into North and South America, Asia, and within Africa itself.[30]

Freshwater herbivores are believed to be the most numerous in Africa, but belong to only six genera: *Labeo, Tilapia, Oreochromis, Sarotherodon, Citharinus,* and *Distichodus.*[32] The pattern of diversity and abundance of these fishes, within fish communities, is similar as it is elsewhere. Herbivores include only 160 species (<7%) of 2500 freshwater fishes in Africa. Despite the small number of species, herbivores commonly constitute a disproportionately large share of the standing fish crop. For example, three herbivorous species comprise 60% of the fish biomass in Lake Georgia and 74% in floodplain pools of the Sakota River, Nigeria; even in the River Niger, where the share of herbivorous fish is only 6% of the species captured, they comprise 19% of the fish biomass.

III. FEEDING BEHAVIOR AND FOOD

Herbivorous fishes can be classified regarding their feeding behavior as browsers, grazers, and filter feeders. Browsers tear or bite rooted plants and macroalgae, whereas grazers feed by scraping, whisking, or sucking sedimentary detritus (dead organic matter, mostly of plant origin) and algae that are affixed to the substratum.[24] Because grazers feed on or near the bottom, they also ingest inorganic materials that are rarely found in browsing species. Grazers are generally nonselective feeders, unlike browsers, which selectively feed on whole or large parts of individual plants.[33] Browsers and grazers, however, do not always fall into mutually exclusive categories within taxonomic units. For example, some herbivorous species from the family Acanthuridae (surgeonfishes) have been observed to feed either as browsers or grazers.

Two feeding behaviors can be distinguished among planktivores: particulate feeding and filter feeding. Particulate feeders visually select a single individual prey item from the water column, whereas filter feeders retain food items by passing volumes of water through their filtering apparatus.[34] Because particulate feeders select their food, they feed primarily on larger zooplankton, whereas filter feeders feed on smaller planktonic algae

Table 5 **Families and major species of freshwater fish that ingest plant material**

Family[a]	Species	Plant material	Ref.
Notopteridae	*Xenomystus nigri*	Algae	678
Mormyridae	*Brienomyrus brachyistius*	Algae	678
Clupeidae	*Dorosoma cepedianum*	Phytoplankton, detritus	34
	Dorosoma petenense	Phytoplankton, detritus	34
Engraulidae[b]	*Anchoviella brevirostris*	Phytoplankton	678
Phractolaemidae	*Phractolaemus ansorgii*	Detritus	678
Cyprinidae	*Abramis brama*	Macrophytes	41
	Alburnus alburnus	Macrophytes	41
	Barbus altianalis	Macrophytes	674
	Barbus carnaticus	Filamentous algae, macrophytes, detritus	454
	Blicca bjoerkna	Macrophytes	41
	Carrassius carrassius	Macrophytes	41
	Catla catla	Algae, macrophytes, detritus	454
	Cirrhinus cirrhosa	Algae, detritus	454
	Cirrhinus mrigala	Algae, detritus	454
	Cirrhinus reba	Phytoplankton, detritus	454
	Crossochilus oblongus	Filamentous algae	40
	Ctenopharyngodon idella	Macrophytes	399
	Cyprinus carpio	Macrophytes, algae	41, 399
	Danio aequipinnatus	Macrophytes, algae	675
	Gobio gobio	Macrophytes	41
	Hypophthalmichthys molitrix	Phytoplankton, detritus	399
	Hypophthalmichthys nobilis	Phytoplankton, detritus	229
	Labeo bata	Detritus, algae	454
	Labeo calbasu	Algae, detritus	454
	Labeo fimbriatus	Algae, macrophytes	454
	Labeo gonius	Macrophytes, algae, detritus	454
	Labeo kontius	Detritus, algae	454
	Labeo rohita	Algae, detritus	454
	Leuciscus cephalus	Macrophytes	41
	Leuciscus idus	Macrophytes	41
	Osteochilus thomassi	Algae, detritus	454
	Puntius amphibius	Macrophytes, algae	675
	Puntius dorsalis	Macrophytes, algae	675
	Rasbora daniconius	Macrophytes	675
	Rhodeus sericeus	Macrophytes	41
	Rutilus rutilus	Macrophytes	41
	Scardinius erythrophthalmus	Macrophytes	41
	Thynnichthys sandkhol	Phytoplankton, detritus	454
	Tinca tinca	Macrophytes	41
Catostomidae	*Pantosteus* sp.	Algae	673

Table 5 (continued) **Families and major species of freshwater fish that ingest plant material**

Family[a]	Species	Plant material	Ref.
Citharinidae	*Distichodus niloticus*	Plant material	677
Hemiodontidae	*Anodus laticeps*	Plant material	678
Curimatidae	*Leporinus maculatus*	Macrophytes	678
Lebiasinidae	*Poecilobrycon trifasciatus*	Algae	678
	Pyrrhulina brevis	Macrophytes, seeds	678
Characidae	*Alestes jacksoni*	Macrophytes	680
	Alestes sadleri	Macrophytes	680
	Colossoma bidens	Fruit	678
	Metynnis roosevelti	Macrophytes	679
Schilbeidae	*Physailia pellucida*	Detritus	678
	Eutropiellus buffei	Detritus	678
Clariidae	*Heterobranchus longifilis*	Detritus	678
Mochokidae	*Synodontis schall*	Detritus	678
Callichthyidae	*Corydoras* sp.	Bottom algae	678
Loricariidae	*Ancistrus spinosus*	Algae	684
Salmonidae	*Salmo gairdneri*	Filamentous algae	681
Cyprinodontidae	*Cyprinodon nevadensis*	Algae, detritus	682
Holocentridae	*Pristilepis fasciatus*	Macrophytes, fruit, filamentous algae	
Cichlidae	*Etroplus suratensus*	Macrophytes	685
	Haplochromis acidens	Macrophytes	680
	Haplochromis cinctus	Phytoplankton, detritus	680
	Haplochromis lividus	Periphyton	680
	Haplochromis nigricans	Periphyton	680
	Haplochromis nuchisquamulatus	Periphyton	680
	Haplochromis obliquidens	Periphyton	680
	Haplochromis paropius	Phytoplankton	680
	Haplochromis phytophagus	Macrophytes, periphyton	680
	Oreochromis alcalicus	Algae	30
	Oreochromis aureus	Phytoplankton	38
	Oreochromis esculentus	Phytoplankton	38
	Oreochromis jipe	Periphyton	38
	Oreochromis leucostictus	Phytoplankton, detritus	38
	Oreochromis macrochir	Phytoplankton	30
	Oreochromis mossambicus	Macrophytes, algae, detritus	38
	Oreochromis niloticus	Phytoplankton	38
	Oreochromis pangani	Periphyton	38
	Oreochromis shiranus	Macrophytes, algae	38
	Oreochromis variabilis	Algae	38
	Petrotilapia tridentiger	Algae	686
	Pseudotropheus zebra	Algae	686
	Sarotherodon galilaeus	Phytoplankton	38
	Sarotherodon melanotheron	Algae, detritus	38
	Tilapia guineensis	Algae, detritus	38

Table 5 (continued) **Families and major species of freshwater fish that ingest plant material**

Family[a]	Species	Plant material	Ref.
	Tilapia kottae	Phytoplankton, detritus	38
	Tilapia mariae	Phytoplankton	38
	Tilapia rendalli	Macrophytes, filamentous algae	30
	Tilapia sparrmanii	Macrophytes, filamentous algae	30
	Tilapia zillii	Macrophytes, filamentous algae	30
Mugilidae[b]	*Liza diadema*	Algae, detritus	683
	Liza dumerili	Filamentous blue algae	73
	Mugil cephalus	Detritus, algae	73,676
	Mugil curema	Filamentous blue algae	73
Osphronemidae	*Osphronemus gouramy*	Macrophytes, fruit, algae	678
Chanidae[b]	*Chanos chanos*	Phytoplankton	676

[a] Order of families follows Eschmeyer.[687]

[b] Primary brackish water or marine fishes.

Table 6 **Contribution of plants to food of lake fishes**

Species	Contribution
Bitterling *(Rhodeus sericeus amarus)*	+
Bleak *(Alburnus alburnus)*	+
Bream *(Abramis brama)*	+++
Carp *(Cyprinus carpio)*	+++
Chub *(Leuciscus cephalus)*	++
Crucian carp *(Carassius carassius)*	++
Gudgeon *(Gobio gobio)*	++
Ide *(Leuciscus idus)*	++++
Perch *(Perca fluviatilis)*	+
Pike *(Esox lucius)*[a]	+
Roach *(Rutilus rutilus)*	++++
Rudd *(Scardinius erythrophthalmus)*	++++
Stickleback *(Gasterosteus aculeatus)*	++
Tench *(Tinca tinca)*	++
White bream *(Blicca bjoerkna)*	++

Note: ++++, high; +++, small; ++, minimal; and +, sporadic occurrence.

[a]Juveniles

From Prejs, A., *Environ. Biol. Fish.*, 10, 281, 1984. With permission.

and detritus suspended in the water column. Filter feeders also ingest zooplankton if the zooplankton are unable to evade the fishes. Furthermore, many of the planktivores can switch from particulate feeding to filter feeding and vice versa, depending on their age, prey composition, and the size and density of the prey. Hence, in many cases it is difficult to distinguish between these two modes of feeding behavior.

Zooplankton are always found in some amount in the diet of planktonic herbivores. Similarly, animal food is also found in the diet of browsers and grazers, indicating that these fishes are not obligate plant eaters, selectively excluding animal food from their diets. For example, the grass carp, a typical browser, while feeding on macrophytes ingests all living organisms associated with plants, including rotifers, oligochaetes, chironomid larvae, and other aquatic insects.[35] In Lake Nyasa (Africa) a large number of cichlids graze on bottom algae and microfauna. This microfauna is very rich in animal life. A 510-cm^2 bottom mat of algae contains 3500 chironomid larvae, over 10,000 ostracods, over 1000 copepods, and a number of other organisms, including mites, mayfly, stonefly, and caddis fly larvae.[36]

IV. STRUCTURAL ADAPTATION TO HERBIVOROUS FEEDING

Feeding modes and food types are associated with morphological and anatomical features such as body form, mouth position and shape, marginal and pharyngeal teeth, gill rakers, and specialization of the alimentary tract.

A. BODY FORM

Among herbivores, either laterally flattened deep bodies or oblong forms with rounded bellies and broad heads prevail. Deep-bodied fishes are well adapted for maneuvering in dense vegetation beds (e.g., tilapia; Figure 4). Moderately deep-bodied forms are also common for open water plankton feeders (e.g., silver carp; Figure 2). This body shape, along with dark dorsal and silvery lateral coloration, makes them less visible to predators approaching from above and below. An oblong body shape is typical of many fishes feeding on aquatic plants both in the water column and near the bottom (e.g., grass carp; Figure 1), and for bottom feeders that eat detritus and algae cells (e.g., mullet; Figure 5).

The mouth position, shape, and size indicates much about feeding habits. Planktonic herbivores feeding in the water column frequently have dorso-terminal mouths (e.g., bighead carp; Figure 3). Most of the macrophyte browsers have a terminal mouth (e.g., grass carp; Figure 1). Grazers scraping algae from substrates have a terminal or ventro-terminal mouth, often situated at the end of a short, blunt snout, as is characteristic for some parrotfishes (e.g., *Scarus sordidus*). They use their powerful jaws to excavate algae from coral reefs. They can take large bites, which leave distinct scars on the coral.[33] In most cases, except for some planktivores, the mouth size of herbivores is small to moderate. Unlike piscivorous fishes, they do not need a large mouth gap to catch and handle large, fast-moving prey.

B. MARGINAL TEETH

Herbivores may have teeth on the marginal bones of the jaw, the tongue, the palatal bones, and the pharyngeal bones. Cyprinids do not have jaw teeth, but have well-developed pharyngeal teeth; in contrast, the Sparidae lack pharyngeal teeth, but have massive jaw teeth.[37] Cichlids have both pharyngeal and jaw teeth. Depending on feeding behavior, teeth may be arranged in sets that form scraping or cropping edges.[33] A highly specialized

Figure 4 Photograph of a blue tilapia *(Tilapia aurea).* (Photograph by C. Cichra.)

Figure 5 Photograph of a striped mullet *(Mugil cephalus).* (Photograph by F. Vose.)

scraping apparatus, consisting of fused jaw teeth, is characteristic of some parrotfishes (Scaridae). Other grazers, such as some Girellidae (e.g., a nibbler sea chub, *Girella nebulosa*), have a single row of massive, chisel-shaped teeth that are used to scrape rock surfaces. In contrast, the jaw teeth of tilapia occur in one to five rows; they are small, unicuspid, bicuspid, or tricuspid, and flattened distally to form blade-like structures that are used as scrapers.[38] The jaw teeth of the striped mullet, *Mugil cephalus,* are minute.

Compared to grazers, browsers often have fewer, smaller, and pointed teeth, which are better suited for biting. Plants caught between the teeth are partly cut and partly torn off by a head jerk or a rapid body spin. Some fishes combine browsing and grazing behavior which is reflected in the structure of their dentition. For example, the luderick *(Girella tricuspidata),* an Australian girellid, browses on large brown algae and grazes on smaller red algal tufts; it has spatulate teeth, straight edged or cuspate, and each is hinged with a ball-and-socket joint. This structure allows the fish to tear large algae for effective browsing and to efficiently graze algae on rock surfaces.[33]

The jaw teeth often undergo morphological changes with age and changes in the diet. For example, the importance of plant material increases in older sparids, and as plant material becomes more important, the canine teeth are replaced by incisors, which are better suited for biting plant material.

C. PHARYNGEAL TEETH

Many fishes possess pharyngeal teeth in the throat. These teeth are well developed in cichlids and consist of two sets of bones: the upper pharyngeal bone in the roof of the throat and the lower pharyngeal bone on the floor. These bones act as a trituration mechanism, breaking food passing through the pharynx into small pieces. The surfaces of these bones have teeth whose shape shows considerable specialization to the diet. They range from fine, thin, hooked structures (e.g., in *Sarotherodon esculentus,* a phytoplanktivore) to the coarse, robust teeth in the redbreast tilapia *(Tilapia rendalli),* a macrophyte consumer.[38]

Pharyngeal teeth are highly specialized in cyprinids, offsetting the lack of teeth on the jaw, vomer, palatines, and tongue. The lower pharyngeal teeth are mounted in sockets and arranged in one, two, or three rows on the fifth branchial arch. The arrangement and number of teeth are important taxonomically and are described by a tooth formula. This formula for the common carp *(Cyprinus carpio),* is 1.1.3.–3.1.1., which stands for three rows of teeth on the left arch with the two outer rows containing one tooth and the inner row containing three teeth (counting the rows from the left lateral side). The teeth on the right are a mirror image of the left arch.

The upper pharyngeal teeth of cyprinids are modified into a thick, horny pad with transverse and longitudinal grooves on the lower masticatory surface.[39] The grass carp has pharyngeal teeth specialized to tear and triturate vegetable material. These teeth are arranged in two rows; the inner row of teeth is more strongly developed than the lateral teeth (Figure 6). In smaller fishes of <30 cm, the inner teeth have a single serrated cutting surface, whereas in large fishes they are thicker and tend to have doubled and flattened serrated cutting and rasping surfaces. These changes, however, are not caused by tooth wear, for in large fishes new replacement teeth also have doubled flattened surfaces. These changes in tooth structure are associated with changes in the food habits of growing fishes; small fishes eat soft, submersed plants, while larger fishes also feed on tough stems of emergent plants and fibrous grasses.

Individual teeth are 2 to 3 mm apart in fishes of 30 to 40 cm length, increasing to 3 to 5 mm apart in fishes 60 to 80 cm. While fishes are feeding, the pharyngeal teeth on both arches are pressed against the horny pad and move toward and away from each other, with some interlocking for about one third of their length. This movement gives a transverse cutting action, and when combined with a fore-and-aft movement creates a rasping and grinding action.[40] The pharyngeal teeth of the planktivorous silver and bighead carp are weakly developed. They have only one row of four teeth on the left and right arches.

The role of the pharyngeal teeth is to prepare food for further digestion by breaking or cutting it into smaller particles. In tilapia the action of the pharyngeal apparatus breaks filamentous algae and large colonial phytoplankton into smaller units *(Sarotherodon esculentus),* breaks detrital aggregate into finer fragments (Mozambique tilapia, *S. mossambicus,* and blackchin tilapia, *S. melanotheron*), or shreds long filamentous periphyton to short units (Sparrman's tilapia, *T. sparrmanii*).[38]

In the rudd (Cyprinidae) the size of the particles bitten off by the pharyngeal teeth depends not only on the space between individual teeth, but also on the kind of plants

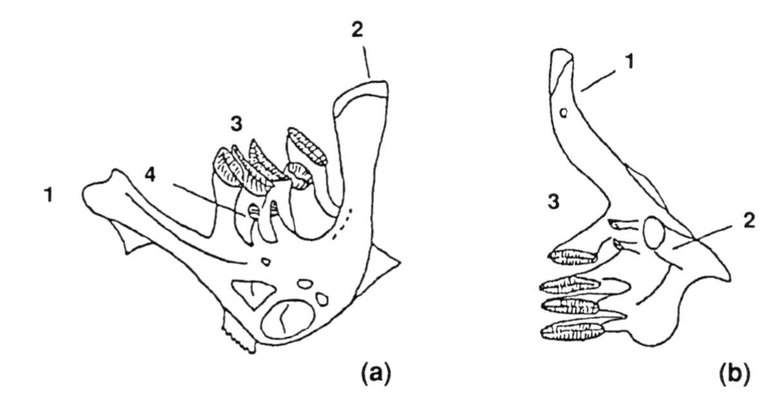

Figure 6 Lower pharyngeal bones of grass carp. (a) Dorso-lateral view of the left pharyngeal bone. (b) Dorsal view of the right pharyngeal bone. 1 = Anterior process, 2 = ascending process, 3 = inner teeth, 4 = lateral teeth. (From Hickling, C. F., *J. Zool.* (London), 148, 408, 1966. With permission.)

consumed.[41] Brittle and stiff plants such as *Potamogeton pectinatus* or *Ceratophyllum demersum* are cut into 2 to 3.5 mm fragments. The size of the particles are positively correlated to tooth spacing. Plants having large elastic leaves are shredded and ground rather than cut, but in such cases considerable reduction in size also takes place. The nature of the plant material also affects the trituration ability of the grass carp.[40] Aquatic plants, having little or no vascular tissue or fibers, are easily broken into small fragments. About 90% of the fragments are <3 mm^2 in area. In the case of land plants with reticulate venation, tearing is irregular and tends to follow the lines of the principal veins.

D. GILL RAKERS

Filter feeding fishes possess a filtering apparatus that allows them to strain microscopic planktonic algae from the water. In some fishes this function is performed by specialized structures called gill rakers and microbranchiospines. The gill rakers are cartilaginous or bony structures that protrude from the inner margins of the gill arches. The size, shape, and number of the gill rakers vary in different fish species, but as a rule in fishes that eat small-sized food the gill rakers are finer, longer, and more closely spaced. The microbranchiospines are smaller than the gill rakers and are located on the gill arches just above the gill filaments (Figure 7).

The gill rakers of gizzard shad *(Dorosoma cepedianum)* are simple comb-like structures with interraker spaces ranging from 1 to 85 μm, with an average space of about 40 to 60 μm.[42] The gill rakers of bighead carp are separated from one another, and their interspace increases from 20 to 60 μm as the fishes grow from 20 to 2000 g;[43] Spataru et al.[44] found higher interspace values ranging from 84 to 87 μm for 150 to 500 g bighead carp. Gill rakers can have more complex structure as, for example, in silver carp, where they are membranous and consist of numerous, elongated filamentous rays. These gill rakers are arranged in two rows on each gill arch and the reticulum conglutinate to the outer surface of each row, making a soft network with a mesh size of 20 to 25 μm.[39,43] Advanced structural specialization of the filtering apparatus occurs in some phytoplanktivorous marine clupeids. For example, *Ethmalosa fimbriata* has an impressive three-dimensional system of ramified branchiospines on the edges of the gill rakers.

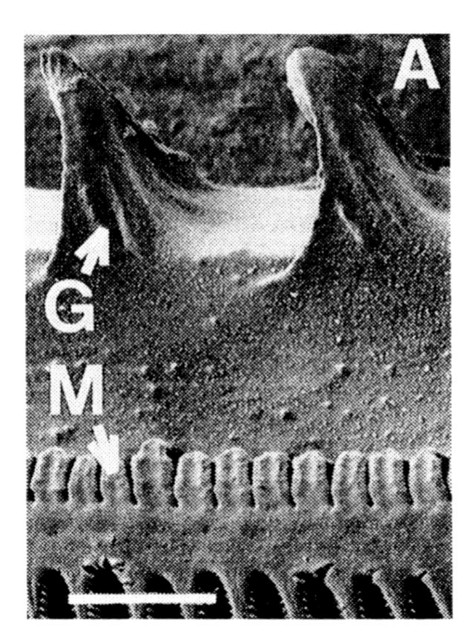

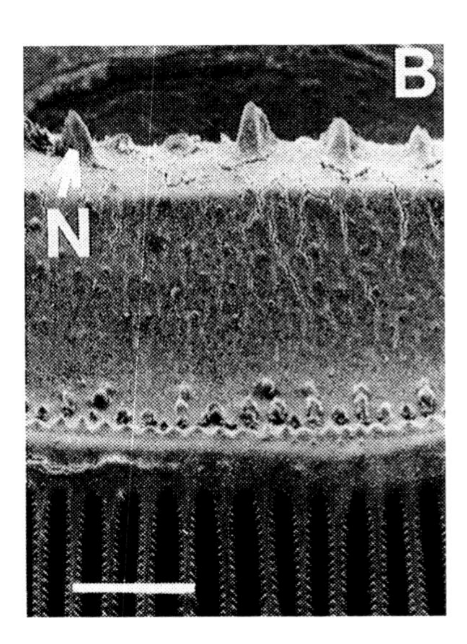

Figure 7 Scanning electron photomicrographs of the filtering apparatus of *Tilapia galilaea.* (A) An untreated fish, (B) a fish whose gill rakers and microbranchiospines had been removed surgically. G = gill rakers, M = microbranchiospines, N = gill raker nubs. Bars = 0.5 mm. (From Drenner, R. W. et al., *Trans. Am. Fish. Soc.,* 116, 272, 1987. With permission.)

Calculation of the filtering efficiency for a particular particle size is frequently made using direct measurements of the spacing between gill rakers. Yet, as stated by Lazzaro,[34] gill raker sieving is only one among several mechanisms involved in particle retention. For example, the presence of mucus, which makes the surface of the filtering apparatus sticky and consolidates food particles, can greatly increase filtering efficiency, making it possible to retain a greater proportion of smaller particles, even smaller than the smallest gill raker space.

Numerous studies were conducted on selective particle ingestion by herbivorous filter feeding fishes. The ingestion rate of gizzard shad increased as a function of particle size, leveling off at 60 μm.[42] Particle ingestion by the Galilee cichlid also increased as a function of particle size, leveling off when particle diameter exceeded 20 μm.[45] Bighead carp filtered food particles within a range of 17 to 3000 μm, with the majority of the phytoplankton 50 to 100 μm in size, whereas particles filtered by silver carp were considerably smaller and ranged from 8 to 100 μm, with the majority between 17 to 50 μm.[46] In another study it was found that silver carp fed selectively on particles larger than 20 μm.[47] Atlantic menhaden juveniles were able to filter phytoplankton cells as small as 1 to 2 μm, whereas the size of the smallest algae filtered by adults was 13 to 16 μm.[48]

Although many studies have shown that a relationship exists between spacing of the gill rakers and the size of particles ingested, the role of the gill rakers in the feeding process still remains unclear. For example, Drenner et al.[45] surgically removed the gill rakers and microbranchiospines of the Galilee cichlid. Fishes that were deprived of these structures (Figure 7) maintained the same particle ingestion rates and particle size selectivity as untreated fishes.

E. ALIMENTARY TRACT

A long alimentary tract is a common feature that distinguishes herbivores from other fishes with different feeding behaviors (Figure 8). Al-Hussaini[49] found that carnivorous, omnivorous, and herbivorous Red Sea fishes had relative gut length ratios (alimentary tract length divided by body length) ranging from 0.5 to 2.4, 1.3 to 4.2, and 3.7 to 6.0, respectively. A similar relationship between gut length and diet existed in African Great Lakes cichlids: carnivores had the shortest relative gut length ratios (about 0.3 to 2.0), followed by omnivores (0.8 to 3.1), and herbivores (1.6 to 8.0).[50] Herbivorous tilapias (redbreast, Mozambique, and blackchin) have relative gut length ratios between 7 and 10[38] and striped mullet between 3.2 and 5.5.[51]

Although gut length is a useful indicator of the feeding habits of fishes, precautions must be taken as there are also known herbivores with relatively short guts. Horn[24] stated that marine stomachless herbivores can have relatively short guts. He cited an Odacid *(Odax pullus)* or a parrotfish *(Scarus rubroviolaceus)* with relative gut lengths of 1.5 and 2.0, respectively. The stomachless grass carp also has a relatively short gut of 1.9 to 2.3 times the length of the fish.[40,52] Other stomachless cyprinids such as the silver carp or the Indian carp *(Labeo horie)* have extremely long guts; the former has a relative gut length of 15 (Figure 8)[53] and the latter of 15 to 21.[54]

As the previous discussion indicates, gut length alone cannot always be used to distinguish herbivores from omnivores and carnivores. The gut is elastic and is subject to considerable variation in length during measurement when different tensions are applied. Gut length also varies, depending on the amount of food and changes in the diet. The relative gut length of the surgeonfish *(Acanthurus nigrofuscus)* decreased by 30 to 50% during 2 d of starvation.[55] The mullet had a longer gut when fed a diet relatively rich in detritus as compared to a diet of algal material.[51] Considerable changes in relative gut length also occur as the fishes grow and change food habits. Larval herbivores feeding on zooplankton have short intestines, as in the larvae of silver carp, whose relative intestine length is <100% of the body length. Relative gut length increases with growth and accompanies the shift to herbivory. This increase, however, is not indefinite, but reaches a maximum length or even declines slightly when the fish attains a certain size. No tendency was found for an increase in relative gut length with increase in body length in 34 to 90 cm grass carp,[40] but the relative gut length increased with body size in *Labeo*.[56]

The alimentary tract of herbivorous fishes shows great structural diversity, which is due to the different digestive mechanisms existing among these fishes. Four types of alimentary tracts can be distinguished depending on the predominant mechanism involved in breaking down plant cell walls:[24,33,57]

1. Type I: Thin-Walled and Highly Acid Stomach

Fishes with this type of digestion have only jaw teeth and depend on acid lysis to break down plant cells in their stomachs, which have low gastric pH levels. These fishes have thin-walled stomachs with moderately long intestines, but otherwise show few morphological specializations for herbivory. Cichlids are representatives of this type (Figure 4). The tilapia digestive tract includes a short esophagus and a small sac-like stomach, the wall of which is composed of several tunicae: mucosa, submucosa, muscularis, and serosa. Gastric glands occupy most of the depth of the mucosa.[58] The stomach is separated from the intestine by a pyloric sphincter. The intestine receives the common bile duct immediately behind the sphincter. The anterior intestinal segment is short, thin walled, and has a diameter greater than the remaining part of the intestine that ends in an anal sphincter.[38] This type of alimentary tract also occurs in Acanthuridae, Chanidae,

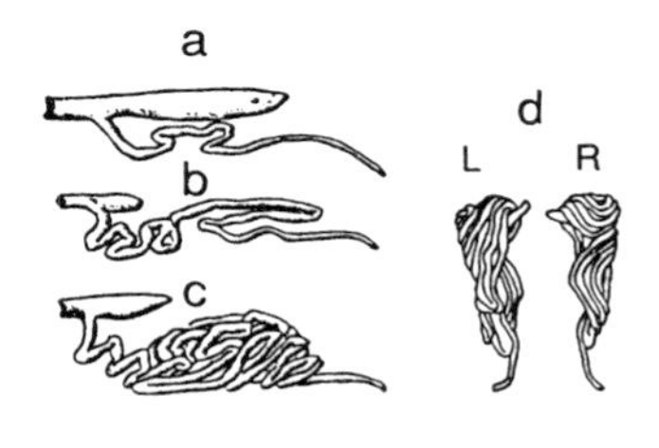

Figure 8 Alimentary tracts of fishes with different food habits. (a) A predator, *Cichla temensis* (Cichlidae); (b) a carnivore, *Geophagus brasiliensis* (Cichlidae); (c) a herbivore, *Tilapia heudelotii* (*S. melanotheron*) (Cichlidae); (d) a herbivore, *Hypophthalmichthys molitrix* (Cyprinidae), viewed from the left (L) and right (R). (a, b, and c from Pellegrin, 1921, data from Verigin, B. V., *Proc. Amur. Ichthyol. Expedition, 1945–1949,* 1, 303, 1950. With permission.)

Pomacanthidae, Pomacentridae, Sparidae, and Stichaeidae. These fishes are browsers rather than grazers, which minimizes sand ingestion during feeding and prevents buffering of the stomach contents by material rich in calcium carbonate.[57]

2. Type II: Gizzard-Like Stomach

The gizzard-like stomach found in fishes is a thick-walled, muscular stomach which serves mainly to grind food, causing cracks and ruptures in the ingested plant cells. Marine fishes with gizzard-like stomachs are found in the families Canthuridae, Plodactylidae, and Mugilidae, and in the freshwater Clupeidae. Fishes with gizzard-like stomachs are mostly grazers, and pick up sand with food and other inorganic material. Sand may comprise >50% of the stomach contents, as was found in striped mullet.[59] It is not clear whether gizzard-like stomachs lack acid-secreting cells and proteolytic enzymes because pepsin-producing glands have been found in a species of mullet.[60] Fishes belonging in this category tend to have shorter intestines than species with thin-walled stomachs, at least in the family Acanthuridae.[24]

3. Type III: Pharyngeal Apparatus

The pharyngeal apparatus is highly developed in herbivores in order to crumble and triturate plant material. The structure and function of this apparatus has already been described. Fish species with pharyngeal teeth do not have other structures for processing food into smaller particles. They have no stomach and their alimentary tract is morphologically poorly defined into a short esophagus, pyloric sphincter, intestinal swelling, intestine, and rectum.[10] The gallbladder is usually large and contains a great volume of bile. For example, the volume of bile in grass carp varied from 2.5 to 53 cm^3 in fishes of 0.5 to 9.5 kg weight, an average of 5.7 cm^3 of bile per kilogram of body weight.[40]

The relative gut lengths of fishes with a pharyngeal apparatus are generally shorter than either those with highly acidic, thin-walled stomachs or gizzard-like stomachs.[24,57] This occurs in marine fishes, but not in freshwater stomachless species. As mentioned previously, the relative gut length of the latter vary from relatively short to extremely long. Herbivorous fishes with a pharyngeal apparatus are either grazers or browsers. Marine species having this apparatus belong to the Hemiramphidae, Odacidae, and Scaridae, whereas the family Cyprinidae is rich in freshwater species (Table 5).

4. Type IV: Hindgut Fermentation Chamber

To date, the hindgut fermentation chamber has been found only in two herbivores, *Kyphosus cornelii* and *K. sydneyanus,* of the family Kyphosidae. These fishes, which are abundant in the temperate and subtropical coastal reef waters of Australia, possess unique digestive tracts and digestive capabilities.[61] Both species have a relatively long and coiled intestine, being 3.3 to 5.8 times the body length in *K. cornelii* and 3.4 to 5.3 times the body length in *K. sydneyanus.* The digestive tract consists of the esophagus, stomach, pyloric ceca (located at the junction of the stomach and intestine), and a thin-walled cecum-like chamber located at the hindgut and rectum. The hindgut cecum is separated by valves from the intestine and the rectum. No structure comparable to the hindgut cecal chamber has been described for any other fishes. This chamber, when distended, could contain approximately 1.5 to 2 times the stomach volume.

These sea chubs are browsers that do not possess a mechanism for trituration of ingested material. Pieces of algae are cleanly bitten off and swallowed intact, and further processing takes place in the hindgut cecal pouch where microbial fermentation occurs.

V. DIGESTIVE MECHANISMS

A. BREAKAGE OF PLANT CELLS: PROBLEMS WITH CELLULOSE DIGESTION

The diets of herbivores are far less digestible than diets of fishes eating animal food because the cell walls and other structural compounds of plants are made from high molecular weight polysaccharides such as cellulose and lignin, which are difficult to decompose enzymatically. It is well known that vertebrates do not produce cellulases (the enzymes hydrolyzing cellulose), and instead digest plant material through intestinal microbial fermentation. However, it still remains unclear, despite numerous papers dealing with this subject, whether fishes possess cellulase enzymes. It is possible that they synthesize cellulolytic enzymes. Although Migita and Hashimoto[62] found cellulase activity in goldfish *(Carassius auratus),* Fish[63] and Barrington[64] believed that fishes do not possess these enzymes. Stickney and Shumway[65] reported that cellulase was produced by the intestinal microflora, whereas other authors related cellulase activity in the gut to the amount of cellulase or cellulolytic bacteria or both ingested with plant detritus or invertebrates.[66-68] Shcherbina and Kazlauskene[69] found both endogenous cellulase and microbial cellulase in the gut of common carp. Cellulose digestion was observed in grass carp, but not in goldfish kept in the same tank.[70] The authors believed that cellulose digestion by grass carp was accomplished through exogenous enzymes taken up with the food which consisted of pellets with high cellulolytic content (18%) and plants. Bacteria having the enzyme allowing cellulolytic activity were not numerous enough in both species under study to ensure cellulose digestion. Das and Tripathi[71] reported a sharp decline in cellulase activity when grass carp were fed a diet containing tetracycline. They suggested that the microbial organisms producing cellulase were killed by the tetracycline treatment, while the remaining cellulase activity still observed after the treatment was derived from endogenous secretion.

In most herbivorous fishes, whatever the source of cellulase activity, the enzymatic processes are not strong enough to break down the plant cell walls to make the cell contents available for further digestion. Therefore, these fishes rely on other mechanisms to destroy the cell wall such as acid hydrolysis or mechanical trituration. Morphological adaptations of the alimentary tract to these mechanisms have been described previously. Breakage of plant cell walls is not only important because fishes must gain access to the nutrients inside the cells, but reduced food particle size also increases enzyme activity due to greater surface to volume ratios and reduced resistance to peristaltic mixing.

1. Acid Lysis

The mechanism for acid lysis of cell walls found in herbivorous fishes is unique among animals. The exact mechanism by which low pH affects the structure of the cell wall is not clear; presumably, increased acidity weakens the hydrogen bonds between cellulose units, weakening the cell wall, and allowing access to the cell contents. Nutrients released by acid lysis are later decomposed by the enzymes present in the alimentary tract.

In the Nile tilapia *(Sarotherodon niloticus)*, changes in stomach pH follow the diurnal feeding cycle.[72] The fishes have empty stomachs at night and begin feeding between 04.00 and 06.00 h and usually stop feeding shortly before sunset. When the stomach is empty hydrochloric acid (HCl) is not secreted, and pH is about 7.0. After the fishes commence feeding, HCl secretion starts, and gastric pH values gradually decrease. In the morning when feeding begins, the feces are brown. After about 4 to 6 h, the feces turn green in color, and later the color gradually reverses to brown. The color of the feces indicates the degree of digestion. Chlorophyll-*a* is present in the green feces, whereas phaeophytin (the product of chlorophyll-*a* decomposition) is present in the brown feces. Therefore, as indicated by feces color, changes in food digestion closely follow the changes in gastric acid secretion (Figure 9). The pH of the stomach fluid in tilapia that are digesting food is frequently as low as 1.25, and values of 1.0 have even been recorded. Such low pH values facilitate lysis and digestion of different food materials, including bacteria, green and blue-green algae, diatoms, and macrophytes.[38,72,73]

2. Mechanical Breakage

Contrary to fishes with thin-walled and highly acid stomachs which utilize chemical digestion, fishes with gizzard-like stomachs and stomachless fishes have developed digestive mechanisms that are primarily mechanical. Payne[73] measured pH values in the alimentary tract of four gray mullets. In *Liza falcipinnis* gastric pH ranged from 2.0 to 5.0, in *L. dumerili* from 7.0 to 8.5, and uniformly 8.5 in striped mullet and white mullet *(Mugil curema)*. The gastric pH was too high in these species to cause appreciable cell lysis; therefore, the mullet relies solely on the grinding action of the muscular stomach. Grinding is particularly effective when abrasives (sand grains) are continually added with the food materials. Wet attrition milling with the use of abrasives is a recognized method of lysing bacterial cells in microbiological studies.

In stomachless fishes the gut is alkaline or nearly alkaline throughout. In grass carp the pH ranges from 7.4 to 8.5 in the anterior part, in the middle section from 7.2 to 7.6, and in the posterior part 6.8.[40] In different species of marine stomachless fishes pH values ranging from 6.1 to 8.6 were found.[24] Hence, the pH in the alimentary tract of stomachless fishes is well above the values needed for chemical lysis of plant material; therefore, these fishes must rely on the efficiency of their pharyngeal apparatus. Prejs[41] argued that trituration by the pharyngeal apparatus was a mechanism that efficiently exposed the contents of higher plant cells and made them more available for digestion than acid lysis; Bitterlich,[74] however, doubted the efficiency of the pharyngeal apparatus in phytoplanktivorous cyprinids.

3. Microbial Fermentation

Although microbial fermentation is found in many terrestrial herbivorous animals, from termites to mammals, this type of digestion was not known in fishes. The two sea chubs discussed previously *(K. cornelii* and *K. sydneyanus)* have this type of digestion. Although they have strongly acid stomachs with pH ranging from 2.8 to 3.9, apparently these values are not low enough to lyse algal cells. In the stomach, algae which remain intact and retain their natural color are softened before passing to the intestine, where pH ranges from 6 to 8. In the anterior intestine these algae become brownish, but retain cellular structure. By the time the algae reach the cecal chamber they are changed to an

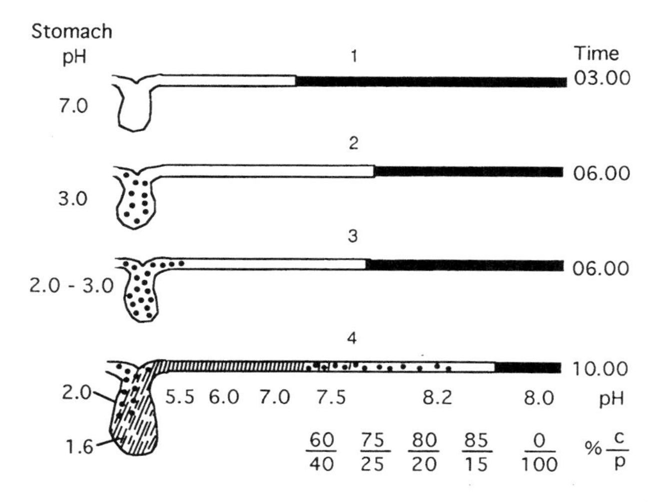

Figure 9 Diurnal changes in stomach fullness, pH values, and food digestion in tilapia *Sarotherodon niloticus*. 1 = No food, 2 = green material (food not digested, chlorophyll present), 3 = greenish-brown material, 4 = brown material (chlorophyll decomposed to phaeophytin). %c/%p = proportion of chlorophyll to phaeophytin. The intestine is drawn to a smaller scale than the stomach. (From Meriarty, D. J. W.; *J. Zool.* (London), 171, 25, 1973. With permission.)

amorphous, yellow-brown paste. The posterior section of the gut, especially the cecal chamber, contains an abundant, diverse microflora of bacteria and flagellated and ciliated protozoans which are undetectable in the foregut.

The occurrence of microbial fermentation in the posterior gut is confirmed by the presence of volatile fatty acids in samples from the cecal chamber and rectum. Volatile fatty acids are end products of anaerobic fermentation and are known to provide a major source of carbon and energy for herbivorous terrestrial animals. Positive evidence for microbial gut fermentation has been provided for only the two kyphosid fishes discussed previously; however, information on complex, symbiotic microfloras in other species is emerging, indicating that microbial fermentation may be more common among herbivorous fishes than was suspected.[33]

B. DIGESTION OF CELL CONTENTS

After the cell contents are exposed, further processing is performed by conventional enzymes. Distribution of these enzymes within the alimentary tract, as well as the strength of their action, differs not only in fishes with different feeding habits, but also among herbivorous fishes.

The protein breakdown enzyme, pepsin, has optimal proteolytic activity at a pH of 2 to 4; therefore, it acts in the low pH environment of the stomach, but is of little importance or has no function in stomachless fishes, the gut contents of which are alkaline. In these fishes proteins are broken down by trypsin- and chymotrypsin-type enzymes secreted by pancreatic tissue, which is either concentrated in a compact organ or scattered in the mesenteric membranes around the intestine and liver. Trypsin may also be secreted from cecal tissue located in the pyloric ceca.

In vitro comparison of protein-hydrolyzing enzyme activity in *Silurus glanis,* a stomach-possessing predator, and two stomachless fishes, common carp (an omnivore) and silver carp (a herbivore) showed that enzymes that are active in both acid and neutral media were present in all three species.[75] However, quantitative distribution of the pepsin- or trypsin- and chymotrypsin-type enzymes varied according to the species and section of the alimentary tract examined. In the neutral to slightly alkaline (7.5) pH range the intestine of silver carp, notably the foregut, yielded highest activity values; common carp ranked second, but showed great variability between the gut sections; and *S. glanis* ranked third. Proteolytic enzyme activity showed a reverse order in acid pH (2.0); it was highest in *S. glanis* and lowest in silver and common carp. Although the proteolytic enzymes of these three fishes were not homogeneous, they consisted mostly of seryl-proteinases.

In herbivorous marine fishes with acid-secreting stomachs, e.g., the luderick, the majority of the proteolytic enzyme activity occurs in the esophagus and stomach, although some protease activity is also found in the posterior portion of the gut.[76] High pepsin activity, however, is not always present in the stomach of herbivores. For example, pepsin was not found in the gastric fluids of Nile and Mozambique tilapia, although a protease with maximum activity around pH 2.1 was present in the gastric mucosa.[38,72] As stated previously, the stomachs of these fishes have very low pH values, which facilitate lysis of algae. These pH values (below 1.6), however, are well below the optimum pH for pepsin activity. Apparently, in these fishes an effective mechanism has evolved for lysis of algae with acid which precludes proteolytic gastric digestion. This is offset by high trypsin and chymotrypsin activities in the intestinal fluids of these tilapia.

Whereas protease is produced in greater amounts in carnivorous fishes, carbohydrases, which break down carbohydrates, are produced in greater quantities in herbivorous and omnivorous fishes. For example, one of the important carbohydrases, amylase, which breaks down starch, is 150 times stronger in its action in carp than in pike (*Esoc* sp.).[77] Generally in herbivorous carps amylolytic activity occurs in the entire intestine, although it is higher toward the proximal end.[40,78] Das and Tripathi[71] found an increase in amylase activity in grass carp from a culture pond compared to fishes fed with *Lemna minor* or leaf protein concentrate. They concluded that grass carp may be better adapted to an omnivorous (plant and invertebrate) diet. This conclusion is supported by Jancarik,[79] who observed improved starch digestion in common carp receiving sufficient supplies of natural food, such as crustaceans, worms, and insect larvae. It was not quite clear, however, whether this improvement resulted from the incorporation of invertebrate endogenous carbohydrases into the digestion process or from other mechanisms.

Strong amylase activity was also confirmed in both marine and freshwater herbivorous fishes having stomachs. Carbohydrates were rapidly digested in the first and second quarters of the Mozambique tilapia intestine, but no evidence of digestion was found in the third and fourth quarters.[38] In *Girella tricuspidata* amylase was present in the pyloric ceca, intestine, and rectum, but not present in the esophagus and stomach.

In spite of the low fat content in plant material, herbivorous fishes possess enzymes that can break down lipids. Lipase was found in the gut of grass carp,[40,71] silver carp,[78] Indian carp (mrigala and rohu), but not in catla.[80] Esterase, but not lipase, activity was detected in the intestinal fluid from Nile tilapia.[72] Lipase was also found in marine herbivorous fishes.[24,76] Generally lipase activity is highly variable in herbivorous fishes, and lower than in carnivorous and omnivorous fishes. The pancreas is the major source of lipase production, even though lipase activity has been demonstrated in various parts of the digestive system.

C. GUT EVACUATION RATE AND NUTRIENT ABSORPTION

As stated previously, herbivorous fishes generally have longer alimentary tracts than the carnivorous and omnivorous. Besides alimentary tract length, the gut evacuation time (the time needed for the alimentary tract to empty after ingestion of a meal) is an important parameter that influences digestion. Gut evacuation rate is not consistent and is dependent on temperature (higher temperature, short evacuation time), meal size (small meals are digested quickly), fish size (larger fishes take a longer time to digest the same relative amount of food), and food type. Fänge and Grove[81] compared the time for gastric or foregut emptying in temperatures ranging from 5 to 30°C for three groups of fishes with different food habits: microphagous (planktivores and insectivores), mesophagous (taking larger invertebrates such as mollusks or shrimps), and macrophagous (eating crabs, fishes, or other vertebrates). Recalculating their data for 20°C and assuming that the time for gastric evacuation is close to 50% for the entire alimentary tract to be cleared, Tanaka and Abe[82] calculated that gut evacuation rates for microphagous, mesophagous, and macrophagous fishes were approximately 10 h, >20 h, and 60 h, respectively. Gastric evacuation rates for herbivorous fishes are usually <10 h, when extreme temperatures are excluded (Table 7). Although great differences exist within and among herbivorous species, evacuation rates are generally higher than in other fishes. Higher evacuation rates indicate shorter digestion and absorption times for herbivorous fishes.

Although interesting, the absorption mechanism and whether the products of digestion are different in herbivorous fishes have not been fully investigated. Carbohydrate (glucose) absorption by the intestine is much higher in herbivores than in carnivores, which is not surprising because the carbohydrate content of plants is higher. In addition, herbivores, unlike carnivores, can regulate the activity of intestinal sugar transporters in relation to dietary carbohydrate levels.[83] Reshkin et al.[84] found no significant differences in basolateral myoinositol (an essential sugar-related cyclohexano vitamin) transport in the intestines of carnivorous and herbivorous fishes, while Vilella et al.[85] found even higher brush-border myoinositol transport in the intestine of a carnivorous fish in comparison to an herbivorous one. Myoinositol deficiency causes poor growth and causes mortality in fishes. These findings, which seemingly contradict the dogma of carbohydrate feeding specialization in herbivores, may be explained by the very different role of myoinositol as a vitamin in fish metabolism. It shows that the dietary role of a compound could be more important than the chemical structure in explaining differences in intestinal absorption patterns among fishes of different trophic levels.

Digested protein products are absorbed from the intestine as amino acids or peptides. Buddington et al.[83] compared the ratio of proline to glucose uptake in different trophic groups of fishes fed the same diet. This ratio decreased from carnivores to omnivores to herbivores mainly due to increased uptake of glucose as herbivory increased. The uptake of proline varied less among these fishes, as species with different natural diets still had similar protein requirements. Because all the species were fed the same diet, this study indicates that species differences in intestinal uptake are genetic adaptations, not phenotypic responses.

D. ASSIMILATION EFFICIENCY

Assimilation efficiency (AE), the amount of food assimilated expressed as a percentage of food ingested, is a measure of digestive and assimilative abilities of fishes. Assimilation efficiency values, however, show great variability depending not only on the fish species studied, but also upon fish age, size, recent feeding history, experimental conditions

Table 7 **Gut evacuation time, daily ration, and food assimilation in herbivorous fishes**

Species Food	Gut evacuation time (h)	Daily ration[a] (%)	Food assimilation[b] (%)
Acanthurus triostegus	2		T 14 (28–29)
Filamentous algae			
Ctenopharyngodon idella	3.5–6.0 (16–25)	102–145 (30–34)	T 20 (22); 31–83
Macrophytes	7.5 (27)		N 45 (27)
Gobius cobitis	22.8 (14)	100[c] (14)	≈C 64 (14) N 89 (14)6
Green alga *Ulva lactuca*	≈4 (20)	≈200[c] (20)	≈C 61 (20) ≈N 85 (20)
	≈4 (25)	300[c] (25)	C 61 (25) ≈N 85 (25)
	2.0 (30)	300[c] (30)	C 66 (30) N 86 (30)
Haplochromis nigripinnis	≈6 (25)	15 (25)	C 66 (25)
Blue-green algae			
Hypophthalmichthys molitrix	3.0–3.1 (20–21)		N 73–93
Phytoplankton	4 (23)		
	6.4 (22–25)		
	10 (23)	6–12 (20–23)	
Hypophthalmichthys nobilis	7.1–12.8 (21–28)	2–6 (21–28)[d]	T 43 (24–28)
Phytoplankton			
Hyporhamphus melanochir	4.4 (17)		O 38 (17)
Seagrass			
Lipoprys pholis	26.8 (11)	100[c] (11)	C 8 (11) N 29 (11)
Green alga *Ulva lactuca*	≈26 (16)	161[c] (16)	C 63 (16) N 73 (16)
	≈13 (21)	161[c] (21)	≈C 54 (21) ≈N 72 (21)
	1.0 (26)	≈146[c] (26)	≈C 8 (26) ≈N 63 (26)
Mugil cephalus	2–6 (20–26)		
Algae, detritus			
Parablennius sanguinolentus	29.5 (14)	100[c] (14)	C 72 (14) N 86 (14)

Green alga *Ulva lactuca*	≈4 (20)	≈197[c] (20)	C 51 (20) N 76 (20)
	≈4 (25)	≈315[c] (25)	≈C 55 (25) ≈N 79 (25)
	2 (30)	≈262[c](30)	≈C 58 (30) ≈N 79 (30)
Plectroglyphidodon lacrymatus	5–6		C 58–78 N 77–92
Algae			P 61
Rutilus rutilus	5.5–9.2 (20)	8–15 (20)	T 30 (20)
Macrophytes			
Sarotherodon niloticus	≈7 (25)	≈10 (25)	C 43
Blue-green algae			
Sarpa salpa	Several (19)	5.6–5.9 (19)[d]	T 61 (19)
Green alga *Ulva lactuca*			N 81 (19)
Scardinius erythrophthalmus	5.5–9.2 (20)	8–15 (20)	T 30 (20)
Macrophytes			
Siganus spinus	2–3		C 6–39
Green alga *Enteromorpha compressa*			
Stegastes lividus	9.7–9.9 (28)		T 72 (28) N 79 (28)
Green alga Enteromorpha clathrata			

Note: Temperatures (°C) given in parenthesis where known. ≈Data recalculated or read from graphs.

[a] In percentage of wet weight per fish wet weight, if not denoted otherwise.

[b] T = total dry weight, O = organic matter, C = carbon, N = nitrogen, P = phosphorus.

[c] Food consumption in percentage of that in the lowest specified temperature (LST); consumption in LST = 100%.

[d] Close to maintenance ration.

(including temperature, kind of food, feeding regimen), and the method used for AE measurements. Contrary to numerous AE measurements for carnivorous fishes, a relatively small number is available for herbivorous species.

The following range of AE values can be quoted for herbivorous fishes (Table 7): 14 to 72% (total dry weight assimilation), 6 to 78% (carbon assimilation), and 29 to 92% (nitrogen assimilation). The influence of temperature on AE is interesting. Whereas significant changes in AE values were found in some species (8 to 63% carbon assimilation in *Lemna pholis* at 11 to 26°C), no significant influence of temperature was reported in others (61 to 66% carbon assimilation in *Girella cobitis* at 14 to 30°C). The reason for this is not clear.

Assimilation efficiency in carnivorous fishes is higher than in herbivorous fishes. For example, crayfish, which has a high proportion of indigestible exoskeleton, was assimilated efficiently (73% food energy) by white bass (*Morone chrysops*).[86] Webb,[87] comparing data obtained by several authors, concluded that approximately 96 to 99% of dietary protein, 80 to 90% of fats, and up to 99% of simple sugars such as glucose and maltose were assimilated; however, complex carbohydrates such as starch were only 38% assimilated in salmonids.

E. PROTEIN REQUIREMENT AND FOOD RATION

White[88] states that the relative shortage of available nitrogen is the major factor influencing the abundance of all herbivores. This also seems to be true in fishes. Protein is the most important dietary component limiting growth of herbivorous fishes. Proteins provide 10 to 13 essential amino acids necessary for fish growth. Because these amino acids are not synthesized by fishes, they must be obtained from the diet. Protein deficiency is not a serious problem in carnivorous fishes, whose diet may contain over 80% protein by dry weight. In contrast, herbivorous fishes eat plants low in protein. In tilapia, the diet may contain $<1\%$ protein. Most values, however, are below 15%.[38] Protein content ranged from 1.6 to 7.0% (dry weight) in 13 species of macrophytes eaten by two marine temperate zone herbivorous fishes.[89] Even though the protein content of plant food is low, the mean protein requirement of herbivorous fishes is equal to that of carnivorous fishes, which amounts to about 49% of the dry food ingested.[90] Therefore, herbivorous fishes evolved morphological, physiological, and behavioral adaptations to cope with the protein shortage in their diet.

The most obvious response to a relative shortage of any nutritional ingredient in the diet is to increase consumption rate. Indeed, plant-eating fishes have higher consumption rates than those of nonherbivorous species. Consumption rates are temperature dependent. At higher water temperatures, daily rations (in wet weight) of grass carp may exceed their body weight. Even in a temperate climate, the daily ration of rudd is 8 to 15% of fish body weight at 20°C (Table 7). High consumption rates are possible because plant food is usually readily available and occurs in excess, which enables the fish to feed continuously.

Morphological and physiological adaptations to high consumption rates are long gut length and high food transition rates. The increased gut length allows ingestion of large quantities of plant material, and the extensive surface area of the intestine facilitates digestion and assimilation of this material. Horn[24] also suggests that another possible function of the long intestine is reabsorption of proteolytic enzymes, which conserves protein.

Selective feeding on plants high in nitrogen (protein) has been demonstrated among terrestrial herbivores. If and how the protein content of plants influences dietary choice in herbivorous fishes remain unclear. While plants with the highest protein content were preferred by fishes in some studies,[91] other studies showed that nitrogen content contributed little to the overall selection of plant species.[89] Horn and Neighbors[91] found that carbohydrate content and energy content rather than the protein content of plants were correlated with increased feeding activity. These examples show the complexity of food selectivity in herbivorous fishes that is apparently driven not only by plant nutritional values, but also by plant morphological and chemical defenses against herbivory.

When the protein requirements of herbivorous fishes are discussed, the contribution of animal food to their diet should not be overlooked; as discussed previously, animal food is ingested along with plant food, irrespective of their feeding behaviors. There is, however, a surprising scarcity of quantitative information on the contribution of animal material to the nutrition of herbivorous fishes. The importance of microorganisms in the diet is essentially unknown. Microorganisms are nitrogen rich and are easier to digest than plant cells.

Fischer and Lyakhnovich[93] found animal food was important when they fed grass carp different quantities of lettuce *(Lactuca sativa)* and animal (Tubificidae) food under laboratory conditions. When grass carp were fed only lettuce, they grew very little. A considerable improvement in growth occurred when the fishes were switched to a diet of tubificid worms. Interesting results were obtained with a mixed diet. When animal food was fed in excess and plant food restricted, there was a clear reduction (almost half) in the amount of animal food eaten. In the reverse situation, however (animal food limited and plant food in excess), the amount of plant food eaten remained unaffected by the amount of animal food available (Figure 10). They concluded that the grass carp needed both animal and plant food for good growth and survival. The animal food ensures good growth and the plant food supplies necessary vitamins and carbohydrates needed for metabolism.

Bitterlich[74] argued that stomachless phytoplanktivorous fishes, including silver carp, *Esomus danrica,* and *Amblypharyngodon meletinus*, were not primarily herbivorous, but rather omnivorous by nature. He found that they had low assimilation efficiencies for phytoplankton, but high digestion rates for zooplankton, which accounted for the apparent lack of zooplankton in their alimentary tracts. Not denying that the diet of the fish species in question may have included animal food, Bitterlich's conclusion may not be valid. There are data showing reasonably high assimilation rates of phytoplankton by stomachless fishes (see Table 7). The assimilation of phytoplankton may depend on the physiological state of the algae. Panov et al.[94] reported much better assimilation rates for dead algae than for living algae of the same species. The lack of zooplankton attributed to digestion by Bitterlich[74] may be due to technique; usually some zooplankton skeletal parts remain undigested and can be found in the intestine.

The most convincing results supporting herbivory or omnivory in fish species could be provided by feeding experiments using all-plant diets. Surprisingly, with the exception of larval fishes, there is an astonishing lack of such experiments and those that exist are contradictory. Herbivorous fishes are not herbivorous when young. Once yolk reserves are exhausted, they depend on high-protein animal diets to survive and grow. They gradually shift to mixed animal-plant food as they grow and develop (Figure 11).

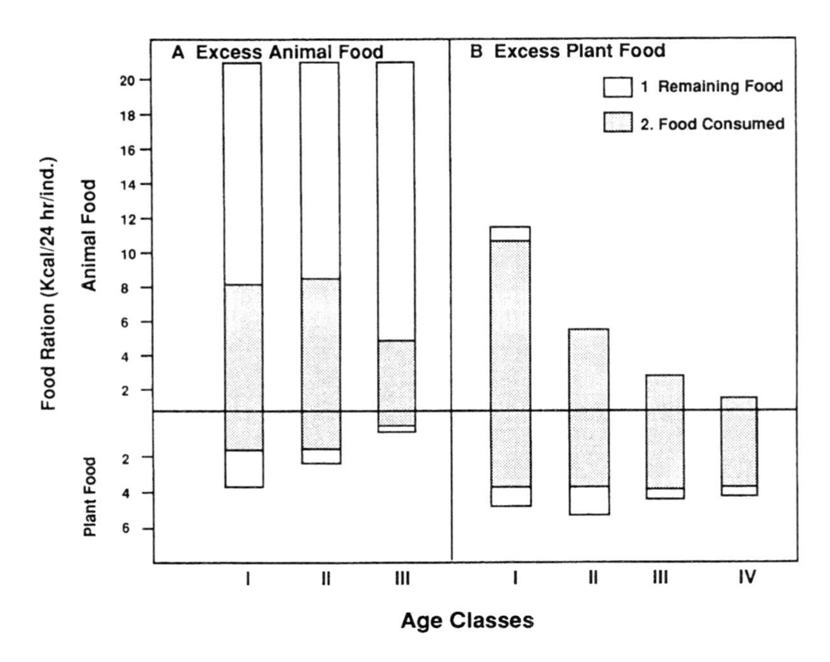

Figure 10 Food ration of grass carp, *Ctenopharyngodon idella*, depending on quantity and quality of food supplied. 1 = Food remaining, 2 = food consumed. A = Excess of animal food + diminishing ration of plant food, B = excess of plant food + diminishing ration of animal food. (From Fischer, Z. and Lyakhnovich, V. P., *Pol. Arch. Hydrobiol.*, 20, 521, 1973. With permission.)

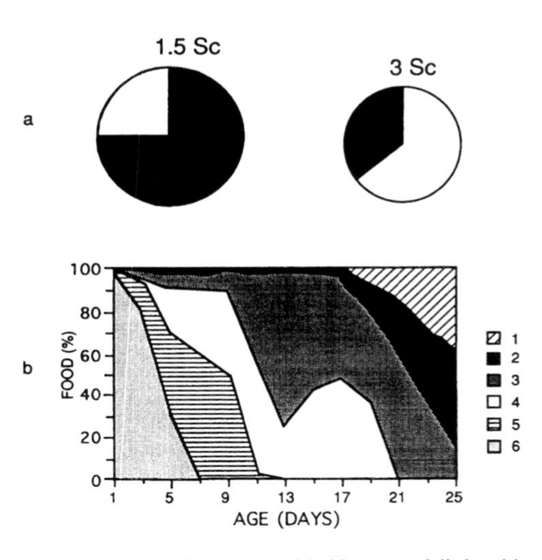

Figure 11a, b Changes in food habits of fishes from the same spawn under different food availabilities (a) growth of fishes, (b) different geographical populations. (a) Silver carp from ponds in Zabieniec, Poland. Stocking density 1500 (1.5 Sc) and 3000 (3 Sc) fishes/ha. Circle size depicts the relative amount of food in the alimentary tracts. Black field = proportion of zooplankton, blank field = proportion of phytoplankton. (From Opuszynski, K., *Aquaculture,* 25, 223, 1981. With permission.) (b) Grass carp fry from a pond in Goslawice, Poland. 1 = Algae and macrophytes, 2 = chironomid larvae, 3 = cladocera, 4 = copepod copepodites, 5 = copepod nauplii, 6 = rotifers. (From Opuszynski, K., unpublished.) (c) Changes in dominant food items with fish age in four different European roach populations. (From Prejs, A., *Environ. Biol. Fish.*, 10, 281, 1984. With permission.)

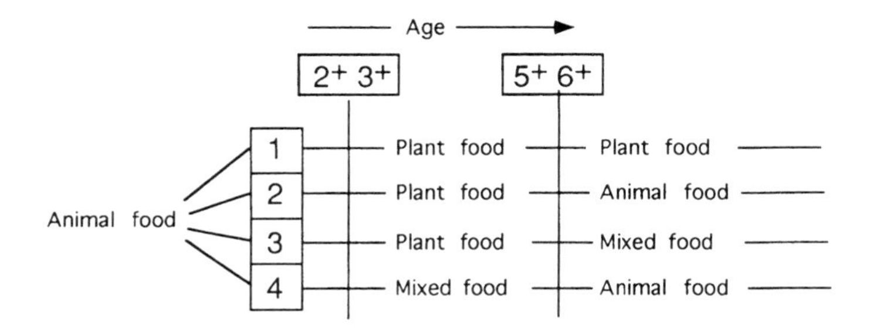

Figure 11c

Few carefully designed experiments have been done on older fishes. Angelfish *(Holacanthus bermudensis)* fed with green algae lost weight when fed with *Enteromorpha* at 19°C, but gained some weight when fed with *Monostroma* at 28°C.[95] The sparids (*Archosargus rhomboidalis*[96] and *Sarpa salpa*[37]) lost weight when fed unialgal diets. Several feeding experiments were done with grass carp because of their commercial value. Krupauer[97] observed poor growth in fishes fed only with macrophytes and suggested that the fishes could not survive on the plant diet alone. When grass carp were fed with lettuce, poor growth rates (50 mg/d in fishes weighing about 40 g) were reported by Fischer and Lyaknovich.[93] In contrast, when Opuszynski[98] fed grass carp with *Chara* sp. and *Fontinalis antipyretica*, he obtained growth rates ten times higher than did Fischer and Lyaknovich.[93] Shireman et al.[99] reported faster growth in fingerlings reared on duckweed *(Lemma minima)* than in those given commercial pellet diets. The above results are difficult to compare because the experiments differed in time of duration (several days to 8 months), fish size, water temperature, and kind of plants.

VI. ENERGY BUDGET: CARNIVORES VS. HERBIVORES

The partitioning of energy consumed with food (C) for metabolism (M), growth (G), and excrement (E) is called the energy budget and can be expressed by the formula:

$$C = M + G + E$$

Assuming that 100 U of energy (energy is measured in calories or joules) was consumed with food, the following equations can generally characterize the energy budgets of carnivorous and herbivorous fishes:[100]

$$\text{Carnivores: } 100C = 44M + 29G + 27E$$

$$\text{Herbivores: } 100C = 37M + 20G + 43E$$

Carnivores need a higher percentage of energy than herbivores for metabolism and growth, but substantial differences between these two categories of fishes exist in the percentage of energy assimilated (A) and excreted. The percentage of energy assimilated $(A = M + G)$ is 73 and 57 and the percentage of energy excreted is 27 and 43 for carnivores and herbivores, respectively. Higher consumption rates, lower food assimilation, and greater fecal production are characteristic features of herbivorous fishes. These factors influence culture methods, use for aquatic weed control, and environmental influences, and are discussed in Chapter 3.

VII. NATURAL DISTRIBUTION OF CHINESE CARP

The Chinese carp are native to eastern Asia (Figure 12). Grass carp and silver carp are found in rivers, lakes, and ponds of Siberia (the Amur River basin), Manchuria, north China, and south China. In the north they occur in the lower and middle reaches of the Amur, Ussuri, and Sungari Rivers, and in Lake Khanka. The southern distribution includes the Liao, Hai, Yellow (Hwang Ho), Hwai, Yangtze, Pearl, East (Tung Chiang), Chientang, and Min Rivers.[11] The natural range of the bighead carp is smaller. This fish is not found in the Amur River basin, but dwells in the lowland rivers of the north China plain and southern China, including the Huai (Huai Ho), Yangtze, Pearl, West (Si Kiang), Han Chiang, and Min Rivers.[22]

A monsoon climate characterizes the area of Chinese carp occurrence. The mean annual air temperature ranges from – 4°C in the Manchurian Plain region to 24°C in the south.[101] Air temperature extremes range from –30° to –16°C during the coolest month (January), and from 20 to 30°C during the warmest month (July). Average annual humidity varies from 70 to 80%. Average annual precipitation is 50 to 100 cm in the Manchurian region, 50 to 70 cm in the north China plain, and 150 to 200 cm in the subtropical and tropical zones of south China. More than 80% of the annual rainfall occurs during the summer monsoon period between May and October.

VIII. HISTORY OF INTRODUCTION AND PRESENT DISTRIBUTION

A. HISTORY OF INTRODUCTION

The first introductions of Chinese carp took place in the Middle Ages when the fishes were transported from Mainland China to Taiwan and the Indochina Peninsula.[9] Apparently unsuccessful introductions were undertaken in Japan in 1878[102] and in the European and central Asian regions of the former U.S.S.R. in 1937.[103] These trials were renewed in the former U.S.S.R. after World War II on a mass scale. Three stages of introductions can be distinguished.

During the first stage (1945 to 1960), extensive studies were conducted on the native population in the Amur River basin to collect more information on the life history and environmental requirements of the fishes. At the same time, introduction of the fishes was begun from the Amur River into various rivers and fish farms in the Ukraine. This introduction was continued every year and the distribution of the fishes was extended into various European and Asiatic republics of the former U.S.S.R. In the late 1950s, the number of introduced fishes reached several million. Initially, the fishes from the Amur River were transported in specially constructed railroad cars. Later, fishes grown at Chinese fish farms were used. These fishes were sent from Beijing to Moscow by air in polyethylene bags filled with water and oxygen.

During the second stage of introduction (1961 to 1963), artificial propagation was begun, and production of larval fishes was started. Simultaneously, pond culture began for stocking fish farms and open waters.

During the third stage, beginning in 1964, mass production of larval fishes and fingerlings was started. In 1964 alone, over 100 million larvae were hatched in large,

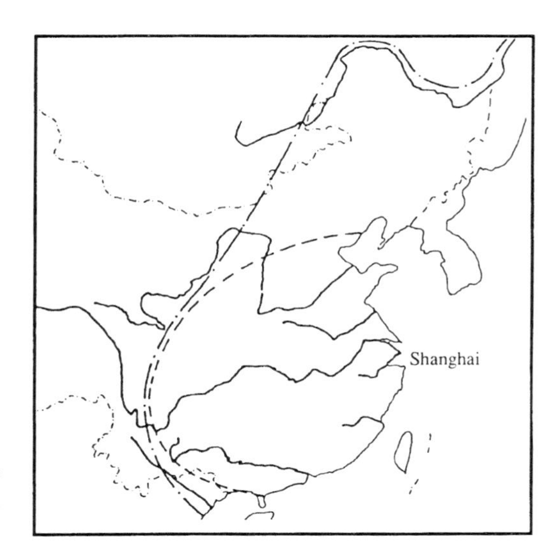

Figure 12 Natural distribution of silver and grass carp (–· –· –· –·) and bighead carp (– – – –).

specialized breeding centers, mainly in the Ukraine and Turkmenskaya. Additionally, mature fishes were distributed by air, often over a distance of a few thousand kilometers, to allow propagation of Chinese carp in areas where maturation was delayed. Consequently, a network of hatcheries was organized, which covered the country's needs for stocking material. At present, Chinese carp are grown in most of the fish farms in the Commonwealth of Independent States. The fishes now occur in great numbers of open waters, and natural spawning populations became established in some water systems.

The introduction of Chinese carp into the former U.S.S.R., followed by development of artificial propagation techniques and mass production of stocking material, allowed further spread of the fishes. In the 1960s they were introduced from the U.S.S.R. to Bulgaria, Cuba, Czechoslovakia, East Germany, Hungary, Iran, Poland, and Yugoslavia; from these countries, in turn, they were exported to Austria, England, West Germany, Italy, Nepal, and Sweden (Table 8).

China also became a distribution point for Chinese carp. From China, fishes were exported to Afghanistan, Hong Kong, Japan, Java, Malaysia, Mexico, Pakistan, Romania, Sri Lanka, Sumatra, and Thailand. Fishes from Japan were exported to Argentina, Ethiopia, India, Indonesia, Iraq, Israel, Laos, and Nepal.

Grass carp were first introduced in the U.S. in 1963.[104] Seventy fishes were imported from Malaysia to the Fish Farming Experiment Station at Stuttgart, AR. In the same year, a shipment of fishes from Taiwan was sent to Auburn University, Auburn, AL. The fishes were artificially spawned at both locations in the spring of 1966. A portion of these fishes were retained for research purposes and some were distributed to other state fish and game agencies and universities. Over the period 1970 to 1975, at least 115 lakes and numerous farm ponds were stocked. The first grass carp introduction in an open system occurred in 1971, when Lake Conway, Arkansas was stocked. Later, grass carp were widely spread throughout the U.S. as a result of research projects, stockings for aquatic weed control, interstate movement from private hatcheries, dispersal from stocking sites, and natural spawning. This fish continues to move northward, and was first captured in the Canadian waters of Lake Erie in 1985.[105]

Table 8 **Introductions of Chinese carp**

Country	Date	Source
Afghanistan	1966–1967	China
Argentina	1970	Japan
Austria	1960	Romania
Bangladesh	1976	?
Belgium	1975	Yugoslavia
Brazil	1979	China
Bulgaria	1964	U.S.S.R.
Burma	1969	India
Cambodia	?	?
Canada	?	?
Costa Rica	1976	Taiwan
Cuba	1966	U.S.S.R.
Cyprus	1976	Israel
Czechoslovakia	1961–1965	U.S.S.R.
Denmark	?	?
Egypt	1976	U.S.
England	1964	Hungary
Ethiopia	1975	Japan
Fiji	1968	Malaysia
France	1975–1976	Hungary
Germany	1964–1965	Hungary, U.S.S.R.
Greece	1980	Poland
Honduras	1976	Taiwan
Hong Kong	?	China
Hungary	1963–1966	China, U.S.S.R.
India	1959	Hong Kong, Japan
Indonesia	1964	Japan
Iran	1966	U.S.S.R.
Iraq	1968	Japan
Israel	1952–1965	Japan
Italy	1972	Yugoslavia
Japan[a]	1878, 1943–1945	China
Java	1949	China
Kenya	1970	?
Korea	1967	Taiwan
Laos	1968	Japan
Malaysia	1800s	China
Mexico[a]	1960	Taiwan, China
Nepal	1966–1972	India, Japan, Hungary
The Netherlands	1968	Taiwan
New Guinea	1965	Hong Kong
New Zealand	1966	Malaysia
Nigeria	1972	?
Pakistan	1964	China

Table 8 (continued) **Introductions of Chinese carp**

Country	Date	Source
Panama	1977–1978	U.S.
Peru	1979	Israel
ThePhilippines[a]	1966–1969	?
Poland	1964–1966	U.S.S.R.
Romania	1959	China
Rwanda	1979	Korea
Sarawak	?	Hong Kong, Taiwan
Singapore	?	?
South Africa	1967	Malaysia
Sri Lanka	1949	China
Sudan	1973	?
Sumatra	1915	China
Sweden	1970	Poland
Taiwan	?	China
Thailand	?	China
Turkey	1972	Romania
United Arab Emirates	1968	Hong Kong
Uruguay	?	?
U.S.[a]	1963	Malaysia, Taiwan
U.S.S.R. (European and Central Asian)[a]	1937, 1950s	U.S.S.R., China
Vietnam	1969	Taiwan
Yugoslavia[a]	?	?

[a] Natural spawning occurred and self-sustaining populations have been established.

Data from References 11, 22, 688, and 689.

Silver and bighead carp were first introduced into the U.S. from Taiwan in 1971.[46] They were imported to Arkansas by a private farmer in an attempt to improve water quality in catfish production ponds and sewage lagoons. In 1974 Auburn University started a research project to evaluate the potential of the fishes in polyculture systems with existing U.S. culture species. Silver and bighead carp have become widely distributed and have been caught from open waters since 1981.[22] Although sterile Chinese carp are required in some areas in the U.S., wild populations have become established.

1. Areas of Natural Spawning Outside the Natural Range

Despite worldwide introduction, Chinese carp naturally reproduce and have established populations in only a few locations. Most of the spawning sites are located in the former U.S.S.R., where they intentionally tried to encourage the establishment of wild populations. Because different climatic zones and numerous river systems with various hydrologic and ecological conditions existed in the former U.S.S.R. greater potential for natural reproduction existed.

At least six spawning sites were found in the former U.S.S.R. (Figure 13). The most eastern location described is in the Ili River.[106,107] The Ili River arises in the Tien Shan

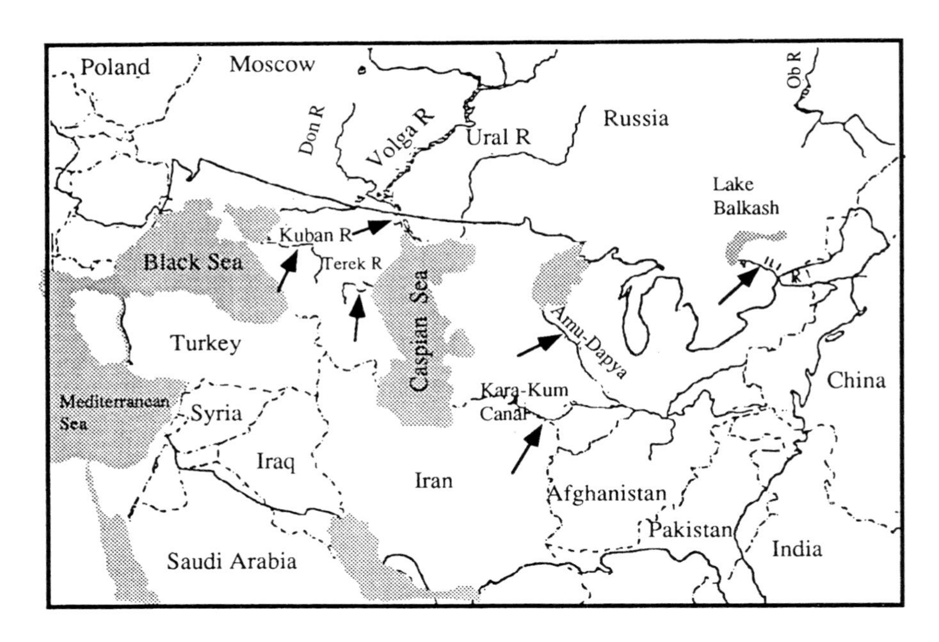

Figure 13 Reported spawning sites within the former U.S.S.R.

Mountains of Sinkiang, China, and flows westward 950 km to Lake Balkhash. Grass carp were introduced to the Ili River basin for the first time in 1958, when they were raised in local ponds. The stocking of grass carp into Lake Balkhash began in 1963. By 1972, a self-reproducing population was established. Chinese carp were also stocked into the Amu Darya River in 1955 and into the Kara Kum Canal in 1958,[108,109] and reproduction was first observed in both sites in 1965. Natural reproduction has occurred in both of the systems.

The most northern grass carp spawning site in the former U.S.S.R. was discovered in the Lower Volga River.[110] Stocking was initiated in 1964 and during the next 8 years about 50 million juvenile fishes were released. In addition, over 1 million yearlings and 2- and 3-year-old fishes were also stocked. Reproduction has occurred since 1970. The two remaining spawning grounds are in the Terek and Kuban Rivers.[111,112] The Terek River flows into the Caspian Sea, whereas the Kuban River empties into the Sea of Azov. Both rivers are typical mountain streams in their upper course, with many gorges and swift water. In the lower course the climate is mild because of the influence of the sea. Chinese carp were stocked in these rivers in the mid-1950s, and they are now established.

Only a few river systems in the southern part of central and western Europe meet the spawning requirements for Chinese carp. One of these, where reproduction may be successful, is the Danube River. Actually, grass carp are reported to reproduce in the Tisa River, a major tributary of the Danube River. According to Djisalov[113] (quoted after Stanley et al.[114]), several thousand juveniles were caught in the floodplain waters of this river.

In Japan, spawning of grass, silver, and bighead carp was confirmed in 1955 in the Tone River.[115,116] Chinese carp have reproduced in the river every year since 1955. The spawning area, however, has been altered by dam construction, extensive herbicide runoff from rice fields, and increased eutrophication. Continuation of these activities make continued spawning in the Tone River uncertain.

In the Philippines, Chinese carp were introduced in 1962 and in 1967, when the United Nations sponsored fish production in ponds located in the Pampanga River system. Grass and silver carp have been caught in the river since these introductions, but not in great numbers. Although spawning has not been observed and verified, the local fisheries staff believe that natural spawning takes place not only in the Pampanga River but also in the Agno River, the other major river of central Luzon.

In Taiwan, spawning of Chinese carp is uncertain. It was reported that grass and silver carp reproduced in the Ah Kung Tien Reservoir.[117,118] It is unknown, however, whether spawning occurred in the reservoir or in a stream from which the eggs drifted into the reservoir and hatched. Apparently spawning was successful in only one year, possibly prompted by unusual climatic events, a drought followed by heavy floods.

The first reproduction of Chinese carp in North America was reported in Mexico.[119,120] Grass carp fry were captured in the Rio Tepalcatepec of the Rio Balsas system in the State of Guerrero in 1974. Reproduction occurred in tributaries during the summer floods and the eggs were washed into the Rio Tepalcatepec where they hatched. The second spawning site was identified in the lake system of Lago Bodegas in the State of Hidalgo. In the U.S., Chinese carp have expanded their range since the first reproduction of grass carp was positively documented in the lower Mississippi River.[121] Besides the lower Mississippi, larval grass carp have been collected in increasing numbers from the lower Red and Atchafalaya Rivers.[122]

Recent reports of larval grass carp in the Missouri and Ohio River drainages and in the Trinity River (Texas) clearly show that this fish continues to expand its range. According to Courtenay et al.,[123] grass carp populations have been established in Arkansas, Louisiana, Minnesota, Mississippi, Missouri, Tennessee, and Texas. In central Missouri, grass carp larvae were captured in four tributaries and at three Missouri River stations between 23 May and 15 July 1987.[124] The bighead carp also must be added to the list of fish species naturalized in the U.S., as larvae and juveniles were collected in the Missouri River drainage.[125] In addition, the silver carp has become so widespread in the central U.S. that the confirmation of a feral breeding population seems imminent. It is difficult to predict further expansion of Chinese carp populations in the U.S. It seems, however, that a population explosion, which would have a destructive environmental effect, is unlikely. Observations to date show that the survival of juvenile fishes is low due to the presence of large numbers of predators in most waters.

2. Areas Where Fish are Introduced but Not Established

No other fishes are known to have spread so rapidly and over such a wide range as Chinese carp. During approximately 30 years, these fishes have spread outside their natural range to other locations in Asia and have been introduced to Africa, North and South America, New Zealand, and Europe. However, in spite of the occurrence of these fishes in over 60 countries (Table 8), self-sustaining populations have been established only in limited locations.

A large portion of the world can be excluded as potential spawning sites because conditions that are needed for natural reproduction are lacking. For example, due to thermal conditions in Europe (Figure 14), natural reproduction is possible only in the southern part, where the average annual temperature exceeds 10°C. Only culture is possible in central Europe, and in areas where temperatures are below 5°C neither

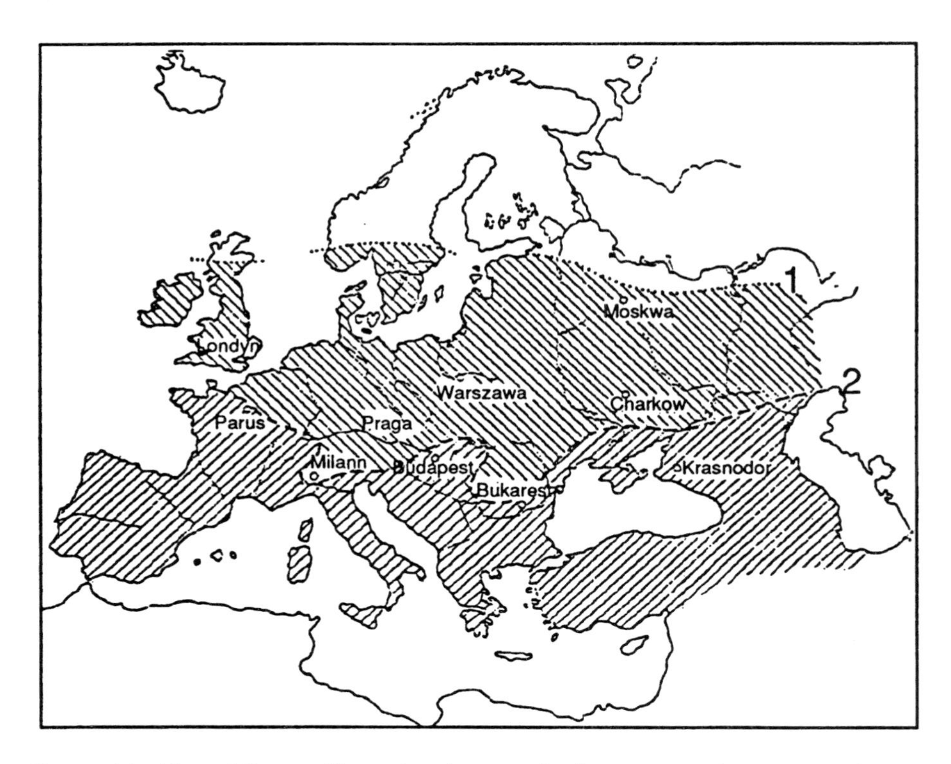

Figure 14 Map of Europe illustrating demarcation between northern range of grass carp (1) and southern spawning area (2).

reproduction nor culture is possible. On the other hand, in extensive subtropical and tropical areas of Asia, Africa, and the Americas, favorable temperatures occur, but the lack of other necessary factors prevents spawning. Therefore, in most locations outside the natural range where Chinese carp can be successfully grown, their reproduction is totally dependent on artificial propagation techniques.

Chapter 3

Life History

I. REPRODUCTION

A. MATURITY

The age and size at which Chinese carp attain maturity varies greatly with climatic and environmental conditions. Males usually mature 1 year earlier than females (Table 9). Among climatic factors, temperature has a pronounced effect on the maturation rate, as maturity occurs earlier in subtropical and tropical regions. In China grass carp and silver carp mature after being exposed to approximately 15,000 degree-days (°D), which is calculated in degrees Celsius as the sum of average daily water temperature for the days when water temperature exceeds 15°C.[126] According to Li and Wang,[127] silver carp need about 20,000°D and bighead carp about 26,000°D. The number of degree-days may differ, however, as temperature acts with other factors, especially feeding conditions, which are very important. An abundant food supply usually accelerates the maturation process because of greater growth rates, but in some cases overfeeding on certain foods, particularly prepared feeds under cultural conditions, may actually retard maturation.[9] Also, Chen et al.[128] suggested that exclusive feeding of grass carp on hydrilla could cause extensive fat accumulation, which may result in the retardation of gonad development.

Shireman and Smith[11] made a comprehensive survey of literature pertaining to sexual maturity of the grass carp in different countries, finding the age of maturity differed at different geographic locations. They reported that 1-year-old males (♂) and 2-year-old females (♀) matured in some locations in India and Malaysia. In the southern U.S. males usually mature at age 2 and females at age 3. In China, grass carp are 3 to 6 years old before they mature, depending on location. In the former U.S.S.R., 2- to 3-year-old males and 3- to 4-year-old females matured in Turkmenskaya, 4-year-old males and 5-year-old females matured in the Krasnodar District (Ukraine), 7- to 8-year-old males and 8- to 9-year-old females reached maturity in the Kiev District (Ukraine), and 9-year-old males and 10-year-old females matured in the Moscow District (Russia). In the Amur River, where they occur naturally, grass carp are 6 to 10 years old when they mature, depending upon location.

Kamilov[129] studied the relationship between the back-calculated growth rate and gonad development of female silver carp in Uzbekistan. He found that growth rate and gonadal development were positively correlated. For instance, fishes at age 1+ with 14 cm increase in body length caught in the spring had oocytes with a diameter of 40 to 44 μm, whereas fishes of the same age but with an increase of 17 cm or more had oocyte diameters of 60 to 73 μm (Figure 15a). The effect of first-year growth rate was more pronounced in older fishes. Two groups were identified among the 40 to 45 cm females of 2+ years caught in the spring and early summer (Figure 15b). The first group had gonads at an earlier developmental stage, with oocytes of 100 to 130 μm in diameter, while the gonads of the second group were more developed, with an oocyte diameter of 190 μm or more. The growth rate study showed that despite similar size, these fishes differed significantly in first-year growth rate: 9 to 14 cm and 18 to 24 cm in the first and the second group, respectively (Figure 15c). Furthermore, it was found that the first-year growth rate of silver carp determined the age of sexual maturity. In Uzbekistan silver carp

Table 9 **Average age and size of Chinese carp at maturity**[139]

Climate	Grass carp	Silver carp	Bighead carp
	Age at First Maturity (years)		
Temperate	4–7	4–6	6–8
Subtropical and tropical	3–5	2–3	3–4
	Size at First Maturity (cm, kg)		
Temperate	50–70 cm	60–80 cm	70–80 cm
	6–8 kg	2–6 kg	5–10 kg
Subtropical and tropical	4–6 kg	2–4 kg	3–7 kg

From Woynarovich, E. and Horváth, L., *FAO Fisheries Papers No. 201*, 1980.

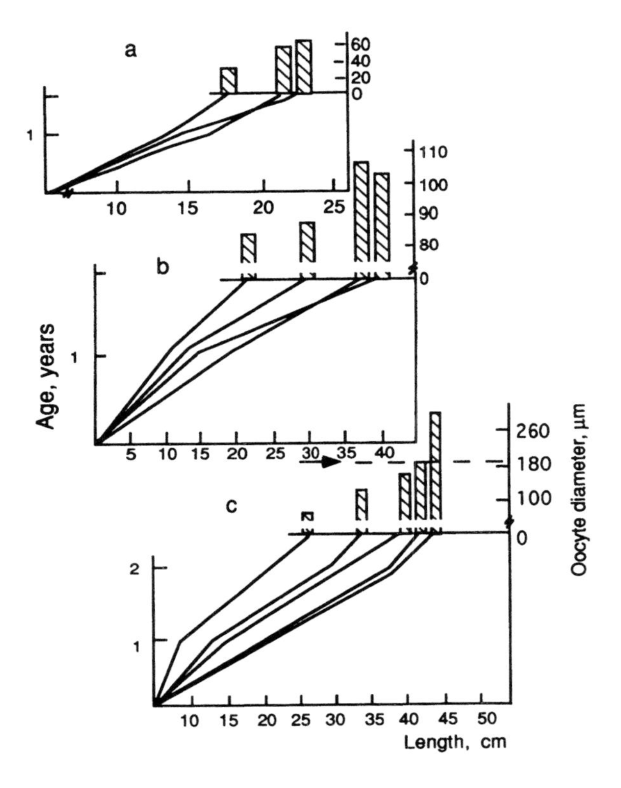

Figure 15 Gonadal development in silver carp females with different growth rates. (a) 1+-year-old fishes in spring, (b) 1+-year-old fishes in autumn, (c) 2+-year-old fishes in spring and early summer. Beginning of oocyte vacuolization is shown by an arrow. (From Kamilov, B. G., *J. Ichthyol.*, 27(2), 135, 1987. With permission.)

females matured at the age of 3 years if they grew to 17 cm and 100 to 120 g or more in the first year. However, age of the fish also affected maturation as the older fishes in the same size classes had more mature gonads.

According to Kuronuma,[130] silver carp mature at 2 to 3 years in south China, 4 to 5 years in central China, and 5 to 6 years in north China. Chang[131] reported that as a rule silver carp mature not only at a younger age, but also at a smaller size in southern China than in northern China. In the Pearl River (southern China) over 80% of the spawners weighed about 1.4 kg and were 3 to 5 years old, whereas in the Heilung River (northern China) females matured at 4 years of age and males at 5 years. In the Amur River silver carp females matured at age 7 when the length was not less than 62 cm.[132] In another river, the Kuban (Ukraine), 5-year-old fishes matured and spawned.[112] Silver carp mature when 6 to 9 years old and 6 to 8 kg in ponds in Romania,[126] and at the age of 7 years in Hungarian ponds.[9]

Bighead carp males matured at 2 to 3 years and females at 3 to 4 years and 5 to 10 kg in south China; males at 3 to 4 years and females at 4 to 5 years and 5 to 10 kg in central China; and males at 5 to 6 years and females at 6 to 7 years and 5 to 15 kg in northeast China.[130] In India 3-year-old bigheads matured when 67 to 70 cm and 4.7 to 6 kg.[133] In the former U.S.S.R. males matured at age 2 to 4 years and females at 3 to 5 years and 10 kg in Turkmenskaya males at 7 to 8 years and females at 8 to 9 years in the Ukraine (Kiev region), and males at 9 years and females at 10 years in the Moscow region.[134,135] In Romanian and Hungarian ponds females matured at 6, 7, and 8 years.[9,136]

B. GONADAL DEVELOPMENT

Anatomical differentiation of the grass carp gonads occurs at a mean total length of 58 mm when fishes are 50 to 60 d old.[137] When first developing, the gonads are slender, milky-white threads, flat in the female and rounded in the male. Different stages of gonadal maturation are distinguished by various authors; usually six stages are recognized by Russian authors[77] and seven by most western authors (Table 10). After stage III of maturation has been reached, further development of the ovary can be very fast at proper temperatures. The onset of ripeness (stage V) depends on the total sum of heat to which the fishes are exposed. In temperate climates, when degree-days are calculated beginning from January 1, silver carp become ripe at 1300° to 1400°D, grass carp at 1350 to 1450°D, and bighead carp at 1400 to 1500°D.[9] These values are approximate indicators, however, because the physiological outcome of thermo-exposure depends not only on the total heat sum, but also on the starting temperature and the distribution of temperatures during the exposure period (see also Chapter 5). According to Jhingran and Pullin,[126] by the time stage V is reached, the gonadosomatic index (the weight of gonads as a percentage of total body weight) may be as high as 20 to 30% for females and 5 to 10% for males. Lower values for grass carp (20% for females and 2.5% for males) were reported by Smith and Shireman.[138]

Histological differentiation of the gonads is a protracted and variable process. Woynarovich and Horváth[139] distinguish the following stages of egg cell development (The first three stages mark the period prior to the accumulation of yolk [lipids] in the developing eggs.):

Stage I. The primitive egg cells (oogonia) are very small, their size being hardly bigger than that of other cells. They multiply by normal mitosis.

Table 10 **Stages of gonad maturation of Chinese carp**

Stage of maturation		Condition of gonad	
		Female	**Male**
I	Immature	A transparent, long, narrow strip	Transparent, thread-like
II	Immature	Thickened strip, translucent	Translucent, thread-like
III	Maturing	Opaque, granular, somewhat grayish, occupying about one third of body cavity	Opaque, pinkish-white, thin, strip-like
IV	Maturing	Dull grayish in color, ova discernible-like granules, occupying about half of body cavity	Thick strip, milky, oozing whitish fluid on applying pressure to belly
V	Mature	Dull gray to greenish, occupying almost entire body cavity, ova distinctly round in shape	Thickened band-like, oozing whitish fluid on applying pressure to belly
VI	Spawning	Having loose eggs in ovarian wall, eggs oozing through genital aperture on applying pressure to belly	Oozing milt freely on applying pressure to belly
VII	Spent	Ovary bloodshot, pinkish-brown mass, but greatly shrunk	Oozing milt freely on applying pressure to belly

From Jhingran, V. G. and Pullin, R. S. V., *A Hatchary Manual for the Common, Chinese, and Indian Major Carps,* Asian Development Bank and International Center for Living Aquatic Resources Management, Manila, 1988, 191 pp. With permission.

Stage II. The egg cells grow and a follicle begins forming around each cell. The follicle functions to nurture and protect the developing egg, and eventually becomes a double layer of cells.

Stage III. During this stage, the egg cell grows significantly larger and becomes enclosed by the follicle.

Stage IV. During this stage production and accumulation of the yolk begins; this process is known as vitellogenesis. The egg continues to grow and drops of lipid accumulate in the cytoplasm.

Stage V. This marks the second phase of vitellogenesis. The cytoplasm is now full of lipid drops and yolk production begins.

Stage VI. This is the third phase of vitellogenesis during which the yolk plates push the lipid drops toward the edge of the cell where two rings begin forming. The nucleoli, which take part in protein synthesis and the accumulation of nutrients, are seen adhering to the nuclear membrane.

Stage VII. The process of vitellogenesis is completed during this stage. Yolk accumulation ends, the nucleoli withdraw into the center of the nucleus, and the micropyle (a small opening on the egg-shell through which the sperm enters the egg) develops during this stage.

Stages IV, V, VI, and VII are stages of vitellogenesis, when yolk is synthesized and accumulated in the egg cell. The egg is now materially ready and fully developed.

Emel'yanova[140] described seasonal changes in the ultrastructure of the oocyte cytoplasm during the previtellogenesis phase in silver carp. Winter and summer oocyte development were described under Uzbekistan conditions, where wintering of silver carp lasts 3 to 4 months. The oocytes studied had diameters of 23 to 240 μm, with nuclei ranging from 19 to 115 μm. In winter the endoplasmic reticulum appeared as small pits with granular-type vesicles, and was often strongly expended, which was never observed in summer. The mitochondria were usually few, spread throughout the cytoplasm, and often arranged in groups. The Golgi apparatus had a large number of vacuoles with an average size of 250 × 200 nm. In summer the Golgi apparatus vacuoles were either few or completely absent. The most distinctive differences between the oocyte cytoplasm in winter and summer were vacuolization of the Golgi complex, a small number and aggregation of mitochondria, swelling of the intermembranal space, and hypertrophy of the endoplasmic reticulum. These features are characteristic for previtellogenic oocytes during periods of suppressed activity at low temperature.

Duvarova[141] detailed the development of silver carp oocytes during the vitellogenesis phase. According to the position of the nucleus in the cells, she distinguished the following subphases of oocyte maturation:

Subphase E_1. The nucleus is in the center of the cell or slightly shifted (by 5 to 10% of the length of the oocyte radius) toward the animal pole where the micropyle is located. The nuclei are mainly round, although sometimes they can be elongated. The nuclear to plasma ratio is 0.11 to 0.15. When eggs are fixed with Serra's fluid, the yolk is almost transparent. Nucleoli of different sizes are located close to the nuclear membrane, with only a few concentrated in the center.
Subphase E_2. The fairly large and dark nuclei are usually of an irregular shape and shifted toward the animal pole by 30 to 70%. They are usually located with their greater diameter perpendicular to the membrane of the oocyte and only rarely parallel to it. The nuclear to plasma ratio is on average 0.3. When eggs are fixed, the yolk is usually clearly transparent. Nucleoli are concentrated in the center of the nucleus. The largest nucleoli are usually three to five times greater than the smaller ones. The small nucleoli are usually the first to concentrate in the center.
Subphase E_3. The nuclei are shifted completely to the animal pole. They are transparent and only the contour of the nuclei are distinctly visible. The nuclear to plasma ratio is 0.2.
Subphase E_3–F. (maximum degree of maturity). In this subphase, disintegration of the nuclear membrane begins.

Observations show that fishes with oocytes in subphase E_1 are not sufficiently mature and do not ovulate after hormonal injection. Successful artificial propagation is possible when most of the oocytes are in subphase E_2 (compare to Chapter 4).

In their native range Chinese carp have a seasonal spawning cycle, with the gonads ripening or recovering in winter and spring for a late spring and early summer spawning. Grass carp spawn intermittently in the Amur River basin, and the fishes that have laid the first batch of eggs predominate among females in stage V maturity toward the end of the spawning season. Although asynchronous development of the oocytes is most common, synchronous development also occurs. Gorbach[142] found two size classes of oocytes in 91% of mature females, while only 9% of mature females had oocytes of one size class. He classified the latter group as one-time spawners during the season.

In the cold climate of Heilungkiang Province, China, mature silver carp females sometimes do not spawn during the season and their oocytes do not show any degeneration until September, when the average temperature is about 13°C. In low winter temperatures only 6 to 11% of the oocytes degenerate and absorb. In May of the following year, when the water temperature rises to 17 to 19°C, both the absorption of degenerated eggs and the development of the new eggs occur. The degenerated eggs are not found after the beginning of June and the new oocytes are developed to mid-stage V.[143] In Malacca, where the temperature is uniformly high at 28 to 32°C year-round, grass carp show no seasonal spawning cycle and they can be ripe at any time of the year.[144] They may undergo more than one ripening and egg degradation within 1 year.

The process of sperm development is far less complex. Primitive sperm cells (spermatogonia) multiply by mitosis in the tubular wall in the testis. Primary spermatocytes develop from the spermatogonia. Each primary spermatocyte divides (meiotic division) into two secondary spermatocytes, and each secondary spermatocyte divides into two spermatozoa or sperm. The sperm remain in the testis in the resting stage until the fishes are ready to spawn.

In the southern U.S., the gonads of female grass carp differentiated histologically when fishes were 94 to 125 d old.[137] Nested oogonia and primary oocytes appeared first. From 180 to 232 d of age, the ovaries were transitional, with oogonia transformed into oocytes. Between the ages of 240 and 405 d, basophilic oocytes occurred in half of the ovaries examined. Most females 675 d old had ovaries dominated by primary oocytes. In males germ cells appeared between 150 and 300 d of age. Spermatogonia dominated in 300- to 675-d-old fishes; primary spermatocytes were not found. In central Russia oocytes began protoplasmatic growth in 3-year-old females, and trophoplasmatic development of oocytes was observed in 6- to 7-year-old fishes.[145] Spermatogonia appeared in the second year for males. Primary and secondary spermatocytes occurred at 37 to 38 months and a few ampullae with mature sperm developed in 39 to 40 months.

C. SEXUAL DIMORPHISM AND SEX COMPOSITION

Males can be distinguished from females by the structure of the pectoral fins. This secondary sex characteristic is not found in small, immature fishes, but develops with the approach of maturity and is most pronounced during the breeding season. Courtenay and Miley[146] described breeding tubercles in grass carp males on the dorsomedial surface of the pectoral fins, scattered along the ridges over the pectoral rays. These structures were easily seen and were rough to the touch (Figure 16). They also found tubercles on the first dorsal fin ray and over the dorsal surface of the caudal peduncle. The presence of breeding tubercles in these regions have not been reported by other authors. Courtenay and Miley[146] did not find tubercles on grass carp females. Chang,[131] however, reported their presence in both sexes, and maintained that in males they spread along the entire length of the fin rays, whereas in females they occurred only in the distal half of the rays in a much smaller area than in males.

Secondary sex characteristics are also manifested on the pectoral fins of silver and bighead carp. In the male silver carp tubercles looking like the row of a fine-tooth comb occur along each of several anterior fin rays, while in the female these fine teeth are present only on the distal half of the fin rays. In the bighead carp the male pectoral fins possess a bony, sharp edge along the dorsal surface of several front rays. According to Chang,[131] these structures are entirely absent in females. He maintains that in bighead the secondary sex characteristics are formed long before maturity, and once formed, they last throughout life. In Chinese carp the pectoral fins of males are usually longer than females, but this feature is not as useful in sexing the fishes as is the presence of tubercles. When females are fully ripe, the abdomen is extended from past the pelvis to the genital aperture, which is pinkish and swollen. The abdomen of males is not generally extended and the vent is pit-like in appearance.

Data are scarce pertaining to the sex composition of natural populations. According to Chang,[131] the sex ratio is close to 1:1 in Chinese rivers, although males always predominate in spawning populations. He reported 22% were females in grass carp spawning populations in the Yangtze River, 25% in the West River, and 20% in the Sungari River. In the spawning populations of silver carp females amounted to only 3, 12, and 4%, respectively, in those rivers. Females, however, predominate (45 to 91%, average 73%) in the catch of grass carp in the Amur River basin.[142,147] The following increase in the percentage of females in spawning populations was observed in consecutive age classes:

Age, years	8	9	10	11	12
Females, %	42	49	81	85	100

D. FECUNDITY

Contrary to numerous data on the fecundity of Chinese carp in cultural conditions, little information pertains to natural populations. One of the most comprehensive studies deals with fecundity of an indigenous grass carp population in the Amur River.[142] The mean absolute fecundity (number of eggs per female) was found to be 820,000, ranging from 237,000 to 1.68 million eggs. Absolute fecundity increased with age and weight of the fishes (Figure 17). A high positive correlation was observed between absolute fecundity and fish age ($r = 0.66$), length ($r = 0.71$), total weight ($r = 0.73$), and body weight without viscera ($r = 0.82$). When body weight increased by 1 kg, average fecundity rose by 167,000 eggs.

Relative fecundity of grass carp in the Amur River ranged from 48,000 to 177,000 eggs. The average number of eggs was 110,000/kg body weight without viscera. The range of relative fecundity was appreciably less (3.7 times) than the absolute fecundity range (7.1 times). Relative fecundity was similar for intermediate age and weight classes, increasing in the oldest and largest fishes (Figure 17). For this reason an insignificant positive correlation was found among relative fecundity and body weight without viscera ($r = 0.02$), total weight ($r = 0.1$), age ($r = 0.14$), and length ($r = 0.24$).

In the Amur River basin fecundity depends strongly on feeding conditions during the year preceding spawning. These conditions are determined by the amount of time the floodplain is inundated and vegetation is available to the grass carp. For example, the floodplain was flooded for almost the entire feeding period (120 to 130 d) in 1966, whereas it was flooded for only 35 to 40 d in 1968. Consequently, all biological indicators, including fish condition factor and fat content as well as absolute and relative fecundity, were higher in 1967 than in 1969. Population fecundity was approximately 2,344 million eggs in 1967, while it was only 540 million in 1969. Although the fourfold reduction in fecundity was due to an exceptionally poor feeding season in 1968, it coincided with a continued trend of decreasing population fecundity in the Amur River basin. The decrease in population fecundity was due to overfishing, reducing the number of spawners and increasing the percentage of young females in the spawning population.

Chang[131] quoted a fecundity of 960,000 eggs for 14.6 kg grass carp in the Yangtze River, a value lower than that for the Amur River. In the Kara Kum Canal absolute fecundity was found to be on the average 1.097 million.[148] Shireman and Smith[11] reviewed grass carp fecundity in different locations. After recalculating their data, the range of relative fecundity was found to be from 13,700 to 110,300 eggs per kilogram of body weight. The number of ovulated eggs is always smaller than that found in the ovaries because not all the eggs are released during spawning. These numbers are especially important for artificial propagation and they are given in Table 11.

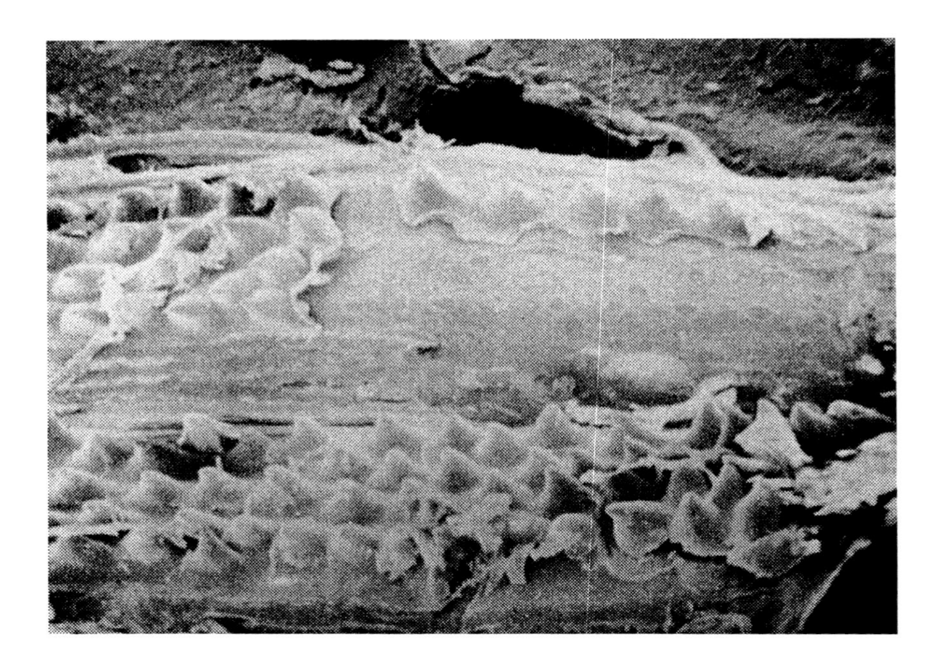

Figure 16a Scanning electron photomicrograph of pearl organs on the medial surface of the pectoral fin rays of a mature grass carp male. (Photo taken from Reference 146.)

Chang[131] reported that fecundity of silver carp was 800,000 eggs for 10.5 kg fishes in the Yangtze River and 158,000 to 650,000 for 1.7 to 5 kg fishes in Kwangtung Province. In the Kara Kum Canal, the average absolute fecundity was found to be 1.525 million.[148] Fecundity of pond-bred silver carp in India ranged from 163,000 to 1.348 million eggs for 0.9 to 7.5 kg fishes, and 164,000 to 180,000/kg body weight, respectively.[149]

Fecundity of bighead carp weighing 18.5 kg was 1.1 million eggs in the Yangtze River.[131] Bighead carp spawning for the first time in the former U.S.S.R. had an average stripped fecundity of 280,000 eggs,[150] while older spawners gave 470,000 to 540,000 eggs.[134] In China the average number of eggs stripped per kilogram of body weight was 59,000 (about 54% of the actual number of eggs in the ovaries). The maximum number of stripped eggs per kilogram was 77,600. These figures are comparable to the best results reported by Santiago et al.[151] They studied fecundity of caged bighead carp in a Philippine lake. Fishes fed a 40% protein diet and fishes feeding only on lake zooplankton were compared. The growth rate was not significantly increased by artificial feeding, but the number of eggs stripped per kilogram of fishes fed a high protein diet was higher than unfed fishes and amounted to 70,000 and 30,000, respectively.

1. Sperm Peculiarities

Not many studies have been conducted on the sperm of Chinese carp. Belova[152] estimated sperm concentrations in the ejaculate of silver and grass carp as 32.6 ± 5.2 and 27.1 ± 3.7 million/mm^3, respectively. The average duration of spermatozoid motility depended on water temperature, and at 20 to 45°C decreased from 53 to 10 s in silver carp and from

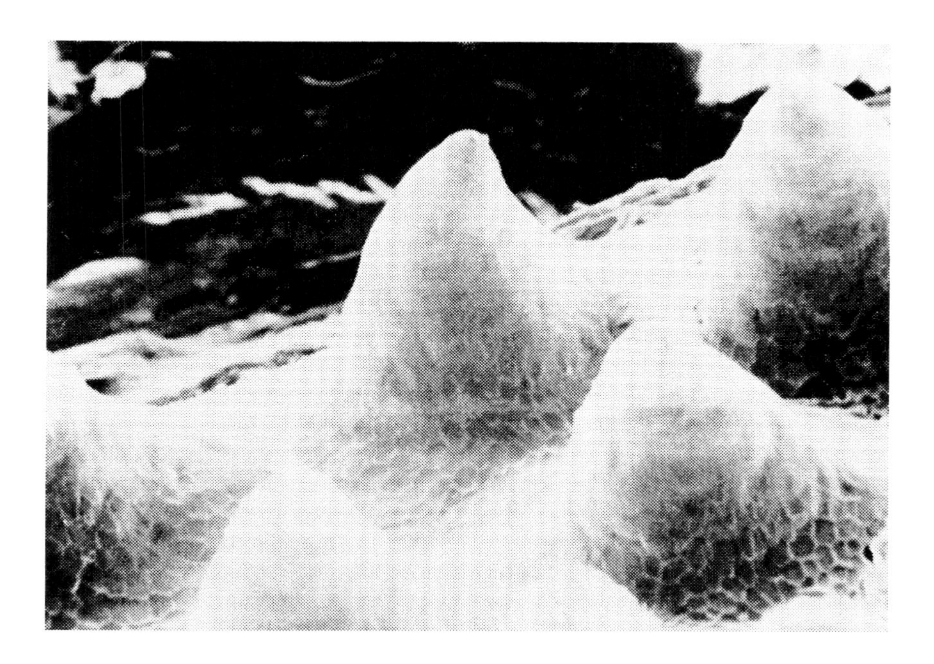

Figure 16b High-magnification scanning electron photomicrograph of pearl organs on the medial surface of the pectoral fin rays of a mature grass carp male. (Photo taken from Reference 146.)

56 to 7 s in grass carp.[153] The spermatozoids move faster in the higher water temperature and their limited energy resources are exhausted more rapidly, which results in reduced motility time. The duration of spermatozoid motility depends also on salinity. The longest motility periods of 128 s for silver carp and 97 s for grass carp were observed in a 0.3% NaCl concentration. With further increases in the NaCl concentration above 0.5%, the motility period decreased rapidly, and at concentrations of 0.9 to 1% sperm were completely inactive. The ability of the sperm to fertilize eggs corresponds to the period in which the spermatozoids remain motile. Extending sperm motility for a longer period is not necessary because egg fertilization sharply declines after 30 to 45 s due to closure of the micropyle.

E. ECOLOGICAL CONDITIONS FOR SPAWNING

Great attention has been paid to the reproductive requirements of Chinese carp because of the worldwide introduction of these fishes. Knowledge of the ecological conditions necessary for spawning allows for evaluation of the likelihood for reproduction and naturalization of Chinese carp in areas beyond their natural range.

In China the spawning grounds of bighead carp are limited to the Yangtze, Pearl, and Hwai River systems. They are not found in northern or central China. The spawning grounds of grass and silver carp are more widely distributed. Besides the above rivers, they also occur in Chekiang and Manchuria. Lin[154,155] and Chang[131] gave detailed descriptions of the conditions existing at the spawning sites in various Chinese rivers. Summarizing their data, the following conditions are conducive to successful spawning:

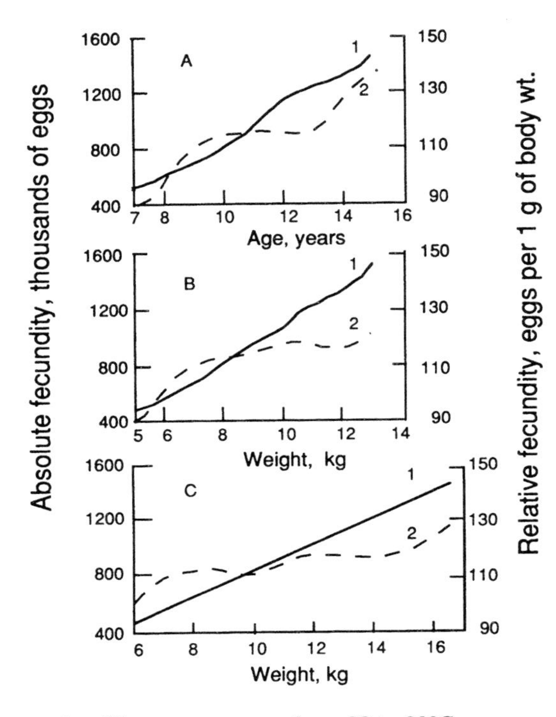

Figure 17 Absolute (1) and relative (2) fecundity of grass carp as a function of age (A), of body weight without viscera (B), and of total weight (C). (From Gorbach, E. I., *J. Ichthyol.*, 12, 616, 1972. With permission.)

1. Water temperature from 20 to 30°C
2. Current velocity from 0.6 to 2.3 m/s
3. Heavy rainfall causing a sudden rise in water level of over at least 0.75 m
4. Bottom consisting of sand and gravel
5. Water generally quite turbid, containing suspended solids, with a visibility of 10 to 15 cm
6. Spawning grounds are situated in places of turbulence, e.g., below confluence of two branch rivers

In addition to China the natural spawning area of the grass and the silver carp is located in the Amur River basin (Figure 12, Chapter 2). Silver carp reproduce in a section of the Amur River, including the tributaries of the Sungari and Ussuri Rivers, approximately 690 km above to 265 km below Khabarovsk. It is also possible that they reproduce higher up in the Amur River. The area where grass carp spawn is approximately half that of silver carp and covers a section of the Amur River from 585 to 95 km above Khabarovsk, including the Sungari and Ussuri Rivers. The rate of flow at the spawning grounds of both fishes is 0.7 to 1.4 m/s. According to Krykhtin and Gorbach,[156] spawning is determined by two factors acting together: water temperature and rise in water level. Spawning begins when the water temperature reaches 17°C and it is most intensive between 21 and 26°C. When the temperature drops below 17°C, spawning ceases, and dead eggs from previous spawns can be found. Spawning time and intensity also depend on water level fluctuations. The fishes spawn only when the water level rises, and stop spawning when the water level drops (Figure 18). However, dependence on a water level rise may not occur when a strong flood surge occurs.

Table 11 **Average number and size of Chinese carp eggs stripped during artifical propagation in Hungary**

Species	No. of eggs: Per kg of body weight (thousands)	Per kg of unswollen eggs (millions)	Per liter of swollen eggs (thousands)	Egg diameter: Unswollen (mm)	Swollen (mm)
Grass carp	60	1	20	1.2	4.2
Silver carp	60	1	25	1.2	4.2
Bighead carp	80[a]	0.82	25	1.3	4.3

[a]50 to 60, according to Reference 139.

Dependence was found between water level fluctuation and the percentage of ovulated females. In favorable seasons, when a substantial water rise occurred, two to three spawnings took place, lasting for one to two weeks each time. The proportion of females with partially or completely unovulated eggs during good spawning years numbers from 4 to 10% for grass carp and from 4 to 20% for silver carp. This proportion increases to 25 to 30% in grass carp and from 50 to 70% in silver carp in unfavorable seasons. The dependence of spawning on rising water level is adaptive because it increases the possibility of fry survival. First, early developmental stages are best protected against predators in turbid water, which is usually caused by rising water. Second, the fry can enter numerous floodplain waters which contain more plankton than the main channel of the river.

Dissolved oxygen in the Amur River was around 10 mg/l during the spawning season, pH 6.2 to 6.4, and free carbon dioxide 3 to 10 mg/l. It is believed that these factors do not influence spawning, as no correlation was found between these values and the spawning time and intensity.

It is interesting to note that the reproductive requirements of Chinese carp may undergo substantial changes when the fishes are introduced to a new environment. The successful spawning of all three Chinese carp (grass, silver, and bighead) in the Kara Kum Canal, Turkmenskaya, contradicts the belief that a rise in the water level is a basic precondition to spawning.[157] The case of the Kara Kum Canal is also interesting because it is probably the only known example of natural reproduction of Chinese carp in a manmade channel. The Kara Kum Canal has a total length of >900 km. It begins from the Amu Darya River and runs northwest, with an extension toward the Caspian Sea (Figure 13, Chapter 2). The water level in the canal is more or less stable and not subjected to substantial fluctuations in the spring–summer period when spawning occurs (Figure 19). The spring rise in water temperature (15 to 17°C) causes a massive upstream migration of sexually mature fishes to the sluice gates which prevent free passage into the Amu Darya River. The length of the section of the canal from the Amu Darya River to its inflow into Kelif Reservoir is around 80 km. The main spawning ground of Chinese carp is located in a 13-km section of the canal between the Amu Darya River and the inflow reservoir. The flow rate is 0.9 to 1.2 m/s. The water is turbid because it carries a heavy load of suspended material from the Amu Darya River.

A case was described in which 3- to 4-d-old larvae were caught in the delta region of the Lower Volga River on 3 June 1972.[110] Taking into account current velocity and

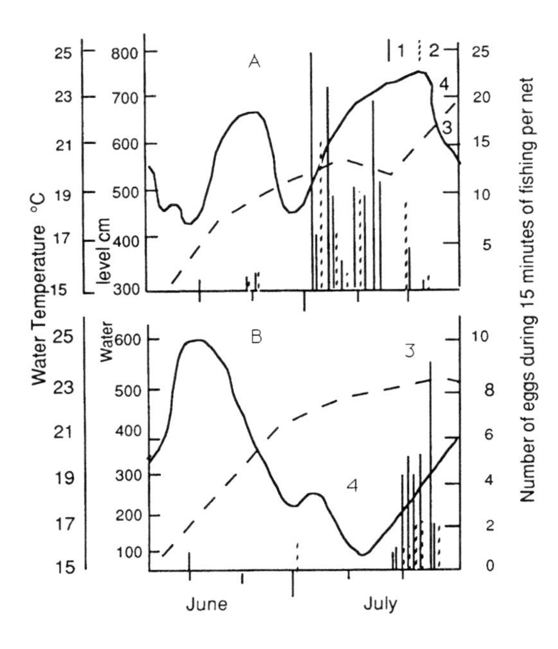

Figure 18 Dynamics of the downstream drift of silver carp eggs in the Amur (1) and Sungari (2) streams of the Amur River near the town of Golovino in 1963 (A) and of Leninskoye in 1967 (B). (3) Mean water temperature and (4) water level. (From Krykhtin, M. L. and Gorbach, E. I., *J. Ichthyol.*, 21, 109, 1981. With permission.)

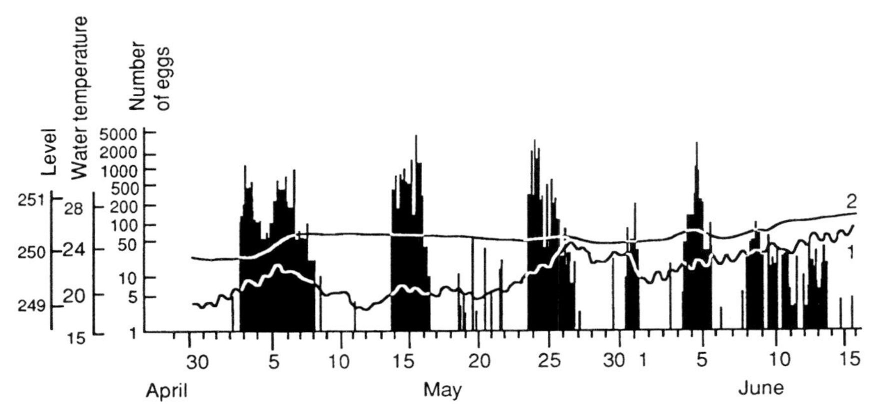

Figure 19 Dynamics of the downstream drift of Chinese carp eggs in relation to water temperature (1) and water level (2). (From Aliev, D. S., *J. Ichthyol.*, 16, 216,1976. With permission.)

incubation rate, the eggs must have been laid 300 to 400 km upstream on 29 or 30 May. The water temperatures in this location were not higher than 14 to 15°C. If these calculations are valid, this is the lowest temperature in which Chinese carp have spawned. Another case indicates that annual temperature fluctuations, which are characteristic in the natural range of Chinese carp, are not necessarily needed for natural reproduction. Temperature does not change appreciably in the Pampanga River basin in the Philippines, where natural spawning apparently occurs (see Chapter 2). The monthly average air temperature ranges from 25.9 to 29.6°C. Heavy rainfall and flooding are rather common in this region.

Among the conditions associated with successful reproduction of Chinese carp is proper flow rate to keep the semipelagic eggs suspended in the water column. Stanley et al.[114] concluded, after reviewing the literature and visiting spawning sites in the former U.S.S.R., that a current above 0.8 m/s was sufficient to suspend the eggs; however, Leslie et al.[158] claimed that a current velocity well below this value was needed. They conducted experiments in a Florida creek with hatchery-derived unfertilized eggs of grass carp and found eggs were transported 3.2 km by a current as slow as 0.23 m/s. This value is close to the minimum flow rate 0.3 m/s determined at the Amur River spawning grounds by Krykhtin and Gorbach.[156] In addition to current velocity the volume of flow also seems to be important. All known spawning grounds are located in large rivers with flow rates of over 350 m^3/s (460 in the Ili River, 350 in the Terek River, 380 in the Kuban River, and 400 m^3/s in the Kara Kum Canal).

The length of the river is also important, as enough length is needed to allow hatching, and the larvae must reach nursery areas where food is available. The length depends on water temperature (incubation time decreases with increased temperature) and current velocity. For example, in temperate climates with cold winters, where fishes spawn when temperatures reach 19 to 20°C, about 180 km are required with an average current velocity of 1.2 m/s. In milder climates, at 28°C and a flow rate of 0.8 m/s, only 50 km of river is needed.[114] It is also possible that a long upstream migration to the spawning grounds may stimulate final sexual maturity and trigger ovulation. Such a mechanism may have adaptive value because it ensures that the eggs have sufficient distance to drift before hatching.

F. SPAWNING BEHAVIOR

The spawning season lasts from April to August in China[131] and from June to July in the Amur River basin.[156] When spawning begins, males actively chase the females — usually more than two males following one female. The fishes swim fast and sometimes leap out of the water. Males occasionally rub the females, pressing their heads against the female's belly, or they butt the female, causing both fishes to turn upside down. When this happens they may swim upside down with their pectoral fins violently trembling and sometimes the eggs and milt are cast into the air. Both sexes usually assume parallel positions, leaning side to side and with their genital openings close as they release gametes.[154] Chang[131] termed these activities as surface spawning and distinguished them from underwater spawning. He claims that bighead carp perform only surface spawns, whereas grass and silver carp perform both surface and underwater spawns. According to Chang,[131] when water levels rise gradually, underwater spawns prevail. If a sudden rise occurs, surface spawning occurs.

II. DEVELOPMENT

A. EGG DEVELOPMENT

The embryonic development and juvenile life history of grass, bighead, and silver carp are very similar.[159,160] Therefore, grass carp development is discussed with only differences noted among the species.

The eggs of the three species are of similar size and structure. The grass carp egg measures 1.21 to 1.36 mm before water hardening. The bighead carp egg is slightly larger,

Table 12 **Differences in the development of grass, silver, and bighead carp**

Characteristics	Grass carp		Silver carp		Bighead carp	
	Mean[a]	Range	Mean[a]	Range	Mean[a]	Range
Eggs (swelling)						
Diameter of chorion (mm)	4.38–5.22	3.80–5.30	3.80–4.50	3.18–4.70	4.82–5.13	4.70–5.22
Diameter of yolk (mm)	1.21–1.36	1.05–1.50	1.10–1.30	1.03–1.30	1.40–1.50	1.25–1.70
Stratification of chorion	Wanting or exceptional		Wanting or exceptional		Chorion stratified as a rule	
Pro-larvae						
Size of hatched embryos (mm)	5.0–5.5	—	4.5–5.5	—	5.5–6.0	—
No. of myotomes						
In trunk	29–31	—	24–26	—	(24)25–26	—
In tail	14–16	—	16–18	—	15–19	—
Location of black pigment on yolk sac	Pigment on anterior part		Pigment on anterior and ventral parts			
Time of appearance of black pigment on head	Pigment appears toward the time that gill-jaw apparatus is in a mobile state and gills begin to function					
Larvae						
No. of myotomes	29–31	—	24–26	—	(24)25–26	—
	14–16	—	14–18	—	15–19	—
Pigmentation of pre-anal fin fold	Pigment absent		Pigment profusely developed		Pigment more weakly developed	
Pigmentation of region between boundary of dorsum and notochord, and between notochord and intestine	Region between boundary of dorsum and notochord pigmented, pigment cells absent or very rare in region between notochord and intestine		Region between dorsum and notochord, and between notochord and intestine profusely pigmented		Region between boundary of dorsum and notochord, and between notochord and intestine less strongly pigmented	

Arrangement of pigment cells at base of pectoral fin	Pigment cells form a distinct semicircle	1–2 pigment cells in upper part of semicircle	
Nature of pigmentation of region at end of notochord	Pigment weakly developed	Pigment weakly developed	Pigment weakly developed
Relationship of length of jaws from time of laying down of pelvic fins	Jaws close on same level	Jaws close on same level	Lower jaw protrudes farther forward than upper jaw
Length of pectoral fins at end of larval development period	Pectoral fins falling far short of base of pelvic fins	Pectoral fins extending only to commencement of base of pelvic fins	Pectoral fins extending beyond base of pelvic fins
Position of the dorsal fin	Dorsal fin ending on level at which base of anal fin commences	Dorsal fin ending on a level midway along base of anal fin	Dorsal fin ending on level of middle or end of base of anal fin
Fingerlings			
Relationship of jaws' length		As in larvae	
Length of pectoral fins		As in larvae	
Position of dorsal fin		As in larvae	
No. of rays in anal fin	8–9	11–14	11–14
Development of ventral keel	Keel absent	Keel developed from base of pectoral fins to anus	
Scales	Scales large	Scales small	Scales small

[a]From Soin, S. G. and Sukhanova, A. I., *J. Ichthyol.*, 12, 61, 1972. With Permission.

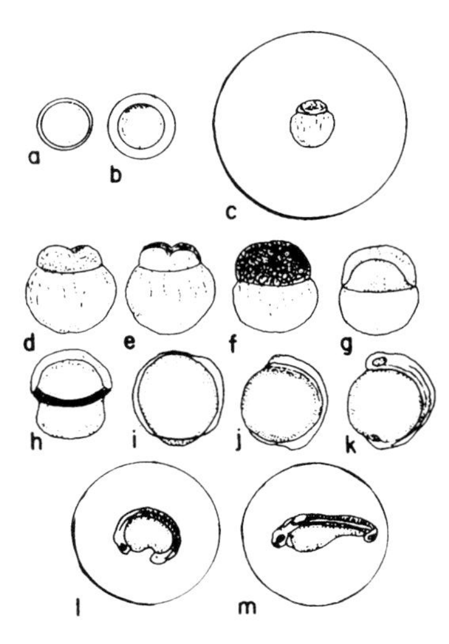

Figure 20 Embryonic development of grass carp (description in text). (From Shireman and Smith, *FAO Fisheries Synopsis*, 135, 1983.)

at 1.40 to 1.50 mm, and the silver carp is smaller, at 1.10 to 1.20 mm (Table 12). The egg has two membranes closely abutting the surface of the egg (Figure 20). The membrane separates within 10 min postfertilization and the cytoplasm collects at the animal pole. Large amounts of water penetrate the chorion, causing the egg to swell to 3.8 to 4.0 mm within 40 min postfertilization. A maximum diameter (4.32 to 5.32 mm) is attained after 1.5 to 2.0 h, which makes the egg buoyant in fresh water.

Development and hatching is dependent upon temperature. Incubation lasts from 16 to 60 h at temperatures ranging from 30 to 17°C.[159] Embryonic development before hatching can be described in six stages according to the Russian *Manual on the Biotechnology of the Propagatin and Rearing of Phytophagous Fishes.*[159] Times and sizes at each stage are predicated on an incubation temperature of 22 to 26°C.

Stage I (0.0 to 0.7 h): During the first few minutes the unhydrated fertilized eggs measures 1.2 to 1.3 mm (Figure 20a). At 10 min membrane separation occurs and the cytoplasm concentrates at the animal pole (Figure 20b). At 40 min the cytoplasm forms into a blastodisc and hydration of the perivitelline space continues.

Stage II (1 to 7 h): The blastodisc divides into two blastomeres at 60 min, four at 80 min, and eight at 100 min (Figure 20d and 20e). The early large cell morula stage begins at 2.5 h (Figure 20f), and the late small cell morula stage occurs at 5 h. The blastula is formed at 6 h (Figure 20g).

Stage III (7 to 13 h): Gastrulation occurs at 7 h, with the blastoderm growing over the yolk toward the vegetative pole (Figure 20h). A node that arises in a section of the blastomere fringe zone rapidly divides and turns inward and under, forming the germ layers of the rudimentary embryonic body, which lengthens and thickens (Figure 20i). Gastrulation concludes at approximately 12 h, when the body of the embryo appears as a thickened spindle, with the broadened head section at the animal pole and the tail section extending to the vegetative pole (Figure 20j).

Stage IV (13 to 24 h): The optic vesicles form at 15 h, segmentation of the mesoderm begins (two somites), and the cerebral vesicles differentiate (Figure 20k). At 21 h eye lenses form, auditory vesicles develop, segmentation continues, a distinct notochord is present, and the embryo occasionally flexes (Figure 20l).

Stage V (24 to 29 h): The tail separates from the yolk sac and the body straightens, and movements of the embryo increase. Numerous small vesicles appear representing hatching glands. Secretions from these glands lead to weakening of the egg membrane (Figure 20m) and hatching occurs.

B. PRO-LARVAL, LARVAL, AND FRY DEVELOPMENT

At hatching the grass carp pro-larvae are 5.0 to 5.2 mm. Pro-larvae of three species are of the same size.[160] Hatchlings possess 28 to 31 trunk and 12 to 16 tail muscle segments. Both bighead and silver carp have fewer trunk segments. Although no differences are found in the number of tail segments, both the silver and bighead carp have larger tail segments. The pro-larvae are unpigmented at this stage. The heart has two chambers, but the vascular system does not contain formed elements. At 30 h posthatching the larvae are about 6.5 mm and all vessels are filled with formed elements. The ducts of Cuvier, located in the anterior portion of the yolk sac, and the caudal vein function as embryonal respiratory organs.

At 3 d the pro-larvae are about 7.5 mm and movable mouthparts develop. Gills are present and gill respiration begins. The eyes are completely pigmented and pigment begins to appear on the body. Pigmentation over the body differs in the three species. In the silver and bighead carp the pigment occurs on the anterior and ventral portions of the yolk sac. In grass carp the pigment is only present on the anterior part of the yolk sac (Table 12). A rudimentary swim bladder forms at this stage and the larvae swim to the surface and swallow air.

The pro-larvae have now reached the larval stage. The swim bladder is filled with air and respiration takes place completely through the gills. The yolk sac is reduced and the larvae enter a mixed feeding stage. Pigmentation increases, and the three species differ according to pigmentation. Generally pigmentation is most intense in the silver carp and varies over the body of the species.

At 7 d the larvae measure 7.5 to 8 mm. The yolk sac has been absorbed and exogenous feeding begins. The larvae swim continuously. Lobes of the unpaired fins differentiate from the common fin fold.

When the larvae are 9 to 18 d old, distinct fin rays appear and the lower caudal fins begin to calcify. At 16 d (9 mm) all unpaired fins are distinctly separated and fin rays are apparent. Both the bighead and silver carp have greater numbers of rays in the anal fins (Table 12). An indentation begins to form in the caudal lobe and the pelvic fins begin to develop. The second anterior portion of the swim bladder begins to form.

At 20 d the larvae measure 11.5 to 18.6 mm. The unpaired fins are well developed and fin rays have appeared in the unpaired fins. The caudal fin is deeply notched. The swim bladder is well developed. The larvae have now reached the fry stage.

The fry measure 2.0 cm at the age of 1 month. Scale formation begins along the midline. The pre-anal fold persists on the abdomen behind the ventral fins. The mouth is pointed and the jaws interlock on one plane. The paired fins are well developed and the pectoral fins do not extend to the base of the ventrals. Internally the pharyngeal teeth are well developed and have the adult formula. The intestine lengthens and approaches the coiled pattern found in adults.

The fry become completely covered with scales at 1.5 months (4 to 5 cm) and lateral-line openings occur along the midline. Differences in scale size occur between the species, and in the bighead and silver carp a keel is developed from the base of the caudal fin to the anus (Table 12). At 55 d (6 to 7 cm) the fingerling appears identical to the adult.

C. PHYSICAL FACTORS INFLUENCING DEVELOPMENT

Chinese carp are highly adaptable, but have strict spawning requirements which limit their distribution.[11] Larval and fry development are limited by temperature. At low incubation temperatures larval development is retarded. For example, Stott and Cross[161] reported that grass carp survival was very low when incubation temperatures fell below 18°C, but 20-h-old larvae tolerated this temperature drop. Opuszynski[162] found that unacclimated fry in Poland had a mean lethal maximum temperature of 40°C and a lethal minimum temperature of 0 to 0.1°C. Yearlings could not withstand temperatures over 35°C, but could survive at 0°C. The *Manual on the Biotechnology of the Propagation and Rearing of Phytophagous Fishes*[159] reported that dropsy occurs in embryos of the three species due to extremely high or low temperatures or oxygen deficiencies or both. Opuszynski[162] reported the mean lethal oxygen levels for acclimated grass carp fry in two experimental groups were 0.45 and 0.41 mg/l for fry and 0.34 and 0.30 mg/l for yearlings. Silver carp means were slightly higher: 0.67 and 0.51 for fry and 0.37 and 0.57 for yearlings. Negonovskaya and Rudenko[163] found lethal oxygen limits to be within the range of the other species.

Indian researchers[164] determined that grass carp fry and fingerlings could survive the following ranges in water conditions: 125 to 215 mg/l turbidity, pH 5.0 to 9.0, dissolved oxygen 1 to 28 mg/l, total alkalinity 88 to 620 mg/l, free ammonia 0 to 3.8 mg/l, salinity 7.5 to 12.0%, free chlorine 0 to 0.2 mg/l, and sulfides 0 to 5.0 mg/l. Other authors have reported on the effects of salinity on growth, survival, and food consumption of grass carp fry and fingerlings. Fry can withstand salinities from 7 to 12%, depending upon the acclimation period and the ionic composition of the water.[165,166] Maceina and Shireman[167,168] found decreasing survival, weight, muscle water content, feeding rate, and growth when fingerlings were exposed to salinities up to 15.7%. They predicted that grass carp could inhabit brackish water up to 9%.

In order for grass carp to survive in natural systems, adequate amounts of macrophytes must be present. Tsuchiya[116] attributed substantial decreases in the population of grass carp in the Tone River (Japan) to the reduction of macrophytes.

III. FEEDING BEHAVIOR

A. GRASS CARP

1. Larval Feeding

Larvae begin feeding 3 or 4 d posthatching. The size of the food items ingested increases along with the size of the mouth gap. Under culture conditions food particles 150 μm or less are ingested after yolk sac resorption to day 3, 150 to 200 μm for 3 days through 5, 250 to 400 μm on days 6 and 7, and up to 0.75 to 1.00 mm at days 10 to 12, when the larvae have reached a mean length of 12 mm.[169] In Polish ponds grass carp larvae initially feed exclusively on animal food consisting primarily of rotifers, planktonic crustaceans, and chironomid larvae (Figure 11b, Chapter 2). The importance of protozoa in the food of early larvae was expressed by several Russian authors.[170–172] Bessmertnaya[170] also states that larvae feed throughout the day and night and that their diet consists mainly of rotifers up to 4 d posthatching; about 50% protozoa (*Arcella* sp.) during the next 5 to 6 d; mostly small chironomid larvaes, copepods, cladocerans, and rotifers from 7 to 10 d; and large cladocerans from 11 to 15 d. She stated that changes in larval diet did not depend on food preference, but rather on size accessibility or on the presence or absence of a given food species in the habitat.

Different feeding habits were shown for grass carp larvae caged in a pond.[173] Phytoplankton and zooplankton were abundant in the cages. The 4-day-old larvae, 6.5 to 7.0 mm in length and 1.6 to 2.0 mg in weight, fed on phytoplankton, including *Ankistrodesmus acicularis, Scenedesmus quadricauda, Pediastrum boryanum, Coelastrum acicularis, Cryptomonas marssonii,* and *Nitzschia* sp. Zooplankton was rarely encountered in the food (one out of ten intestines). At 5 d, when the larvae were 7.2 to 7.4 mm long and weighed 1.8 to 2.0 mg, their diet changed to predominantly zooplankton. The most abundant zooplankters in the pond, *Keratella vulga* and *Moina rectirostris,* constituted the main dietary components. Two-week-old grass carp, 12 to 17 mm long, fed selectively on zooplankton.[174] The preferred species were *Daphnia longispinna, Polyphemus pediculus, Bosmina longirostris,* and *Scapholeberis mucronata,* while copepods, *Chydorus,* and *Ceriodaphnia* were avoided. Linchevskaya[175] reported that algae in fry ponds were important in the diet. The food of fishes 5 to 30 d posthatching consisted of animal organisms, accompanied by almost every algal species occurring in the ponds. After 20 d, the role of algae increased, and filamentous algae along with macrophytes appeared in the diet. Grass carp started feeding almost exclusively on macrophytes at 1 to 1.5 months posthatching.

Disagreement exists as to the exact size or age at which grass carp become herbivorous. Gaevskaya[176] maintains that herbivory depends on the development of the digestive tract. When the larvae feed on small invertebrates and algae, the length of the intestine is about 57% of the total body length. As the fishes grow, the length of the intestine also increases, as does the importance of macrophytes in their food. According to De Silva and Weerakoon[177] grass carp change from a carnivorous to a herbivorous diet at a length of 25 to 30 mm. In their studies, however, the plant diet consisted of phytoplankton rather than macrophytes. The importance of the large colonial alga *Pediastrum* sp. increased gradually, accounting for 40% of the diet in 30 to 40 mm fishes (Figure 21).

Opuszynski[178] reported the transition from animal to plant food at a weight of 1.1 to 1.8 g and body length 36 to 43 mm or total length of 49 mm. Sobolev[174] showed that grass carp converted completely to feeding on macrophytes at an age of 36 to 40 d and a length of 50 mm. Watkins et al.[179] reported that hydrilla *(Hydrilla verticillata)* and grasses composed 85% of the diet when fishes were 87 mm in length. Some data indicate that the transition depends, apart from the size or age of the fishes, upon water temperature. Scheer et al.[180] consider that young fishes feed on zooplankton up to a water temperature of 15°C, and on plants above this temperature. Also, Fedorenko and Frazer[181] maintain that grass carp become herbivorous sooner in warm water than in cool water.

2. Preferred Plants

Hundreds of plant species, aquatic and terrestrial, eaten by grass carp are documented in the literature. Many studies pertaining to plant selection also have been conducted. The results of these studies, however, are often confusing because different studies state that the same plant species is eaten well, poorly, or not eaten. There are many reasons for these discrepancies including fish size, water temperatures during the studies, and the experimental methods used. Some studies were conducted under natural conditions, while others were conducted in laboratories. Results were also influenced by the plant species and the number of plants from which the fishes were able to select simultaneously, and by the methods used to calculate food preference.

All authors agree that grass carp food selectivity decreases with increasing fish size and water temperature. For example, small fishes weighing around 14 g did not eat *Azolla*

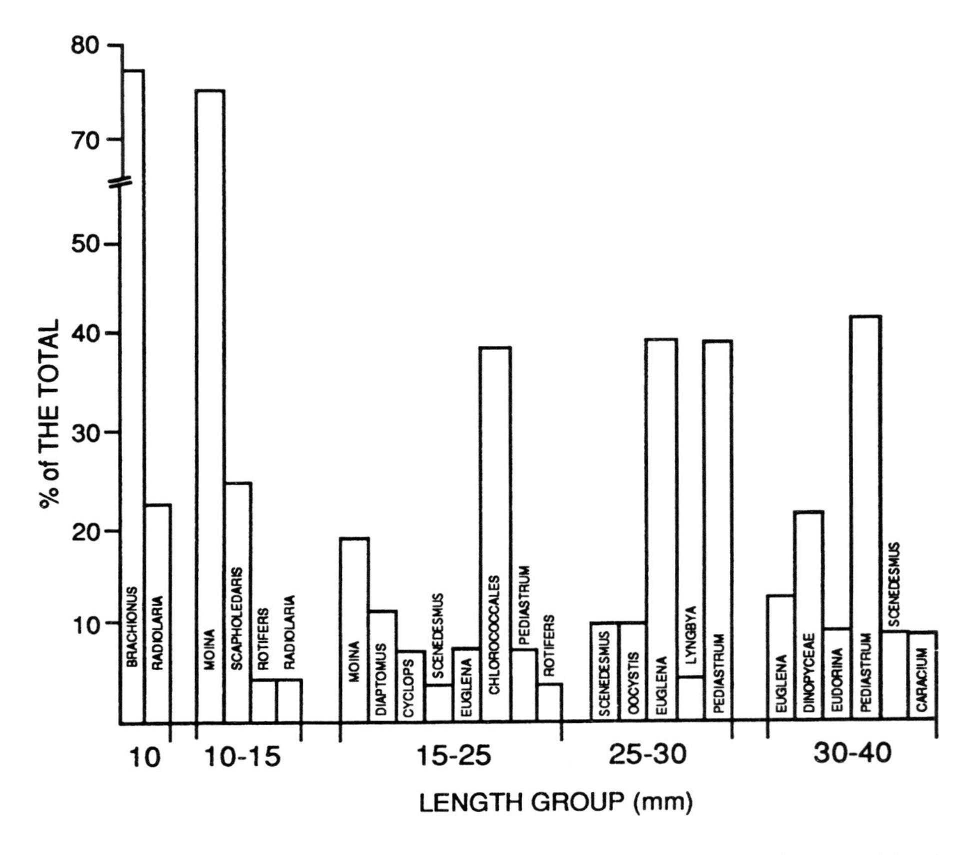

Figure 21 The dominant food items, expressed as a percentage of the total food ingested, in five length groups of grass carp fry. (From De Silva, S. S. and Weerakoon, D. E. M., *Aquaculture,* 25, 67, 1981. With permission.)

rubra, Myriophyllum propinquum, and *Lagarosiphon major* at a temperature of 21°C, but larger fishes weighing around 500 g ate considerable quantities of these weeds at the same temperature in New Zealand.[182] Also, smaller fishes eat fewer plant parts than larger fishes. Small fishes usually take only the leaves and reject the stems, while larger fishes eat the entire plant. A spirited discussion took place among some scientists in the former U.S.S.R. as to whether grass carp ingested large emergent plants having fibrous or woody stems, such as cattail *(Typha latifolia),* or phragmites *(Phragmites communis).* According to Stroganov,[183] cattail and phragmites higher than 1 m were not consumed. On the other hand, Aliev[184] and Verigin et al.[185] stated that these plants were readily eaten. Gaevskaya[176] believed this apparent controversy arose because Stroganov's studies were conducted at low water temperatures in the Moscow District, whereas the other studies were conducted in Turkmenskaya or in ponds receiving heated effluents from a power station.

It is interesting to note that the palatability of the same plants from different locations may differ depending on chemical content.[186] In this respect, calcium and cellulose were identified as being most important (Figure 22). Plants high in calcium may supply grass

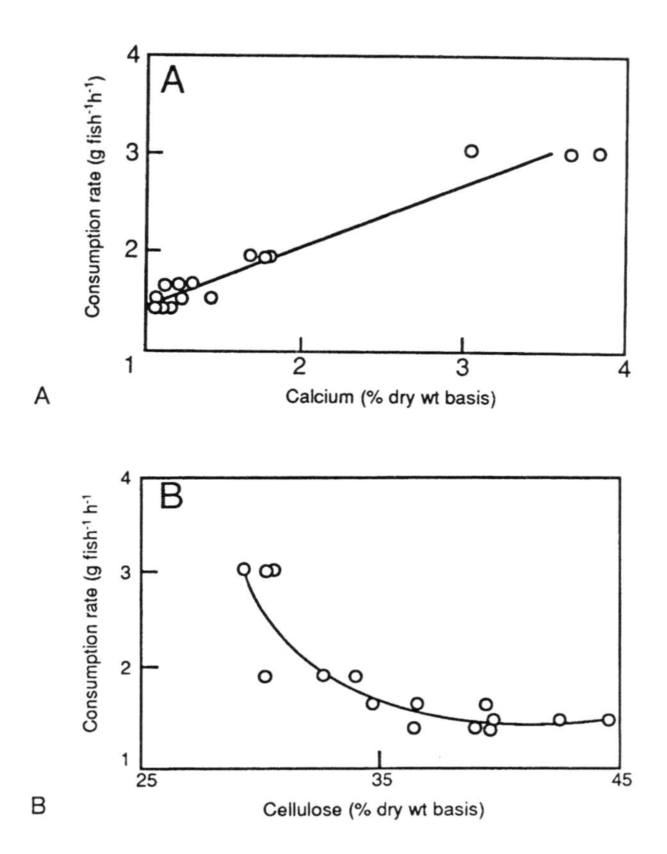

Figure 22 Relation between Elodea sp. chemical content and consumption by grass carp. (From Bonar et al. 1990) (A) Calcium: $Y = 0.64X + 0.78$, $r = 0.976$, $p < 0.001$. (B) Cellulose: $Y = -0.0028X^2 + 0.2246X - 3.8790$, $r = -0.922$, $p < 0.01$. (From Bonar, S. A., Sehgal, H. S., Pauley, G. B., and Thomas, G. L., *J. Fish Biol.*, 36, 149, 1990. With permission.)

carp with an element necessary for growth, and, therefore, may be selectively eaten. Plants high in cellulose, however, must be masticated by fishes before the digestive process begins; thus the high cellulose content may reduce the consumption rate.

Differences in plant preferences by triploid grass carp were found in static and in flowing water.[187] In the spring, fishes consumed a significantly higher amount of sago weed *(Potamogeton pectinatus)* and American *(P. nodosus)* pondweed in static water. Higher consumption of Eurasian watermilfoil *(Myriophyllum spicatum)* and American pondweed also occurred in static rather than flowing water in the summer. The authors concluded that observed differences in shoot length and nutritional content of plants in static vs. flowing water may have altered the consumption rate and preference of the fishes.

Generally, the foods most preferred are soft and succulent submersed plants (Table 13), but even the preference for the same plant may differ depending on the age of the plant. Young plants that are soft and tender are eaten more readily than old, fibrous, woody plants. The status of filamentous algae as a food is uncertain. They are mentioned as the most attractive food for the grass carp,[188,189] or as not eaten willingly,[183] or

Table 13 **Preferences of plant food by grass carp and hybrid grass carp: *Cyprinus carpio* ♂ × *Ctenopharyngodonidella* (HC) and *Ctenopharyngodon idella* ♀ × *Hypophthalmichthys nobilis* ♂ (HH)**

Plants Eaten Readily or Moderately	
Alisma plantagoaquatica[505]	*Myriophyllum* sp.[698,699,702]
Anacharis sp.[a] [192,695]	*M. spicatum*[187,196,520,566]
Aponogeton distachyus[699]	*M. brasiliense*[188]
Azolla caroliniana[703] HH[690]	*M. pinnatum*[520] HH[690]
A. rubra[182,691,699]	*M. propinquum*[182,691]
Azolla spp.[698]	*M. verticillatum*[505]
Brasenia schreberi[520]	*Najas flexillis*[606,616,698,703]
Callitriche sp.[203,505,696]	*N. guadalupensis*[188,217,273,520] HH[690] HC[700]
C. stagnalis[182,691,699]	*N. minor*[606]
Carex nigra[702]	*Najas* spp.[695]
Ceratophyllum demersum[a] [188,505,698,699,701,702]	*Nasturtium officinale*[a] [182,691,702]
Chara sp.[a] [196,217,273,566,606,694,695,698,699,702]	*Nitella hookeri*[182,691,699]
Chara vulgaris HC[700]	*Nitella* sp.[701]
Cladophora spp.[699,703]	*Paspalum notatum*[695]
Egeria densa[a] [698,699]	*P. distichum*[699]
Eleocharis acicularis[188] HC[700]	*Phalarus arundinacea*[488]
Eleocharis sp.[488,616,695,699]	*Phragmites communis*[a] [205,488,702]
Elodea canadensis[188,606,693,698,699,701,702]	*Pithophora* sp.[188,698] HC[700]
E. densa[182,191,205,488,691,693,698,701]	*Polygonum* spp.[695]
Elodea spp.[192]	*Potomoyeton* spp.[694,695]
Eremochlea ophivroides[196,566]	*P. amphibium*[488]
Fimbristylis acuminata[192]	*P. cappillaceus* HC[700]
Fontinalis spp.[191,205]	*P. cheesemanii*[701]
Fuirena umbrellata[192]	*P. crispus*[a] [182,606,698,701]
Glyceria fluitans[a] [699]	*P. diversifolius*[188,196,566]
G. aquatica[488]	*P. follosus*[606,698]
G. maxima[191,205,699]	*P. illinoensis*[273,520,616]
Hydrilla spp.[694]	*P. lucens*[a] [182,488,691,702]
H. verticillata[a] [192,273,694,699] HH[690]	*P. natans*[a] [182,488,691,702]
Hydrocharis morsus-ranae[702]	*P. nodosus*[187]
Hydrochloa carlineusis HC[700]	*P. ochreatus*[701]
Iscetes kirkii[701]	*P. pectinatus*[187,191,205,208,606,696,702,703] HH[208]
Juncus effusus[702]	*P. pusillus*[698] HC[700]
Lagarosiphon major[182,691,699]	*P.* spp.[694,699]
Lagarosiphon sp.[192,701]	*Ranunculus curcinatus*[696]
Lemna gibba[217] HH[208]	*R. fluitans*[488]
L. minor[182,217,691,692,698,702] HC[700]	*Rhynchospora aurea*[192]
Lemna sp.[191,205,699,701,703] HH[690]	*Sagittaria graminea*[488]
L. trisulca[505,702]	*S. sagittifolia*[205]
Limosella lineata[701]	*Salvinia herzogii*[701]
Lyngbya spp.[698,701]	*S. lacustris*[191,205]
Ludwigia palustris[699]	*Schoenoplectus jacustris*[191,205]

Table 13 (continued) **Preferences of plant food by grass carp and hybrid grass carp: *Cyprinus carpio* ♂ × *Ctenopharyngodonidella* (HC) and *Ctenopharyngodon idella* ♀ × *Hypophthalmichthys nobilis* ♂ (HH)**

Plants Eaten Readily or Moderately	
Scleria poaeformis[192]	*T. latifolia*[a 488,505,702]
Sirogonium spp.[698]	*Utricularia foliosa* HH[690]
Spargunium erectum[505]	*U. gibba*[520] HC[700]
Spirodella sp.[188,699]	*Vallisneria americana*[188,273,520]
S. punctata[699]	*V. gigantea*[701]
Spirogyra polyrhiza[a 188]	*Vallisneria* spp.[694]
Spirogyra sp.[703]	*Wolfia australiana*[699]
Trapa natans[488]	*W. columbiana*[698] HH[690] HC[700]
Typha angustifolia[a 488]	
Plants Eaten Reluctantly or Not At All	
Alternanthera philoxeroides[a 188] HC[700]	*Nelumbo lutea* HC[700]
Anacharis sp.[188]	*Nitella hookeri*[182,691]
Azolla rubra[699]	*Nymphaea alba*[699,701]
Brasenia schreberi HC[700]	*N. odorata*[520] HC[700]
Cabomba caroliniana[703] HC[700]	*Panicum repens*[520]
C. hudsonii[702]	Pasture grasses[701]
Ceratsphyllum demersum[a 606,703] HH[208]	*Phalarus arundinacea*[488]
Chara sp.[a 192]	*Phragmites communis*[a 191,205,488]
Cladium jamaicense[520]	*Pistia stratiotes*[703] HH[690] HC[700]
Egeria densa[a 701] HH[690] HC[700]	*Polygonum amphibium*[a 488]
Eichhornia crassipes[692,703] HH[690] HC[700]	*P. crispus*[a 182,691]
Eleodea sp.[703]	*P. decipiens*[699]
Heteranthera dubia HC[700]	*P. diversifolius*[196,566]
Hydrilla verticillata[a] HC[700]	*P. illinoensis* HH[690]
Hydrocelys nympaeoides[699]	*P. lucens*[a 488]
J. repens HC[700]	*P. natans*[a 488]
Lagarosiphon major[182,691]	*P. nodosus* HC[700]
Limnobium spongia[703] HC[700]	*P. pectinatus*[191,205]
Lydwigia peploides[699]	*R. fluitans*[488]
L. repens HC[700]	*Ruppia martinima*[703]
Lyngbya sp.[698]	*Sagittania graminea*[616]
Myriophyllum brasiliense[701] HH[208] HC[700]	*S. sagittifolia*[191,205]
M. elatinoides[701]	*S. subulata* HC[700]
M. heterophyllum HC[700]	*Salvinia natans*[699]
M. pinnatum[703]	*S. rotundifolia* HH[690]
M. propinquum[701]	*Spirogyra* sp.[a 192]
M. spicatum[a 182,196,566,691] HC[700]	*Typha angustifolia*[a 701]
Myriophyllum sp.[606]	*T. orientalis*[699]
N. graminea[192]	*U. inflata* HC[700]
Nasturtium officinale[a 182,691]	*Vallisneria* spp.[694]

[a] Plant is also included in the other preference category by another author.

completely neglected.[190] Opuszynski[191] reported that in ponds in which massive development of filamentous algae occurred, the algae accounted for only 9% of the total food intake, while macrophytes constituted 72%. He stated that when grass carp had a choice, they clearly preferred macrophytes to algae. The filamentous alga, *Pithophora,* was freely taken by grass carp in Malacca, but *Spirogyra* and *Mougeotia* were ignored.[192]

3. Animal Food

Considerable attention has been paid to the proportion of animal food in the grass carp diet (see Chapter 2) because this information is important in fish culture and aquatic weed control. Numerous data collected in different regions and bodies of water indicate that the proportion of animal food in the diet is very low if the plants preferred by grass carp are abundant. Boruckij[35] studied the food of grass carp 33 to 58 cm long in natural conditions in the Amur River. He found the alimentary tracts filled with macroflora and a negligible number of whole and the biomass of animal organisms. Animal food represented <1% of the diet in 2-year-old fishes raised in ponds in Poland.[191] The digestive tracts of 0.5 to 10 kg grass carp from a lake in Arkansas (U.S.) were filled with smartweed *(Polygonum fluitans),* but did not contain any form of animal remains.[193] Grass carp 6 to 22 cm long were strictly herbivorous and ate animal material only in trace quantities (<0.1%) in a vegetated pond in Florida (U.S.).[194]

In an Alabama (U.S.) pond,[195] grass carp rarely ate anything except plant material, which comprised on the average 88% of the food by volume; crustaceans and insect larvae accounted for 1%, and other materials consisted mainly of supplemental feed. In other Alabama ponds[196] grass carp stocked alone consumed mainly macrophytes and algae (75 to 95% by volume), and only a small amount of mature insects (0 to 18%). When stocked with other fishes their diet consisted of 84% macrophytes and only 9% insects.

As reported by many authors, adult grass carp may switch to alternate foods when the supply of macrophytes is low. In ponds in Louisiana (U.S.) plant material occupied the majority of the gut volume in grass carp collected at the beginning of the season, when vegetation covered the bottom, whereas animal remains accounted for 80 to 100% of the gut contents after weeds had been eliminated.[197] Terrestrial macrophyte fragments were the most abundant food item in grass carp from a Georgia (U.S.) pond devoid of vegetation. Animal remains, which accounted for only 0.05% of the intestine contents, consisted of 99% earthworms (Oligochaeta) and 1% dipteran pupae, ants, and unidentified fragments.[198]

Grass carp weighing between 18 and 40 g ate most of the common invertebrates from streams and ponds which were offered to them in aquarium experiments in the presence of palatable plants.[199] The fishes were attracted by movements of the prey, but they also ate slow-moving snails. Oligochaetes were usually the first to be eaten, and soon afterward mayflies, caddises, amphipods, chironomids, and *Archichauliodes* were ingested. Rainbow trout *(Oncorhynchus mykiss)* eggs were ignored, but fry were taken. When stones were provided as a cover, many of the invertebrates survived. It may be expected that under natural conditions, where prey animals stay hidden in the substrate, grass carp predation would be substantially lower. The piscivorous habits of young grass carp 7 to 12 cm in length were also confirmed in other laboratory experiments.[200] The fishes fed actively on the larvae of common carp even in the presence of preferred plants. The eggs of common carp, although taken into the mouth, were later rejected, but the developing eggs were damaged. Larger grass carp (20 to 25 cm) preferred to starve rather than feed upon the larvae of common carp. Singh et al. stated, however, that contrary to experimental results no distinct piscivorous tendency was noted either in fry and fingerlings or in adult grass carp kept in ponds.

4. Consumption Rate

Considerable data have been collected on the amount of food eaten by grass carp. Unfortunately, these data are difficult to compare because food ingestion is often quantified in different ways: as a ratio of wet or dry weight of food consumed to wet weight of fishes, or as a percentage of the body calorific content, or as calories consumed per unit of time, or simply as wet weight of food consumed. Furthermore, the sizes of fishes examined differed as did experimental conditions. Great variability in consumption rates exists when daily consumption as a percentage of fish body weight is compared. Relative consumption rates ranged from 3 to 216% and 1 to 10% as calculated as fresh and dry weight of the plants eaten, respectively (Table 14). As expected, preferred plants are consumed in the greatest quantity (Figure 23).

Fischer[201] presented a regression equation of $\log C = \log 0.30 + 0.81 \log W$ for the relationship between food consumption (C in calories per day) and grass carp body weight (W in grams). This regression, however, covers a relatively small range of fish weights (20 to 100 g) and was calculated using experimental data when fishes were fed with a homogenous diet of lettuce. As a general rule, relative food consumption decreases as fishes grow. However, no statistically significant difference in relative consumption rates was detected in either diploid or hybrid grass carp in size classes up to 1.5 kg.[202] Visual inspection of these data, however, suggests some reduction in consumption rate by the largest size classes of both fishes (Figure 24). Relative consumption apparently decreases when broader size ranges of fishes are compared. Chapman and Coffey[203] found the daily consumption rate dropped from 63% of body weight in 3.2 kg fishes to 32% in those weighing 10.2 kg. Grass carp over 6 kg ate 26 to 28% of their body weight per day in a Florida (U.S.) lake.[204] The food of the fishes consisted of hydrilla, which was the dominant plant in the lake.

Temperature greatly affects food consumption. Grass carp take relatively little food and select aquatic invertebrates rather than plant food at temperatures below 10°C.[181] The consumption of aquatic plants commences at 12°C and the fishes begin intensive feeding at 22 to 23°C.[205] Colle et al.[194] report that grass carp (94 to 186 mm) grew rapidly until temperatures fell below 14°C. When fingerlings were fed with *Chara* sp. and *Fontinalis antipyretica,* daily rations increased more than twofold in temperatures from 16 to 28°C. At temperatures from 22 to 28°C, the increase in food consumption was similar for fingerling (8 to 10 g) and 2-year-old fishes (91 to 109 g) and amounted to 29 and 30%, respectively.[162] Above 33°C feeding activities decrease.[192]

Attempts were made to express the relationship between temperature and consumption rate by mathematical formulas. A linear regression equation of $Y = 2089 + 187x$ was established for the amount of hydrilla consumed by grass carp (Y) and water temperature.[206] This regression explained 62% of the observed variation in the amount of weeds consumed and appeared to be a fairly good predictor of hydrilla consumption by this fish. Wiley and Wike[202] fitted daily consumption of diploids, triploids, and hybrid grass carp to an asymptotic log function of daily mean temperature (T). Fishes smaller than 800 g were used for calculations. The relative consumption rates C (proportion of the rate at 25°C) were expressed by the following regression equations:

$$\text{Diploids: } C = -2.4596 + 1.07480 \log_e T$$

Table 14 **Daily consumption of plants as a percentage of body weight by different size grass carp at different water temperatures**

Plant species	Consumption (%/fish/d) FW[a]	DW[b]	Weight (g) or length (mm)	Temp (°C)	Ref.
Carex nigra	31		170–260 g	30–34	185
Ceratophyllum sp.	114		170–260 g	30–34	185
	104–193				705
C. demersum	15	1	46 g[c]	20	216
		1	52 g	20–23	214
	22	2	>450 mm	24	605
Chara sp.	36	6	>450 mm	24	605
	168		165 g		540
Egeria densa	27	2	890–1300 g	22–24	212
Elodea sp.	109		170–260 g	30–34	185
E. canadensis	42	10	>450 mm	24	605
Elodea densa	41	4	46 g[c]	20	216
Hydrilla sp.	127–216				705
H. verticillata	27		76 g		540
	140		1208 g		540
Hydrocharis morsus-ranae	145		170–260 g	30–34	185
Lemna sp.	102		170–260 g	30–34	185
L. minima		7	3 g	25	209
		4	35 g	25	99, 337
L. minor		6	90–130 mm	21	168
Myriophyllum sp.	27	3	46 g[c]	20	216
	36		170–260 g	30–34	185
	10	2	>450 mm	24	605
Najas flexilis	41	6	>450 mm	24	605
N. foveolata	146		96 g		540
	131		930 g		540
	99		1837 g		540
N. minor	38	4	>450 mm	24	605
Phragmites sp.	17		170–260 g	30–34	185
Polygonum amphibium	34		170–260 g	30–34	185
Potamogeton filiformis	145		170–260 g	30–34	185
P. pectinatus	30	3	>450 mm	24	605
P. perfoliatus	15		205 g		540
	18		2800 g		540
Scirpus sp.	17		170–260 g	30–34	185
Typha sp.	31		170–260 g	30–34	185
T. angustata	3		200 g		540
	14		4000 g		540
Vallisneria sp.	36		170–260 g	30–34	185

Table 14 (continued) **Daily consumption of plants as a percentage of body weight by different size grass carp at different water temperatures**

Plant species	Consumption (%/fish/d) FW[a]	DW[b]	Weight (g) or length (mm)	Temp (°C)	Ref.
Mixture of water plants	4–61		22–38 g	20–22	704
	5		186 g	13	207
	24		178 g	18–29	207

[a] FW = fresh weight.

[b] DW = dry weight.

[c] Triploid grass carp.

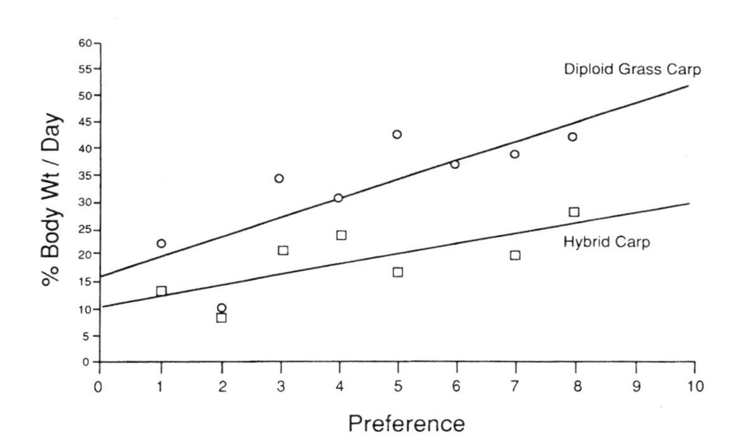

Figure 23 Consumption rate plotted against increasing plant preference for grass carp and hybrid grass carp, *Ctenopharyngodon idella* x *Hypophthalmichthys nobilis.* The following order of preference was determined, from least to most preferred plants: (1) *Ceratophyllum demersum,* (2) *Myriophyllum* sp., (3) *Potamogeton crispus,* (4) *P. pectinatus,* (5) *Elodea canadensis,* (6) *Chara* sp., (7) *Najas minor,* (8) *N. flexilis.* (From Wiley and Gorden, 1984.)

$$\text{Triploids: } C = -2.8591 + 1.19889 \log_e T$$

$$\text{Hybrids: } C = -4.0873 + 1.58047 \log_e T$$

On the basis of these regressions diploid fishes fed most actively at lower temperatures, triploids occupied an intermediate position, and hybrids fed the least actively. Also, as estimated from these equations, the threshold temperatures at which feeding began were 10, 11, and 13°C for diploid, triploid, and hybrid grass carp, respectively. It can also be concluded that hybrid grass carp fed at 66% of the rate of diploid grass carp. Triploid grass carp were feeding much better than the hybrids, but still at only about 90% of the diploid rate.

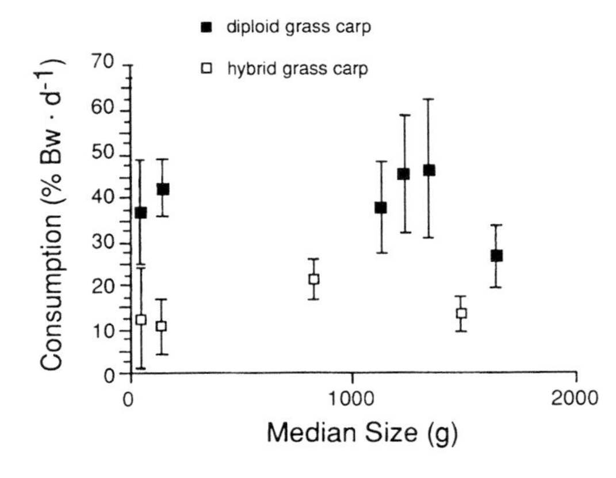

Figure 24 Mean food consumption by size class for diploid and hybrid grass carp. Bars = ± standard deviation, Bw = live body weight. (From Wiley, M. J. and Wike, L. D., *Transactions of the American Fisheries Society*, 115, 857, 1986. With permission.)

A positive correlation between food consumption and water temperature was not always found. For example, Kilambi and Robison,[207] feeding grass carp at 12.8, 18.3, 23.9, and 29.4°C, determined daily food consumption rates were 5, 24, 23, and 24%, respectively. Food consumption was lowest at 12.8°C, but did not differ significantly from the consumption rates obtained at the other temperatures. Uninhibited feeding at low temperatures was also reported in another study. Hybrid grass carp consumed as much plant material and in the same order of species preference at 12 to 15, 17 to 20, and 25 to 28°C temperature ranges.[208] The exceptions to temperature-dependent consumption for the hybrid grass carp probably result because of the overall low consumption of aquatic plants by this fish.

Among other environmental factors, oxygen concentration influences consumption rate. Observations in Malacca ponds showed that grass carp consumed less food at low oxygen levels.[192] Fingerling consumption of duckweed in tank experiments was significantly correlated ($r = 0.77, p < 0.01$) with oxygen levels, and when oxygen concentration dropped below 4 mg/l consumption was reduced by 40%.[209] Stanley[210] also found that oxygen levels below 4 mg/l drastically affected the amount of food consumed. Increasing water salinity beginning from 9‰ caused reduced consumption rates.[168] The presence of herbicides may also influence food intake. Grass carp stopped feeding when exposed to sublethal concentrations of some herbicides. At concentrations four to five times lower than 96-h LC_{50} values, feeding was substantially suppressed.[211]

The effect of light on feeding has yet to be investigated. Some data indicate, however, that increasing or decreasing photoperiod may affect feeding activity.[208] Increased food consumption at lower stocking densities was found under cultural conditions in which food was provided in excess.[207,209]

5. Food Conversion Ratio, Assimilation, and Energy Budget

Numerous data have been collected on grass carp food conversion ratios, not only because of the practical importance for fish culture, but also because it is relatively simple to determine. The food conversion ratio (FCR) indicates how many units of food are needed to produce one unit of fishes. Recalculating the data in Table 15 (FCR = $1/K_1$), the calorific FCR values of plant food between 1.3[212] and 53[213] were obtained. However, the exceptionally low FCR value calculated by Stanley[212] may be questionable.[202]

The FCR of 25 to 35, expressed as wet weight (w.w.) of food per live weight (l.w.) of fish, were determined in ponds where 3-year-old grass carp were fed with terrestrial

Table 15 **Grass carp published values for U^{-1} (assimilation/consumption), K_1(growth/consumption), and K_2 (growth/assimilation)**

Av. fish weight (g) or age (d)	Food	Temp (°C)	U^{-1}	K_1	K_2	Ref.
10–16 g	Algae and chara	19–20	0.005–0.009			219
	Lettuce	19–20	0.042			219
	Elodea sp.	19–20	0.016, 0.025			219
	Potamogeton sp.	19–20	0.022			219
3–6 d	Animal	21	0.255	0.085	0.340	706
12–22 d	Animal	21	0.810	0.385	0.480	706
19–124 d	Animal	21	0.430	0.180	0.400	706
	Plant	21	0.385	0.080	0.210	706
	Mixed	21	0.380	0.185	0.480	706
365–545 d	Animal	21	0.340	0.135	0.400	706
	Plant	21	0.165	0.025	0.140	706
	Mixed	21	0.385	0.290	0.760	706
23 g	Plant	23	0.122	0.020	0.165	213
52 g	Plant	23	0.158	0.028	0.210	213
23–52 g	Plant	23	0.123	0.019	0.165	213
3–100 g	Animal protein	22	0.261	0.152	0.601	707
	Animal carbohydrate	22	0.923	0.003	0.003	707
	Plant protein	22	0.410	0.065	0.159	707
	Plant carbohydrate	22	0.754	<0.001	<0.001	707
86–107 g	Animal	22	0.395	0.125	0.404	93
22–55 g	Plant	22	0.172	0.022	0.145	93
890–1300 g	*Egeria densa*	22–24	0.58	0.77	0.90	212
52–1809 g	Plant	19–29	0.20–0.50	0.02–0.20	0.10–0.57	202
50–135 g	Plant	17–26	0.30–0.45	0.09–0.17	0.24–0.50	202(T)[a]
106–1037 g	Plant	16–34	0.11–0.46	0.0–0.02	0.0–0.08	202(H)[b]

Note: All parameter values are expressed in calories.

[a] T = triploids.

[b] H = hybrids.

From Wiley, M. J. and Wike, L. D., *Transactions of the American Fisheries Society*, 115, 860, 1986. With permission.

plants.[191] The availability of invertebrates in those ponds could have accounted for the relatively low FCR values. In aquarium experiments, in which fishes were fed with aquatic weeds *(Chara* sp. and *Fontinalis antipyretica),* FCR values were 78 to 92 w.w./l.w. for 8 to 10 g fishes, and 35 to 36 w.w./l.w. for 91 to 100 g fishes. The high FCRs for fingerlings indicate that they still require considerable amounts of animal food for rapid growth.[191] Shireman et al.[99] reported FCRs of 1.6 d.w./l.w. for 3 g fishes and 2.7 d.w./l.w. for 63 g fishes. A reason for this discrepancy may be the high protein (31%) and low moisture content (7%) food *(Lemna minima)* used in the experiments of Shireman et al.[99] Other FCR values expressed as d.w./l.w. for grass carp fed with aquatic plants are 4,[214] 4 to 7 (hybrid),[215] 6 to 12,[216] and 21 to 39,[206] or, expressed as w.w./l.w.: 39 to 51[217] and

57 to 94.[216] When grass carp were fed with lettuce, FCRs for calorific values and for dry and wet weights were 29, 42, and 223, respectively.[218] For the same plant, Fischer and Lyakhnovich[93] reported a FCR equal to 45 cal.

Assimilability of plant food recalculated from Table 15 (U^{-1} 100) ranged from 0.5 to 58%.[212,219] Panov et al.[94] reported that grass carp weighing 5 g assimilated 69% of the animal food *(Daphnia)* on a calorific basis and only 15 to 18% of the plant food (*Potamogeton* sp.). Cai and Curtis[216] determined dry matter assimilation as 13 to 38% for different plants and Hajra[214] gave values of 49 to 51%.

Wiley and Wike[202] computed an energy balance for grass carp fed with a mixture of *Potamogeton crispus* and *Lactuca* sp. at 25°C. This balance was $100C = 12M + 74E + 14G$, where C = consumption, M = metabolism, E = excretion, and G = growth. It is interesting to compare this balance with general balances for carnivorous and herbivorous fishes given by Brett and Groves[100] (see Chapter 2). The proportion of food energy assimilated by grass carp ($M + G$) is lower, and that excreted (E) is much higher than the respective average values for herbivorous fishes. Grass carp energy balances must be examined carefully because fishes feeding freely in the field may perform better than those feeding under laboratory conditions, where they are restricted to undiversified exclusively plant diets. The optimum feeding strategy for this fish is clear: it must minimize metabolic costs and maximize consumption rate. Growth rate data (see Section IV below) show that this strategy can be very successful provided that the fishes have an unlimited supply of preferred plants for food.

B. SILVER CARP

1. Larval Feeding

The silver carp feeds on zooplankton during early life stages. The importance of rotifers in the diet has been emphasized by many authors. According to Panov et al.,[94] silver carp larvae can ingest only small zooplankton during the transition period to active feeding, and their principal food is rotifers. Even the small-sized cladoceran, *B. longirostris*, which is willingly consumed by the larvae of other freshwater fishes, is generally too large for the larvae of silver carp. Lupaceva[220] showed that 5- to 7-d-old silver carp fed mainly on the rotifer *K. cochlearis*. Seventy-nine individuals of this species were found in one alimentary tract. Also, Sobolev and Abramovitch[221] reported that 99% of the food consumed by 4- to 7-d-old larvae, 6 to 7 mm long and 1.6 to 3.8 mg in weight, consisted of rotifers. *Bosmina* and *Sida* appeared in the food of 8-d-old and 7 mg larvae, but rotifers still constituted up to 91% of all the organisms eaten.

In ponds in Belorussia silver carp began feeding on phytoplankton at an age of 14 d and at a length of 12 to 14 mm. Zooplankton consisting of small individuals of *Bosmina* and *Chydorus* accounted for 25% of the food consumed. Phytoplankton feeding occurred at an age of 18 d and a length of 20 to 22 mm. Only single zooplankters were then found in fish intestines.[174] In Polish ponds, phytoplankton became the essential food component when fry reached a standard length of over 32 mm and 0.7 to 0.8 g.[191] Quantities of zooplankton, however, were still found in the food.

2. Food Quality

The composition of the food eaten by silver carp varies. Detritus made up 90 to 99% of the food in the spring and 60 to 100% in autumn in the Amur River. In summer, during periods of algal blooms, detritus dropped to 5 to 15%; zooplankton comprised 1%, and phytoplankton the remaining part of the food.[53,222] A high proportion of detritus in the

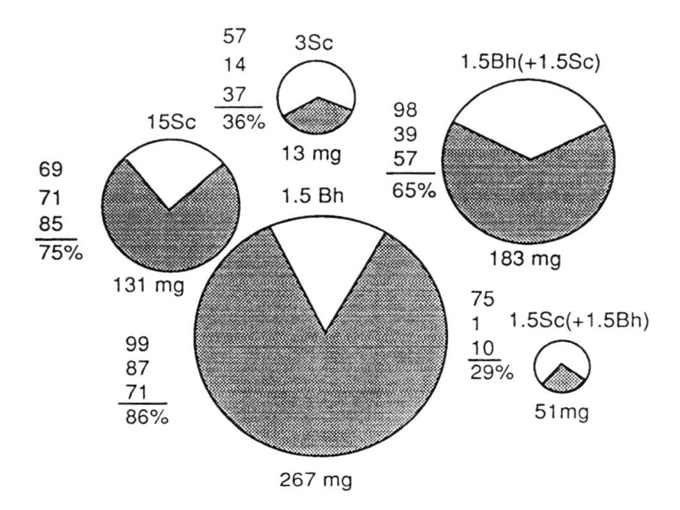

Figure 25 Amounts and proportions of zooplankton (screen field) to phytoplankton in the food biomass of silver carp (Sc) and bighead carp (Bh). 1.5 and 3 = stocking densities in thousands of fishes per hectare. Numbers at left of circles show the proportion of zooplankton in three consecutive samples (June, July, and August) and the average. Numbers under circles show the food biomass in milligrams of wet weight per fish. (From Opuszynski, K., *Aquaculture*, 25, 233, 1981. With permission.)

food of silver carp was also noted in ponds: 89 to 94%,[223] >99%,[224] and 90 to 99%.[225] It should be noted, however, that other authors did not find detritus in the food,[226] or found it in very small amounts.[227]

Opuszynski[228] reported an increase in the proportion of detritus and a decrease in phytoplankton and zooplankton as fish density in ponds increased from 1500 to 12,000/ha. The combined percentages of phytoplankton and zooplankton in consecutive densities of silver carp was 10, 4, 3, 2, and 1. In another study Opuszynski[229] showed that not only did the share of phytoplankton and zooplankton in the food drop as the stocking density increased, but the proportion between the amount of zooplankton and phytoplankton also reversed. Zooplankton was most abundant at lower fish densities, and phytoplankton was more abundant in the higher fish densities (Figure 25). According to Adamek and Spittler,[230] phytoplankton constituted 97%, detritus only 3%, and zooplankton was not important in the food of 3-year-old silver carp in ponds in the former Czechoslovakia. Also, Cremer and Smitherman[46] reported that phytoplankton was the bulk of food of free-swimming and caged silver carp in ponds, detritus accounted for only 13 to 15%, and zooplankton 0.0% of the food.

In Polish lakes, when silver carp were released in net cages and polyethylene enclosures, phytoplankton accounted for an average of 48% (10 to 70%), zooplankton 14% (10 to 30%), and other food, mainly detritus, for 38% (20 to 70%) of the diet.[231] The food of silver carp of different sizes (0.12 to 17 kg) was studied in Lake Kinneret, Israel, from 1973 to 1981.[232] Fishes fed mostly on phytoplankton from February through August and predominantly on zooplankton from September to January. Another study conducted in this lake between 1982 and 1984 showed that phytoplankton was the most important food of large fishes (a planktonic alga, *Peridinium cinctum*, constituted up to 92% of the total food volume), while smaller fishes ingested a large quantity of zooplankton. As the fishes grew from 1 to 20 kg, zooplankton was less important, and *P. cinctum* increased.[233] The food of silver carp, in a eutrophic German lake, was comprised of about 5% zooplankton, mainly *Bosmina, Cyclops, Keratella,* and *Brachionus*.[234] Detritus seems also to be unimportant in the diet of silver carp in Chinese lakes. In Qingling Lake[235] and Dong Hu-Lake[236] the fishes fed on plankton. In Dong Hu-Lake a blue-green alga, *Microcystis* sp., constituted about 90% of the food during the summer.

It may be concluded from the above review that detritus is an important food constituent in rivers and some ponds, but is relatively unimportant as food of silver carp in lakes. This inference appears to be consistent with the abundance of detritus in rivers, where it is kept suspended by the water current, as well as in ponds, where it is stirred up by bottom feeding fishes. The pond situation, however, is not uniform because cases were reported where common carp, an effective bottom stirrer, was present, but the proportion of detritus in the silver carp diet was negligible. For example, in intensive polyculture ponds in Israel, the vast majority of the silver carp food consisted of phytoplankton, while detritus was seldom found.[44]

3. Food Selection

Food selectivity by silver carp is a controversial issue. Some authors (e.g., Lupaceva,[226] Salar,[227] and Tarasova[237]) are of the opinion that the dominant phytoplankton in the environment constitutes their basic food. They compare the feeding behavior of silver carp to a "natural plankton net". This opinion is also supported by some recent studies. Spataru and Gophen[232] did not find selective feeding of silver carp for phytoplankton or zooplankton in Lake Kinneret and concluded that this fish collected suspended food items from the water mechanically rather than selectively. Similarly, Zhou and Lin[235] found the plankton composition in Lake Qingling in keeping with the food composition of silver carp and suggested no selectivity.

Distinctive food selectivity was found in other studies. Kajak et al.[231] reported selectivity for unicellular or colonial algae larger than 30 μm, which are more or less globular in shape, and also filamentous algae with a diameter >5 μm and a length of 200 to 300 μm. They argued that food selection was not only mechanical, but other mechanisms were involved. For example, in one of the investigated lakes, a blue-green alga *(Oscillatoria redeckei)* with trichomes longer than 200 μm was a dominant form. This form, however, was infrequently eaten by the fishes. A similar species *(O. agardhii)* dominated in the food, although it was relatively rare in the lake. Also, Barthelmes and Janichen[238] maintained that silver carp are able to distinguish between different seston qualities and that they feed in localities where a concentration of preferred seston mixture is concentrated. In Polish ponds silver carp fed on 108 species of phytoplankton, but only a few species in each sample made up the bulk of the food biomass. In 1968 the first three species dominating in the food constituted an average of 71% (42 to 91%) of the phytoplankton food biomass and 65% (45 to 90%) in 1974. The food of the silver carp did not reflect the phytoplankton composition in the ponds (Table 16). Of 54 samples in 1968, only in 10 did the same phytoplankton species dominate in the food and in the ponds, whereas in 1974 in a total of 63 samples, 30 of the same species were dominant in both the food and the ponds.

A more precise analysis of the above data was made using Ivlev's selectivity index (Table 17). This index confirmed higher food selectivity in 1968 (average +0.65) than in

Table 16 **Dominant phytoplankton (3 taxons showing highest percentages in the biomass of particular samples) in the food of silver carp and in the ponds of the Inland Fisheries Institute, Zabieniec, Poland, 1968 and 1974**

Taxon	1968		1974	
	No. of cases of dominance in food	No. of cases of dominance in food, phytoplankton simultaneously	No. of cases of dominance in food	No. of cases of dominance in food, phytoplankton simultaneously
A				
Pediastrum boryanum Menegh.	11	—	—	—
P. duplex Meyen	6	—	—	—
Gomphosphaeria naegeliana Lemm.	2	—	—	—
Protococcus viridis C.A. Agardh	2	—	—	—
Bacillariophyta	1	—	—	—
Chlamydomonas sp.	1	—	—	—
Dinobryon divergens Brunth	1	—	—	—
Microcystis aeruginosa Kützing	1	—	—	—
Navicula sp.	1	—	—	—
B				
Melosira granulata (Ehrbg.) Ralfs	11	9	7	—
Coelastrum microporum Naeg.	7	1	4	2
Stephanodiscus Hantzschii Grun	4	—	16	14
Fragilaria capucina Desm.	2	—	2	—
Pandorina morum Müller (Bory.)	1	—	4	3
Scenedesmus quadricauda (Trup.)Breb.	1	—	5	3
Cymatopleura solea (Breb.) W. Sm.	1	—	2	—
Dictyosphaerium pulchellum Wood.	1	—	1	—

Table 16 (continued) **Dominant phytoplankton (3 taxons showing highest percentages in the biomass of particular samples) in the food of silver carp and in the ponds of the Inland Fisheries Institute, Zabieniec, Poland, 1968 and 1974**

	1968		1974	
Taxon	**No. of cases of dominance in food**	**No. of cases of dominance in food, phytoplankton simultaneously**	**No. of cases of dominance in food**	**No. of cases of dominance in food, phytoplankton simultaneously**
C				
Cryptomonas rostrata Troitzkaja	—	—	7	4
Eudorina elegans Ehrbg.	—	—	5	—
Nitzschia palea (Kütz) W. Sm.	—	—	4	1
Euglena sp.	—	—	2	2
Trachelomonas volvocina Ehrbg.	—	—	1	1
Phacotus sp.	—	—	1	—
Synedra ulna (Nitzsch) Ehrbg.	—	—	1	—
Nitzschia sigmoidea (Ehrbg.) W. Sm.	—	—	1	—
Total	54	10	63	30

Note: Dominant taxons in the food: A = in 1968 only, B = in 1968 and 1974 , C = in 1974 only.

From Opuszynski, K., *Ekol. Pol.*, 27, 93, 1979. With permission.

Table 17 **Selectivity of the first 3 plankton dominants[a] in the food of silver carp at various stocking densities, determined by Ivlev's selectivity index,[b] in ponds of the Inland Fisheries Institute, Zabieniec, Poland**

	Silver carp stocking density (thousands of individuals/ha)				
	1968		1974		
Month	**1.5**	**3**	**4**	**8**	**12**
June	−0.78	−0.77	−0.70	−0.32	−0.38
July	−0.54	−0.54	−0.60	−0.55	−0.49
August	−0.59	−0.67	−0.28	−0.22	−0.36

[a] For dominants, see Table 16

[b] Ivlev's (1961) selectivity index = $r - P/r + P$, where r = percentage of a given item in the food, and P = percentage of this item in the pond. The index values range from −1 to +1; −1 indicates total avoidance and +1 indicates maximal preference.

From Opuszynski, K., *Ekol. Pol.*, 27, 93, 1979. With permission.

1974 (average +0.45). In 1974 the decrease in food selectivity was apparent throughout the growing season and also in the higher stocking densities. The decrease was probably caused by increased food competition as fish biomass increased. It is noteworthy that in many cases food selectivity for the same algal species was different in different samples. Only in 1968 did the dominant species in the food show positive index values. For example, *Peridinium boryanum* averaged +0.81 (from +0.22 to +1.0); *P. duplex*, +0.86 (+0.68 to +0.93); *Gomphosphaeria nageliana*, +1.0; and *Protococcus viridis*, +0.61. Of the algae dominant in 1974, negative index value ranges were shown for *Cyclops rostrata*, +0.48 (−65.0 to +1.0), *N. palea*, +0.50 (−0.08 to +0.89); *Euglena* sp., −0.61 (−0.78 to −0.44); and *Trachelomonas volvocina*, −0.13. Spataru et al.[44] also determined food selectivity of silver carp using Ivlev's[239] index. They found average index values for phytoplankton were >+0.5 and that Chlorophytes (*Chlamydomonas* sp., *P. simplex*, *Tetraedron* sp., *S. quadricauda*, and *Eudorina elegans*) were ingested most often.

Few data exist regarding silver carp selectivity for zooplankton. In Israeli ponds a strong negative selection (index values about −0.8) was observed. The frequency of occurrence for zooplankton was much lower than for that of phytoplankton. Cladocerans were found in 11% and rotifers and copepods in only 7% of the fish guts examined. A more complex situation was reported by Opuszynski.[228] He found that fishes definitely preferred a small cladoceran (*B. longirostris*) whose preference initially increased with increasing stocking densities. However, the preference for *B. longirostris* declined in the highest fish densities as the season progressed (Table 18). In general, cladocerans were treated indifferently or were avoided by fishes, whereas copepods (excluding nauplii) were usually avoided. As regards nauplii, they were strongly avoided (selectivity index usually −1). Despite an ambiguous situation for rotifers, which were either strongly selected for or avoided, they sometimes along with *B. longirostris* constituted an important component of animal food (Figure 26).

4. Consumption Rate and Food Utilization

The silver carp is a filter feeding fish whose consumption rate depends on filtering rate (a volume of water pumped through the filtering apparatus in a unit of time), density of

Table 18 **Ivlev food selectivity indices[a] of silver carp for certain zooplankton taxons in ponds of Inland Fisheries Institute, Zabieniec, Poland**

		Silver carp stocking density (thousands of individuals/ha)				
		1968		1974		
Month	**Zooplankton taxon**	**1.5**	**3**	**4**	**8**	**12**
June	*Bosmina longirostris*	–0.47	+0.82	+0.72	+1.0	–0.84
	Cladocera (without *B. longirostris*)	+0.06	–0.56	–0.01	–0.40	–0.15
	Copepoda (without nauplii)	–0.05	–0.64	–0.43	–1.0	–1.0
July	*Bosmina longirostris*	+0.59	+0.80	+0.87	+0.15	+0.08
	Cladocera (without *B. longirostris*)	–0.42	–0.21	+0.18	–0.57	+0.21
	Copepoda (without nauplii)	–0.15	–0.56	–1.0	+0.68	–0.99
August	*Bosmina longirostris*	+0.36	+0.73	+0.51	–1.0	—
	Cladocera (without *B. longirostris*)	–0.86	+0.04	–0.23	–0.33	–0.19
	Copepoda (without nauplii)	–0.17	+0.62	–0.27	+0.11	–1.0

[a] See Table 17 for selectivity index.

From Opuszynski, K., *Ekol. Pol.*, 27, 93, 1979. With permission.

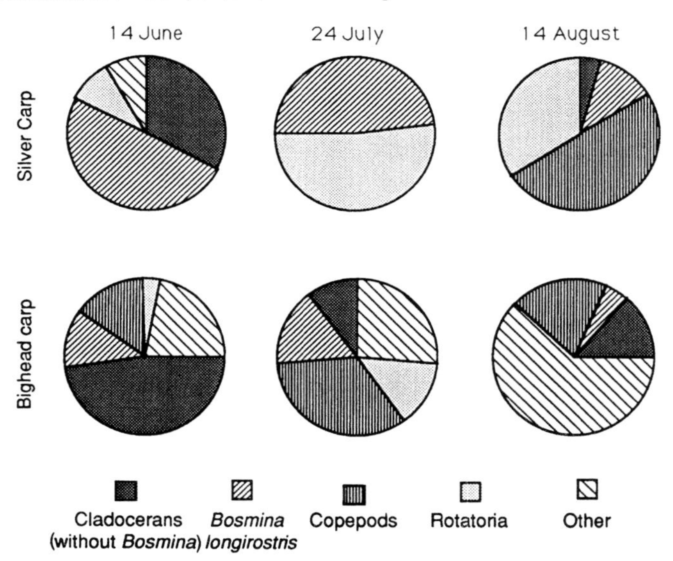

Figure 26 Composition (percentage) of the zooplankton food biomass of silver carp and bighead carp grown in the same ponds (average from three ponds). (From Opuszynski, K., *Aquaculture*, 25, 223, 1981. With permission.)

the food particles in the water, and filtering efficiency (the proportion of food particles that is retained by the filtering apparatus).

A filtering rate equal to 241 ± 139 ml/h was reported by Spittler[240] for fishes weighing 1.39 g, and Herodek et al.[241] determined that the filtering rate was 108 to 137 ml/h/g body weight for fishes weighing 1 to 5 g. Reliable results for filtering rates of larger fishes were

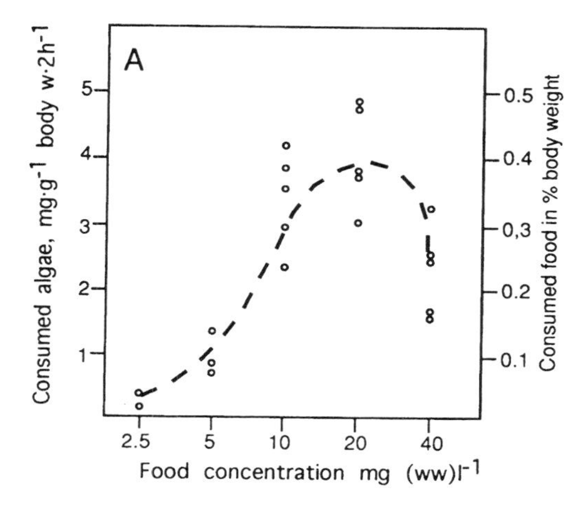

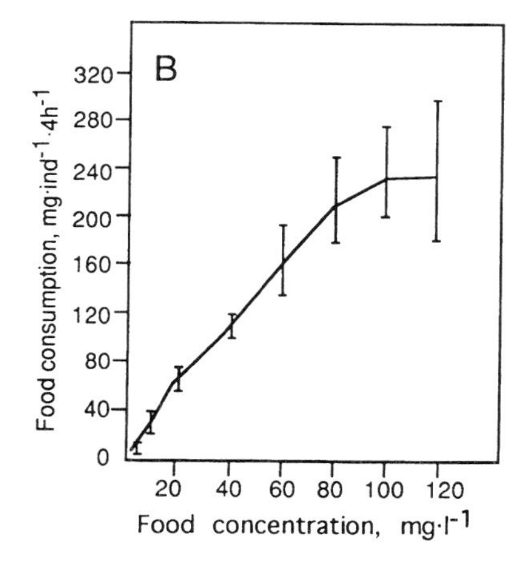

Figure 27 Consumption rates of silver carp fed with different concentrations of algae. (A) *Scenedesmus ellipsoideus,* (B) *Anabaena flos-aquae.* (From Herodek et al., *Aquaculture,* 83, 331, 1989. With permission.)

not found. Food consumption initially increases with increasing algal density and at a certain density levels off or even decreases. When small silver carp (average weight 1.1 g) were fed *Scenedesmus* the maximum consumption was at a food concentration of 20 mg/l, whereas the maximum consumption of *Anabaena* was observed at a concentration range of 100 to 120 mg/l (Figure 27). Adamek and Spittler[230] found that silver carp selected food particles larger than 11.2 μm, medium size particles (>6.25 μm) were equally abundant in the phytoplankton and in the fish food, and the smallest particles (<1 to 6.25 μm) were slightly avoided.

Wang et al.[242] determined daily consumption rates of silver carp from fry to fingerling periods. The relative food consumption of 1.6 to 3.4 mg larvae was 139% (expressed as wet weight of food and live weight of fishes), whereas the consumption rate of fingerlings 18 to 40 g was 16%. However, a steady decrease in food consumption was not observed for all the weight classes compared. Actually, the food consumption of 69 to 165 mg fishes was higher (63%) than that of 23 to 63 mg fishes (31%), but decreased with a further increase in fish weight. The increased relative food consumption rate for 69 to 165 mg fishes coincided with the change from a zooplankton to a phytoplankton diet. Relative food consumption rates reported by other authors were 17 and 12% for 1.4 and 5.8 g fishes,[243] 4 to 6% (dry weight) for 0.3 to 2.4 g fishes,[244] 12% for 16 g fishes,[245] 17% for 320 to 370 g fishes,[246] 10 to 16%,[238] and 6 to 12% for 653 to 1493 g fishes at temperatures 19 to 23°C in a lake during the summer,[247] and 13 to 23% for 5 to 34 g fishes.[248] High consumption rates of 46% in July and 27% in August were reported for fingerlings whose diet was comprised of 89 to 94% detritus,[223] and even higher consumption rates of 65 to 80% for 20 g fingerlings feeding on the blue-green alga (*Microcystis* sp.) in Dong Hu-Lake, China.[236]

The food efficiency indices (K_1 = growth/consumption) were low for silver carp feeding mainly on detritus and decreased from 6 to 2% as stocking density in ponds increased.[191] The K_1 value in the highest stocking density was similar to that reported by Fischer and Lyakhnovich[93] for grass carp fed with lettuce. Interestingly, as the stocking density of silver carp increased from 4000 to 12,000/ha, food rations also increased from 16 to 23 g per gut. Nevertheless, the growth rate of fishes decreased. Food analysis showed that the percentage of detritus increased and that of phytoplankton and zooplankton decreased as stocking density increased. Apparently, the fishes ate more detritus to offset the lack of plankton, but the nutritional value of detritus must have been lower as fish growth decreased.

Opinions vary, however, regarding the nutritional value of detritus. Savina[249] maintained that silver carp feed on detritus only under conditions of hunger. Lin et al.[250] fed silver carp with artificially prepared detritus from *Microcystis* and *Daphnia*. The detritus-fed fishes lost weight, whereas the control group fishes gained weight when fed with a commercial food. The authors concluded that the nutritional significance of detritus for silver carp was grossly overestimated. On the other hand, Panov et al.[94] demonstrated that decomposing algae were assimilated better than live algae. Kopylova[223] found satisfactory fish growth, with over 50% of the detrital nitrogen transformed into fish protein when detritus was abundant in the water.

Data pertaining to food assimilation by silver carp are scarce. When fed with blue-green algae, these fishes assimilate 40% of the protein when daily rations are 1.6%, and only 6.6% when daily rations are increased to 9.8%.[78] Assimilation rates are 35 to 48% when fishes are fed with *M. aeruginosa* and 17 to 36% when fed with *Euglena* sp.[251] Ekpo and Bender[252] compared digestibility of a commercial fish feed and wet and dry algae in juvenile fishes weighing 5 to 10 g. They found the digestibility of organic matter was 63% commercial fish feed, 60% wet algae, and 44% (dry algae). The respective figures for crude protein were 80 (commercial fish feed), 75 (wet algae), and 60% (dry algae). Herodek et al.[241] reported that 17% of ingested *Anabaena* was used for growth, whereas Hamada et al.[253] reported that 15, 11, and 1.5% was used for growth when fishes were fed with *Anabaena*, diatoms, and *Microcystis*, respectively.

C. BIGHEAD CARP

1. LARVAL FEEDING

First food of bighead carp comprises a wider range of organisms than that of silver carp. The most suitable particle size for larvae when they begin active feeding is 150 to 200 μm,[254] but as they grow from 9 mm to 6 mg they ingest most zooplankton, including rotifers, nauplii, and copepodite stages of cyclopoids, and smaller cladocerans such as *Bosmina* or young *Moina*. Early larvae also feed on phytoplankton and infusoria,[171] but their growth is slower. When zooplankton is abundant, larvae between 9 and 15 mg eat only zooplankton; the proportion decreases to 69% for larvae 10 to 45 mg and to 39%

(with small chironomid larvae) for larvae 14 to 125 mg. The remaining proportion of the food consists of phytoplankton.[255]

2. Food Quality

Everyone studying the food habits of Chinese carp agrees that zooplankton is of greater significance as food for bighead than it is for silver carp. However, the bighead carp is very adaptable and its food may differ widely depending on feeding conditions. Many authors agree that differences in the food habits between these two carp result because of differences in the structure of the filtering apparatus (see Chapter 2).

In Florida (U.S.) ponds bighead carp acted as escape-selective zooplankton predators and size-selective phytoplankton filterers.[256] All zooplankton species in the ponds were selected for by the fishes (Ivlev's selectivity index: >+0.9), except for adult copepods (+0.1). Copepods, despite their large size, can evade the fishes because of their swimming ability. *Botryococcus braunii,* a large phytoplankter, was always selected for (>+0.9), not only because of its large size (80 ± 51 μm), but also because of its spherical shape. The filamentous algae (*Lyngbya lagerheimii* and *L. limnetica*) with long trichomes (125 to 130 μm) but small diameter cells (1 to 2 μm) were not efficiently retained by the filtering apparatus (-0.1 to +0.1). In general algal selectivity increased with the size of colonies described as the product of the two largest dimensions. The size of a colony, however, seemed not to be the only criterion. For example, 294 μm^2 *Navicula* sp. had a selectivity index of +0.8, whereas 375 μm^2 *Microcystis incerta* had an index of -0.2. It is possible that algae having a rigid structure (for example, *Navicula*) are more easily retained than those that are gelatinous and flexible, such as *Microcystis.* However, the possibility that selectivity is also affected by algal palatability or nutritional value or both cannot be excluded.

Although the structure of the filtering apparatus certainly affects the quality of ingested food, it does not explain how bighead carp can feed on small phytoplankton and detritus when large phytoplankton and zooplankton are scarce. Opuszynski et al.[257] showed that the dominant phytoplankton species in the food of bighead carp caged in a pond were much smaller than the gill raker space reported for this fish. However, filter-feeding efficiency was low. It increased as algal size increased. This was shown distinctly when *Chlorella* and *Scenedesmus* were compared, even though both algae were smaller (6 and 12 μm, respectively) than the space between the gill rakers. The filtering efficiency for *Chlorella* was 0.1% and for *Scenedesmus* 13% of that found for rotifers having a mean size of 140 μm. The mechanism used by bighead carp for small-particle food capture is not clear. Possibly, small particles are embedded in the mucus secreted by the fishes and form aggregates that can be ingested. Whatever this mechanism may be, filter feeding on small algae is inefficient as bigheads feeding on small alga exhibit slow growth.

The proportions of zooplankton, phytoplankton, and detritus in the diet varies depending on feeding conditions. In the Kara Kum Canal (former U.S.S.R.), bighead carp fed predominantly on zooplankton when it was abundant during spring and early summer. During late summer and autumn the fishes fed mainly on phytoplankton.[258] Also, Danchenko[259] observed that bighead switched to phytoplankton and detritus when zooplankton declined. Lazareva et al.[255] found that while phytoplankton usually constituted a small percentage of the diet, algal blooms resulted in increased consumption of algae, amounting to 70% of the food ration.

In Alabama (U.S.) catfish ponds bighead fed primarily on zooplankton in May, detritus in June, and switched to colonial algae in July and August, when the biomass of algae was very high and the biomass of zooplankton low.[260] Also in Alabama, Cremer and Smitherman[46] compared the food of bighead swimming freely or caged in ponds. Detritus averaged 69% and zooplankton 24% of the food of pond fish, while in caged fish detritus declined to 25%, zooplankton to 5%, and phytoplankton constituted the remaining food content.

Detritus made up the bulk of bighead food in ponds in Poland. Plankton constituted 9 to 15% of the food, with zooplankton in greater amounts than phytoplankton.[229] When bighead were grown alone, zooplankton amounted to 86% of the diet, but when bighead and silver carp were grown together, the proportion of zooplankton in the food of both species was 65 and 29%, respectively (Figure 25). Considerable differences were found between these species as to the species composition of the zooplankton eaten (Figure 26). Bighead carp ate cladocerans, including *Alona quadrangularis, A. rectangula, Daphnia longispinna,* and *Polyphemus pediculatus,* and two species of copepods (*Cyclops* sp. and *Diaptomus* sp.). Other food consisted almost exclusively of small chironomid larvae. Green algae, diatoms, and blue-green algae were also present in the food. Large algae such as *Melosira granulata, Cymatopleura solea,* and *P. boryanum* were frequently dominant.

In Israeli ponds bigheads ate primarily zooplankton, including rotifers *(Botryococcus calyciflorus, Filinia longiseta,* and *Hexarthra fennica),* cladocerans *(Melosira rectirostris),* and copepods *(Mesocyclops dybowskii).*[44] Phytoplankton was consumed in much lower quantities, with large colonial chlorophytes *(Pandorina morum* and *Coelastrum microsporum)* constituting the most frequent forms. In Qingling Lake (China) zooplankton was the dominant food,[235] whereas in Dong Hu-Lake (China) bigheads ingested cladocerans (30%) in addition to phytoplankton.[236] In ponds phytoplankton and zooplankton were found to be of equal importance.[261]

3. Consumption Rate and Food Utilization

Data pertaining to quantitative feeding and food utilization by bighead carp are limited. Opuszynski et al.[257] determined that the filtration rate ranged from 185 to 256 ml/h/g for 34 to 2242 g fishes. According to Sifa et al.,[262] the consumption rate is influenced mainly by light intensity, water temperature, and dissolved oxygen. They found that fishes fed most intensely during July and August and that the highest daily consumption rate was between 12 p.m. and 8 p.m. The daily ration amounted to 6.6%. A similar daily ration was reported by Opuszynski and Shireman[263] for 13 to 2850 g fishes. These results, however, may have been underestimated because the fishes were caged in a pond in which zooplankton were not abundant and small algae dominated. In Chinese studies the daily rations were found to be 9 to 15% for 2 to 28 g fishes[248] or even 47 to 57% for 20 g fishes.[236]

Panov et al.[264] determined food assimilation rates for 24 mm and 400 mg juvenile bighead. Blue-green algae *Anabaena spiroides* (37%), *Aphanizomenon flos-aquae* (41%), and a diatom, *Nitzschia* sp. (49%), were assimilated at the highest rates. However, green algae, including *Chlamydomonas*, *Ankistrodesmus*, and *Chlorella,* were poorly assimilated (11, 15, and 27%, respectively). The assimilability of cladocerans was 59% for *D. pulex* and 61% for *C. quadrangula*. It is interesting to note that the assimilability of algae increased when the algae were killed and given to the fishes as detritus. This contradicts the statement of Lin et al.,[250] who reported that detritus was a low nutritive food for bighead carp. The estimated mean nitrogen assimilation efficiency for *Microcystis* and *Daphnia* was 57%.[236]

IV. GROWTH

A. GRASS CARP

Under conditions for which food supply is unlimited, grass carp, when compared to fishes of similar growth potential, exhibit an intensive growth rate, probably as great as any other species.

1. Diet

The foods eaten by grass carp are influenced primarily by the size of the fishes. For example, larval and small fingerling grass carp feed on zooplankton and algae until they exceed 30 mm, at which time they begin to feed on small, succulent macrophytes. The types of foods ingested, consumption rate, and assimilation are dependent upon temperature.

A number of researchers have investigated the dietary requirements of larval grass carp in order to develop commercially processed foods that would provide the necessary ingredients for growth and survival. The availability of these diets is advantageous because they are more easily fed, consistent in essential nutrients, minerals, and vitamins, and culture facilities for growing live foods are not needed.

Shireman and Smith[11] reviewed much of the early literature pertaining to larval grass carp foods and growth. Many natural and prepared foods were tried. The prepared foods were compared usually to results obtained with live foods, including cultured zooplankton and brine shrimp *(Artemia salina)* nauplii. For example, Sharma and Kulshrestha[265] found yeast and a vitamin B complex to be superior to groundnut oil cake, rice bran, hydrilla, and pondweed for fry and fingerlings. Appelbaum and Uland's[169] data supported this work as they found Alkan yeast with a vitamin and protein supplement to be superior to the other diets they tested. More recent studies report similar results. Dabrowski and Poczyczynski,[266] in a 25-d feeding experiment, tested several compound diets and live zooplankton on growth and survival of larval grass carp. Larval grass carp grew significantly faster (380 mg) on a diet containing yeast without fish tissues. Slowest growth occurred when fish tissues were added to the diet. Dabrowski[267] determined the optimum protein content of a cesium diet was $52.6 \pm -1.93\%$ for the greatest growth and conversion efficiency in fry. Dabrowski and Kozak[268] reported that optimum growth of 0.49 g fry was obtained with a diet of 40% fish meal, which yielded a 209% relative weight increase in 70 d. Different combinations of green algae, trout feed, and whey powder soybean diet were tested by Meske and Pfeffer.[269] They found that pure trout food and algae diets caused dorsal lordosis. More recently, Opuszynski et al.[270] fed grass carp larvae zooplankton and four dry diets for 14 days. He found that fishes grew best when zooplankton was added to the EWOS C10 diet at a 1 to 10 *ad libitum* rate. During the experiment the quality of zooplankton available for the grass carp was deficient because the zooplankton available was too large. Rottmann et al.[271] compared live foods (freshwater rotifers [*B. rubens*] brine shrimp, and nematodes [*Panagrellus* sp.] and two commercial foods. Grass carp larvae fed freshwater rotifers grew best, and growth of the fishes fed the commercial diets was less than those fed live foods. It appears that commercial foods for larval grass carp have not been developed that equal live foods for both growth and survival.

For fingerling grass carp and larger fishes, growth is determined by the amount of food available, temperature, and the palatability of the food. Consumption, food conversion ratio, assimilation, and energy budgets have been discussed previously concerning various plant and animal materials.

A number of plants have been investigated as pertains to their influence on growth. Tan[272] found that grass carp fed hydrilla grew faster than those fed napier grass *(Pennisetum purpereum)* and tapioca leaves. Other authors have reported good growth when fishes were fed or had the opportunity to consume hydrilla;[204,206,273,274] however, Venkatesh and Shetty[275] reported that grass carp fingerlings fed hybrid napier grass grew three times better than those fed hydrilla. Duckweed (*Lemna* sp.) is also a plant preferred by grass carp, and because of its small size it is a superior diet for fingerling grass carp.[217] Sutton[217] reported 20 to 22 g/d growth in Florida, from April through June, when consumption was 50% of body weight. Better growth was also obtained with duckweed when compared to pelleted food.[99,276,277] In Florida lakes where grass carp had an unlimited hydrilla food supply, growth was very rapid. Gasaway[278] reported that growth was fast in Florida lakes when hydrilla was abundant. Growth was reduced in one lake after hydrilla was treated chemically. Growth during the first year after stocking was 15.3 g/d and 5.7 g/d during the next 7 months.

Grass carp growth patterns, as described by weight-length relationships, generally follow the cubic relationship. Weight-length relationships have been reported for culture ponds,[279-281] two lakes in Florida,[168,282] and for a northern U.S. lake.[283]

Chow's[279] calculated weight-length relationship for fishes 27 to 66 cm from Hong Kong culture ponds was $\log_{10}W = -0.556 \times 10^{-5}L^{3.108}$. Comparisons of weight-length relationships between diploid and triploid grass carp fingerlings were made by Cassani and Caton.[281] They found that diploid fishes had significantly higher condition factors (1.32 ± 0.16 vs. 1.19 ± 0.17), which was supported by weight-length formulas (triploid: $\log_{10}W = -5.311 + 3.205 \log_{10}L$, diploid: $\log_{10}W = -5.827 + 3.468 \log_{10}L$). Shireman[282] reported weight-length relationships for grass carp of 29 to 252 mm total length (TL) in Lake Wales, Florida, to be $\log_{10}W = -4.916 = 3.02 \log_{10}L$. For larger fishes in Lake Baldwin, Florida, Maceina and Shireman[168] reported formulas of $\log_{10}W = -4.821 + 3.005 \log_{10}L$ for fishes 450 to 700 mm TL, and $\log_{10}W = -5.239 + 3.127 \log_{10}L$ for fishes 700 to 1111 mm TL. For fishes over 650 mm TL females were significantly heavier than males at the same length. In a northern U.S. lake Mitzner[283] calculated a relationship of $\log_{10}W = -3.484 + 2.477 \log_{10}L$ for fishes 60 to 75 cm TL. The low exponent (2.477) indicated that the northern fishes weighed less for their size than the southern fishes. This resulted because measurements were taken in the winter. Shireman and Hoyer[284] reported condition factors K (TL) for Lake Baldwin grass carp after vegetation was unavailable. Between 1978 and 1980, when hydrilla was abundant, K values ranged from 1.20 to 1.50, but by 1983, 1.5 years after vegetation disappeared, K (TL) averaged 0.97.

Growth rates for grass carp were reported by Gorbach[147] for fishes living in the Amur River, which is their native habitat. He found that fishes increased in length 9 to 10 cm annually during the first 4 or 5 years. Growth rates began to decline in subsequent years, reaching 6 to 7 cm in years 6 and 7, and 2.5 cm after year 8. Weight increased with age, especially from years 5 to 7. During year 5 growth averaged 1.8 g/d and increased to a maximum of 7.4 g/d in year 8, then declined rapidly in years 9 and 10 (2.8 and 3.2 g/d, respectively). Gorbach reported no difference between males and females in growth. Growth in more southern latitudes is more rapid. For example, Shireman et al.[274] and Shireman and Maceina[204] reported that the growth of grass carp having an unlimited food supply in two central Florida lakes ranged from 10.1 to 10.4 g/d from 1 to 4 years after stocking. Shireman and Hoyer[284] reported, however, that after vegetation was removed from one of the lakes growth declined, as indicated by K (TL) values. Other authors have reported that growth in subtropical and tropical lakes ranged from 10 to 22 g/d (Table 19). Growth rates vary considerably according to different environmental conditions, but are usually less under culture conditions (Table 19); therefore, in order to make definite

Table 19 **Growth of grass carp in different countries under varying conditions**[11]

Country and culture conditions	Size	Growth (g/d)	Age or time period	Ref.
China, high density polyculture	30–100 g	0.08–0.27	1 year	308
	280–300 g	0.38–0.41	2 years	
	1.8–2.4 kg	1.64–2.19	3 years	
China, pond culture	25.5 cm, 0.68 kg	1.86	1 year	708
	1.8–2.3 kg	1.64–3.15	2 years	
	4.5 kg	3.08	4 years	
Fiji, pond culture	6.5 g		—[a]	709
	1341 g	5.50	243 d	
	2174 g	5.96	365 d	
	3069 g	6.31	486 d	
India, pond culture (Cuttack)	17.2–20.5 cm		4.5 months	708
	280 g	0.93	9.5 months	
	60 cm, 2.7 kg	4.74	19 months	
India, polyculture (Cuttack)	16–37 g		—[a]	710
	1.3–2.6 kg	3.52–7.02	1 year	
India, pond culture (Kalyani)	31 g		—[a]	711
	2.53 kg	13.66	6 months	
India, intensive polyculture (Kalyani)	45 g		—[a]	712
	1.59 kg	12.88	4 months	
	45 g		—[a]	
	2.03 kg	13.23	5 months	
	82 g		—[a]	
	5.04 kg	13.58	1 year	
India, pond culture (Karnal)	2.00–2.13 kg	5.48–5.84	1 year	694
India, pond culture (Tamilnadu)	1.5 kg, 50–55 cm	4.11	1 year	694
	4.0 kg, 66.5–70 cm	5.48	2 years	
	7.0 kg, 85–88 cm	6.38	3 years	
	8.0 kg, 90 cm	5.48	4 years	
India, pond culture (Uttar Pradesh)	1.67–1.95 kg	9.13–10.66	6 months	712
Israel, pond culture	113 g		—[a]	713
	2.86 kg	15.26	180 d	
Malaysia, pond culture (Malacca)	3.3 kg	12.36	267 d	714
	4.24 kg	10.27	413 d	
South Africa, pond culture	0.96 kg	2.63	1 year	715
	3.1 kg	4.25	2 years	
	6.4 kg	5.84	3 years	
	9.8 kg	6.71	4 years	
U.S., culture ponds (Alabama)	13–14 cm, 33–46 g	0.41–0.51	80–90 d	716
	0.8 kg		—[a]	
	5.3 kg	5.29	2.4 years	

Table 19 (continued) **Growth of grass carp in different countries under varying conditions**[11]

Country and culture conditions	Size	Growth (g/d)	Age or time period	Ref.
U.S., polyculture	6.0 g		—[a]	717
(Alabama)	1.35 kg	7.34	6 months	
U.S., vegetated lake (Arkansas)	4 g	0.02	6 months	695
	21 g	1.9	9 months	
	372 g	3.9	1 year	
	1271 g	9.9	15 months	
	1816 g	6.0	18 months	
U.S., vegetated lake	3.8 kg, 661 mm	10.2	1 year	274
(Lake Wales, FL)	7.5 kg, 812 mm	10.1	2 years	
	11.2 kg, 900 mm	10.1	3 years	
	15.0 kg, 962 mm	10.4	4 years	
U.S., vegetated lake (Florida)	3 kg, 592 mm	12.1	6 months	204
	5.24 kg, 720 mm	12.3	1 year	
	7.27 kg, 807 mm	11.1	18 months	
	9.17 kg, 876 mm	10.4	2 years	
U.S., vegetated pools (Florida)	40.5 g		—[a]	206, 273
	531.5 g	4.38	112 d	
	55 g		—[a]	
	195 g	1.25	112 d	
	304 g		—[a]	
	740 g	7.79	56 d	
	410 g		—[a]	
	593 g	3.27	56 d	
U.S., vegetated ponds (Florida)	48 mm		—[a]	194
	186 mm		5.5 months	
U.S., vegetated lake (Iowa)	1.82 kg, 47.2 cm		—[a]	283
	4.26 kg, 70.2 cm	16.26	5 months	
U.S.S.R., vegetated lake	148 g, 20.0 cm		—[a]	184
	975 g, 38.5 cm	6.78	122 d	
U.S.S.R., sown rice fields	29 g		—[a]	718
	460 g	5.39	80 d	
U.S.S.R., fallow rice fields	250 g	3.12	80 d	718
U.S.S.R., drainage ditch	230 g	2.88	80 d	718

[a] Age not reported but initializes time period for values following.

From Shireman, J. V. and Smith, C. R., *FAO Fisheries Synopsis*, 135, 86, 1983.

statements about growth rate the environmental and culture conditions must be known. Environmental factors that affect growth include temperature, stocking density, oxygen level, and salinity.

Low temperature affects growth indirectly by causing reduced consumption and directly through reduced metabolism. Faster growth is evident in the tropics and during warmer seasons in temperate regions (Table 19). Colle et al.[194] observed slower growth in fingerlings 48 to 186 mm in a vegetated pond during the winter months. Growth was especially slow below 14°C when consumption fell drastically. Seasonal growth was also reported in New Zealand studies. Fingerlings weighing 6 g grew slowly from April to October, when the temperature was below 14°C, but grew rapidly from October to February.[182] In another New Zealand study Chapman and Coffey[203] reported winter growth rates of 4.8 g/d for 8.3 kg fishes, and summer growth rates of 8.1 and 26 g/d for 3.3 and 10.2 kg fishes, respectively. Other authors have reported that grass carp growth rates increase along with temperature.[98,202,216] Kilambi and Robison[207] found that growth was not different at temperatures of 18.3, 23.9, and 29.4°C, but was significantly less at 12.8°C. It appears that smaller fishes are also more sensitive to changes in temperature than larger fish. In a study conducted in South Florida,[206] a water temperature increase from 23 to 29°C was correlated with a faster growth rate in 0.1 kg fishes, but not in 1 kg fishes. The effects of temperature on growth can be generally summarized by saying that consumption and growth are reduced when the environmental temperature falls below 14°C, and growth is fairly uniform through a temperature range of 18 to 32°C, with best growth occurring between 25 to 32°C.[285]

Fish density and oxygen levels are often interrelated as to their effects on grass carp growth. Stanley[286] reported that feeding stops when oxygen levels reach 2.5 mg/l. In a tank experiment Shireman et al.[209] reported that stocking density did not have an apparent effect on the growth of 2.7-g fingerlings until oxygen levels fell below 4 mg/l, when the consumption rate decreased 45%. Kilambi and Robison[207] reported reduced consumption when the environmental temperature in tanks was below 12.8°C.

Density-dependent growth has been reported by a number of authors. Tank stocking densities of 0.53, 1.06, 1.59, and 2.11 fishes per liter were investigated by Shireman et al.,[209] who found that density did not significantly affect growth until day 35. After 88 d, fingerlings in the lowest density were almost double the size of fishes in the highest density. In a pool experiment Blackburn and Sutton[287] reported that 154 to 159 g grass carp grew 13.5 g/d at 950/ha, 10.9 g/d at 190/ha, and 3.6 g/d at 3800/ha. Cage culture of grass carp has been practiced in The Netherlands in power plant cooling water. Grass carp were stocked at 2500, 12,500, and 500 fishes per 6 m^3 cage. In a vegetated pond study Kilgen and Smitherman[196] reported significant differences in growth when fishes were stocked from 49 to 395 fishes per hectare. In this experiment growth ranged from 0.8 g/d (395 fishes per hectare) to 4.9 g/d (49 fishes per hectare). Shelton et al.,[288] in another pond study in Alabama, reported that first-year growth was density related, and average size was related to stocking density. The stocking densities ranged from 14,000 to 470,000 fishes per hectare. It was more difficult to evaluate second-year growth as grass carp were stocked with other species and densities at draining ranging from 5 to 395/ha. They reported an average growth rate of 11 g/d, with best growth in a pond containing vegetation. Data pertaining to density relationships and growth have been gathered primarily from culture situations. Fry and fingerling grass carp are often stocked in excess of 75,000 fishes per hectare[289,290] and raised in these densities for stocking material. As Van Zon[291] points out, they are seldom raised in monoculture, but are either stocked in polyculture with other Chinese carp or used for weed management purposes in which growth would be optimal until vegetation is removed.

The effects of salinity on growth and survival have been investigated primarily to determine the effects grass carp may have if they entered estuarine areas. Normal fingerling growth was observed by Doroshev[165] at salinities ranging between 7 and 9‰; however, Maceina and Shireman[168] reported significant reductions in fingerling growth at 3 to 6‰ salinity. These data agree with Kilambi,[292] as he also reported less growth than controls at salinities from 3 to 9‰, but growth was not significantly different at these ranges. Mortality usually occurred when salinities reached 10‰.[165,168]

B. SILVER CARP

The first foods of silver carp, like the other larval cyprinids, are small zooplankton. Larvae can feed on phytoplankton, but if sufficient zooplankton are unavailable, growth rate is slowed.[249,293] Opuszynski[162] found that survival and growth of silver carp larvae were influenced by temperature, food quality, and density. Fishes in ponds in which rotifers predominated grew faster than those feeding on crustaceans. In a pond treated to enhance rotifer production larval fishes reached 285 mg vs. 57 mg in the control pond after 31 d. Larvae also showed superior growth and survival rates when grown in heated effluent. Density became a factor at later stages when larvae were large enough to feed on crustaceans as fishes in low densities grew faster.

A feeding experiment was conducted by Prinsloo and Schoonbee[294] to compare live food (rotifers) and a prepared food (EWOS C10 Larvestart). Five hundred 4-day-old larvae were stocked in 150-l tanks and fed for 10 d. Best growth and survival was obtained with the live food diet. In 10 d rotifer-fed fishes grew to 103 mg and EWOS-fed fishes grew to 27.5 mg. Dabrowski[295] reported that fry grew better when fed only a dry diet (23.5 mg) as compared to live zooplankton (15.8 mg) for 15 d, but that slow growth on zooplankton was due to an insufficient supply of zooplankton. The growth and survival of silver carp larvae were greater than other Indian carp when they were stocked together. Silver carp grew 40.08 mg/d for 28 d in small ponds receiving fertilization and supplementary feeding.[296] Konradt[297] reports that the silver carp is a fast-growing fish in the Amur River, reaching weights of >16 kg. In Romania fishes 6 to 9 years old weigh 6 to 8 kg. Growth is slow during the first year, but increases in subsequent years, depending upon available food (Table 20).

In the former U.S.S.R. where silver carp have been stocked, growth is greatest in the southern region (Table 21). Karamchandani and Mishra[298] stated that silver carp grew rapidly in Kulgarhi Reservoir, India, up to 25 months. Growth declined during the next 25 to 50 months when the amount of rotifers declined in the zooplankton. In Germany 3509 silver carp grew rapidly when stocked into 10 to 50 eutrophic lakes (stocking rate = 180). Seven years after stocking they reached a weight of 5.5 kg.[299] Growth rates are faster in Israeli lakes and reservoirs. Silver carp stocked in Tsalmon Reservoir grew 5 kg in 4 years[300] and fishes in Lake Kinneret, Israel, reached 5 to 6 kg in their second year and 6 to 10 kg at year 3.[301]

Silver carp growth rates appear to be influenced primarily by temperature and food availability. A study conducted by Li et al.[302] indicates that genetics is also an important factor. They studied four populations of silver carp and found that both wild and hatchery populations from the Changjiang River gained about 10% more body weight than those from the Zhujiang River. Wild Changjiang River fishes grew 3 to 5% more than hatchery stock from both sources. They recommend therefore, that broodfish procurement should be reevaluated.

The silver carp has been used in polyculture situations. Opuszynski[229] stated that although bighead grew faster than silver carp in common carp ponds, silver carp were more suited as a species for carp ponds due to the low level of food competition.

The growth rates of silver carp from polyculture situations are difficult to compare due to the conditions of each experiment. In two cages in which silver carp were stocked with

Table 20 **Weight increase[9] of silver carp from different geographical areas**

Location	Age group 0	I	II
China	12–22	500–1500	2000–3000
Moscow area	3–5	120–800	417–1205
Turkmenskaya	21–55	1100–1850	3290–4250
Romania	164	2030	2500–4000
Hungary	57–66		

From Woynarovich, E., *FAO Fisheries Report No. 5,* 162, 1968.

Table 21 **Rate of increase of silver carp in different bodies of water, former U.S.S.R.**

Body of water	Age 1+	2+	3+	4+	5+	6+
North, Kuybyshev Reservoir			1.6			
North, Volgograd Reservoir	0.11	0.5	0.9			
North, Lower Dnieper		1.5				
North, Tsimlyansk Reservoir	0.38	1.4	2.3	3.7		
North, Proletarskoye Reservoir	0.3	1.4	1.7	4.6	5.2	6.1
North, Shendzhiy Reservoir		1.5	2.6	5.7	6.0	
	1.2	2.6				
South, Khauzkhan Reservoir	1.0		6.4			
	3.1		10.1	13.0		

From Negonovskaya, I. T., *J. Ichthyol.,* 20, 101, 1980. With permission.

Mozambique tilapia and suspended in human sewage and pig feedlot waste treatment systems, silver carp (17/m^3) grew from 15 to 260 g in 190 d (1.29 g/d).[303]

In Israel silver carp are stocked at 10,000 fishes per hectare and grown for 1 year, reaching a size of approximately 100 g. These fishes are grown a second year, and stocked with other fishes at a stocking rate of 800 to 1000 fishes per hectare. At this stocking rate they reach a weight of 1.5 to 2 kg (3.8 to 5.29 g/d). When silver carp are stocked at 300 to 500 fishes per hectare they reach 2.5 kg (6.6 g/d) and do not influence the growth of the other species.[276,277]

Moav et al.[304] tested nine combinations of stocking densities. Silver carp were stocked with common carp, blue tilapia, and grass carp. In 1974 total stocking combinations ranged from 19,700 to 6130 fishes per hectare. Silver carp growth was 9.0, 8.2, and 8.7 g/d at the lower stocking densities (1250 silver carp) and 6.1 and 6.5 g/d at the higher densities (2500 silver carp). They concluded that silver carp responded to crowding, but not to food availability. Other authors reported growth rates for silver carp grown in polycultures, but growth rates were less than those reported by Moav et al.[304] In U.S. ponds supplemented with pig wastes, silver carp grew 3.97 g/d.[305] Dimitrov[306] reported growth rates of 3.5 g/d in Hungary and in a later study[307] reported slightly higher growth rates of 4.86 and 4.11 g/d.

C. BIGHEAD CARP

The first foods of bighead carp larvae are small zooplankton, but in experiments larvae have been fed a number of different foods. In a 14-d feeding experiment Opuszynski et al.,[270] starting with 3-day-old larvae, determined growth rates when larvae were fed zooplankton, a combination of zooplankton and EWOS C10 Larvestart, two diets prepared by Polish scientists, and a trout starter diet made commercially in Poland. The diets were evaluated by average fish growth if the survival rate for a diet was ≥50%. Fishes grew best when fed zooplankton and poorly when fed EWOS C10 only. Satisfactory growth was obtained when fishes were fed a mixture of zooplankton and dry food. The data of Opuszynski et al. supported data presented by Dabrowski,[295] who found zooplankton to be superior to an artificial compound diet after 15 d. Rottmann et al.[271] compared the growth rates of bighead carp when fed freshwater rotifers, nematodes, brine shrimp nauplii, and two commercial diets (EWOS C10 Larvestart and Fry Feed Kyowa). Larvae were stocked in a 76-l aquarium at 57 and 28 fishes per liter in two experiments. Freshwater rotifers proved to be a significantly better food than the other diets. Fishes fed the Kyowa diet were significantly larger than those fed brine shrimp, nematodes, and EWOS Larvestart. The EWOS Larvestart diet fishes were significantly smaller than fishes grown on the other diets.

Most authors agree that larval bighead carp grow better when fed zooplankton alone or in combination with prepared foods. The results obtained with prepared foods may have occurred because bighead larvae and fry were fed diets containing improper amounts of protein. Santiago and Reyes[307] found that the growth of 3.8 g bighead carp fry, when fed isocaloric diets (290 kcal digestible energy per 100 g), was best when protein levels were between 20 to 30%. Growth decreased when protein levels were increased past 30%. These protein requirements are low when compared to other carp fry; therefore, diets containing protein levels above 30% may inhibit growth. It is likely, however, that when larval cyprinids are shown to grow best on commercial diets, the zooplankton diet fed to these fishes was probably inadequate.[270]

Bighead carp generally reach 0.75 to 1.5 kg by age 2 and 3 to 4 kg in their third year.[308] Baltadgi[309] reported a maximum size of 40 kg for a 9-year-old bighead in the Ukraine. Growth rates, however, were shown to vary according to location in the former U.S.S.R. Vinogradov[135] reported that 2-year-old bighead reached 500 to 600 g in the middle zones and 1 to 1.5 kg in the southern zones. Older fishes (5 years and older) reared in Kuban estuaries reached 15 to 16 kg. Bighead grew 2.5 kg/year in ponds in the Ukraine, but grew faster in power plant reservoirs (6 kg/year),[309] indicating the positive influence of temperature. Negonovskaya[310] reported that bighead attained 21.0 kg by age 5+ in cooling ponds. These fishes grew 57.54 g/d, which is very rapid growth.

Growth rates in Lake Dgal Wielki, Poland, ranged from 3.66 to 6.31 g/d from ages 4 to 7. These fishes were stocked at age 3, prior to which time they were grown in ponds in which growth rates ranged from 0.1 to 2.33 g/d.[311] Leventer[300] reported that bighead carp grew 10 to 20% faster than silver carp in Tsalmon Reservoir, Israel. Growth after a few years was estimated to be 5.5 to 6 kg (3.8 to 4.1 g/d).

Several authors reported that bighead grow well in culture systems and outperform both grass carp and silver carp.[229,312,313] Results from culture systems, however, are difficult to compare due not only to different climates, culture conditions, species of fishes stocked, and stocking rates, but also because of genetic differences. Experiments in China, for example, indicate that bighead carp taken from two river systems and grown under identical conditions in ponds exhibited significantly different growth rates.[314]

These authors recommended that selective breeding programs should be carried out to select broodfishes for faster growth.

Most growth data reported for bighead carp come from polyculture situations, but in cases in which bighead are stocked alone, stocking density can cause reduced growth of fry. Tal and Ziv[276,277] found that stocking density in Israeli ponds affected growth. Bighead carp fry stocked at 10,086 fishes per hectare gained an average 1.68 g/d after 120 d and 1-g fry stocked at 570 fishes per hectare gained 9.3 g/d after 126 d. In polyculture systems growth was also faster at lower stocking densities. At densities of 300 to 500/ha bighead carp gained 10 to 17 g/d and at 100 to 2910 fishes per hectare they gained 3 to 10 g/d. Green and Smitherman[315] used two stocking rates for 0.9 g fry: 49,400 and 98,800 fishes per hectare. The weight of fingerlings averaged 15.2 and 13.3 g, respectively, after 60 d. These differences, however, were not statistically significant.

Bighead carp fingerlings have been cultured in pens with other species.[285] In 1985 bighead were stocked with common carp, crucian carp, silver carp, grass carp, and blue tilapia. The maximum daily growth increment for bighead carp during 1 month was 6.6 g/d, which was greater than for the other species stocked. They recommend that bighead should comprise from 7 to 10% of the population. Cremer and Smitherman[46] reported on cage culturing bighead carp fingerlings (13.2 g) in U.S. ponds. Best growth was obtained for free swimming fishes in a fertilized pond with supplemental feed (3.6 g/d). Growth was considerably less in cages, but best in cages in which fishes were receiving supplemental feed.

Several authors have reported using bighead carp in polyculture with other species in systems enriched with animal effluents.[305,316,317] In most cases bighead carp comprised a minor portion of the population. For example, Maddox et al.[316] stocked 500 silver carp and only 50 bighead carp, and reported that bighead carp weight increased 115% in 52 d.

Chapter 4

Culture of Chinese Carp

I. ARTIFICIAL PROPAGATION

Eggs and larvae of Chinese carp have been collected from the great rivers in China and grown as food fishes for more than 2000 years. The development of Chinese fish culture and the successful spread of Chinese carp throughout the world became possible only after methods for artificial propagation were established. Artificial propagation techniques include capture or raising of broodfish, inducing ovulation, and fertilizing and incubating eggs. These techniques were first used in China and the former U.S.S.R. in the early 1960s and were further developed, modified, and adapted to local conditions in different parts of the world.

This chapter discusses the biological bases for artificial propagation techniques and the latest developments in the field. Technical details can be found in hatchery manuals that are available in English.[11,126,139,318,319]

A. HORMONE-INDUCED MATURATION

1. Reproductive Control Mechanisms

The physiological mechanisms that control fish reproduction processes are relatively well known. Because these processes involve both the nervous and endocrine systems, the mechanisms are singly called the neurohormonal mechanism of fish reproduction. Environmental stimuli of reproductive importance (see Chapter 3) or genetically imprinted internal cycles stimulate a portion of the brain called the hypothalamus (Figure 28). The hypothalamus produces gonadotropin-releasing hormone (GnRH) and gonadotropin release-inhibiting factor (GRIF) (dopamine). Gonadotropin-releasing hormone stimulates the pituitary (hypophysis), a small gland located beneath the brain, to produce and release gonadotropic hormones (GtH), which target the gonads (ovaries and testes). Elevated blood levels of GtH cause final maturation of the sex products through local action of the steroid hormones (progesterone stimulates final maturation of the eggs and testosterone stimulates spermiation). Prostaglandins act as local ovarian mediators of the ovulatory action of GtH. They play a role in the final stages of ovulation, including rupture of the follicle and expulsion of the mature oocyte.

The internal mechanisms that regulate spawning are similar for most fishes. The external environmental regulators are complex, vary considerably among fish species living in different environments, and are difficult, if not nearly impossible, to simulate under hatchery conditions. For these reasons, hormone-induced spawning is often the only reliable method for reproducing fishes in captivity or outside their natural range of occurrence. The successful worldwide introduction of Chinese carp would not have been possible without development of hormone-induced spawning techniques.

The hormones and other substances used for induced spawning influence the internal mechanism of fish reproduction at several levels by either promoting or inhibiting the processes (Figure 28). Hormones used for these purposes include pituitary extracts, purified gonadotropins, and luteinizing hormone-releasing hormone analogues (LHRH-a), injected alone or together with dopamine blockers.

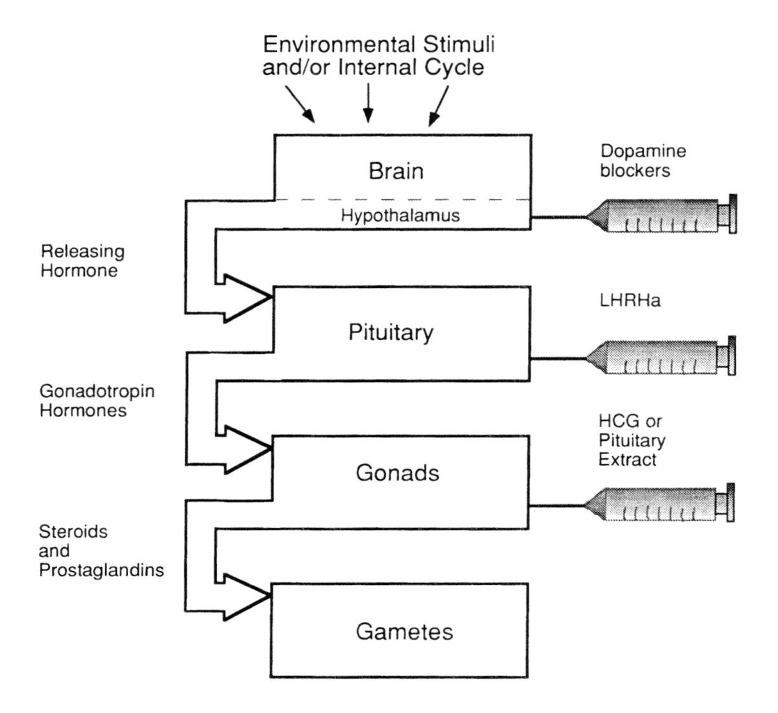

Figure 28 Neurohormonal mechanism that regulates reproduction in fishes (right side) and mode of influence by hormone-induced spawning techniques (left side). LHRH = Luteinizing hormone-releasing hormone, HCG = human chorionic gonadotropin.

2. Hormones for Induced Spawning

The first material used for hormone-induced spawning of Chinese carp was pituitary glands collected from fishes of the species to be spawned. This method underwent numerous modifications before being used worldwide. First, pituitary GtH was found to not be species specific and pituitaries collected from one species could be used to ovulate other species. Common carp pituitary glands are the most commonly used, but pituitaries from other fishes, even phylogenetically remote species, such as salmon, tilapia, catfish, and some marine species, are used, depending on local availability. For example, the pituitary from the sharptooth catfish *(Clarias gariepinus),* which is generally available in South Africa, is suitable for grass carp spawning under local conditions.[320] Pituitary glands can be used fresh or they can be dried in acetone and stored in sealed vials, preserving their activity for several years.[321] Powdered pituitary materials are also available.

Hormones must be administered in proper amounts. If the dose of hypophysis given is to be determined precisely, not only the weight of the fish, but also the size of its gonads must be known. The more eggs a female has, the greater the amount of hypophysis required for ovulation. Gonad size or the number of eggs is estimated by measuring fish body girth. In Europe the dose of hypophysis (per kilogram of fish body weight) is determined from empirically developed nomograms of the dependence of pituitary dose on body girth.[9,318] To calculate the total dose, the nomogram value is multiplied by fish body weight. It is probably not necessary to be this precise if the potency of the pituitary material used is not known. Potency depends mainly on the maturity of the donor (the hormone content in the pituitary is greatest just prior to spawning and the lowest in immature fishes and fishes after spawning) and on the methods of hypophysis collection, processing, and storage.

In practice, the amount of pituitary used is determined by the success rate of the initial spawning attempts. The amount of pituitary used in subsequent spawnings can then be adjusted. The pituitary dose varies within 3 to 6 mg/kg of the female's weight and is usually administered in two injections (0.1 and 0.9 of the total dose), 12 to 24 h apart. In countries in which donor fishes are readily available or commercial hormonal preparations are expensive, or both, this method is still widely used with good results: e.g., the Commonwealth of Independent States, eastern Europe, China, Malaysia,[322] Sri Lanka,[323] and Syria.[324]

Purified human chorionic gonadotropin (HCG) is used to replace GtH produced by the pituitary. Human chorionic gonadotropin is used to spawn fishes because it is reasonably priced, and the concentration of the hormone is known. The concentration is measured by the biological activity of the hormone and is expressed in international units (IU). Human chorionic gonadotropin does not seem to work in all cases, however, and is not equally good for all Chinese carp. Generally, the preovulation phase can be obtained easily by administering HCG, but it is difficult to achieve ovulation when broodfishes are not in good condition or not fully mature prior to injection or both. The dose varies, but usually ranges between 800 to 1000 IU/kg of body weight. Human chorionic gonadotropin is administered in two doses at 10-h intervals. To enhance its effect, HCG is sometimes used in conjunction with fish pituitary GtH.

Human chorionic gonadotropin did not induce ovulation in grass carp when used alone;[319] however, fishes responded favorably to two injections of HCG (400 and 1600 IU/kg, respectively) followed by a resolving dose of carp pituitary (6 mg/kg). Injections were administered 12 to 24 h apart. There are reports on using HCG alone to spawn grass carp in China, but high doses were needed to induce ovulation.[126]

Silver carp have been successfully spawned using HCG. At present, in most hatcheries in the Commonwealth of Independent States, induced spawning of silver carp is carried out via HCG.[325] In India HCG is used alone or in a mixture with fish pituitary.[326] Some studies revealed that HCG with pituitary extract was more effective,[327] but others showed successful spawning when only HCG was administrated in low doses of 300 IU/kg body weight for females and 100 IU/kg body weight for males.[328] For mass breeding in India, HCG proved to be twice as inexpensive as pituitary gland.[329] In China higher doses were used, amounting to 800 to 900 IU/kg for silver carp and 500 to 2200 IU/kg for bighead carp.[126] Human chorionic gonadotropin at a dose of 1800 IU/kg was found to be ineffective in inducing female bighead carp to spawn in the Philippines.[330]

Recently, synthetic LHRH-a (or GnRH-a) have been used for induced spawning of Chinese carp. Luteinizing hormone-releasing hormone analogues stimulate the fish's pituitary to produce the GtH necessary for ovulation or spermiation, rather than circumventing the endocrine system with pituitary extract from donor fishes or GtH (e.g., HCG) derived from other species. Luteinizing hormone-releasing hormone analogues proved to be very effective; they are easier to use and less costly (in some countries) than other hormone preparations. Rottmann and Shireman[331] reported that after a single injection of LHRH-a (10 μg/kg body weight) 92% of grass carp broodfishes successfully ovulated in Florida (U.S.). Higher doses were used to spawn grass carp in other countries (e.g., reaching 200 μm/kg in Czechoslovakia[332]), or LHRH-a was administered in two injections, or both.[323] Two injections of LHRH-a are used for inducing spawning of silver and bighead carp.[323] In Thailand successful spawning of these fishes was reported after injections of 5 and 15 μg/kg given 18 to 20 h apart.[333]

Luteinizing hormone-releasing hormone analogue action can be improved by combined injection with a dopamine antagonist (blocker). Dopamine inhibits the release of GtH from the pituitary, in many cases effectively blocking the pituitary's response to injected LHRH-a. There is a family of drugs (e.g., reserpine, pimozide, domperidone, and haloperidol) that acts as dopamine blockers, either by preventing the release or by inhibiting the synthesis of dopamine.

In combination with pimozide, LHRH-a was used to spawn silver carp in China.[334] In the Philippines, domperidone in combination with LHRH-a lowered the cost of induced spawning of bighead carp.[330] In Poland combinations of pimozide and LHRH-a (50 μg/kg LHRH-a + 10 mg/kg pimozide) allowed induced spawning of grass carp after a single injection.[335]

Milt usually can be stripped from male broodfishes without hormonal inducement. In most cases, however, males are also injected with a fraction of the dose given to females to enhance milt production.

3. Coordination of Hormone Application with Fish Maturity

If broodfish ovaries have not reached a certain stage of development, injected hormones will not produce ovulation, or if eggs are ovulated they will be of low quality. Negative results will also be obtained when eggs are overmature. Therefore, if induced spawning is to be successful, it must be performed when the ovaries of broodfishes are in the preovulation stage.

A general indicator of fish readiness to spawn is the number of degree-days (°D) to which a fish has been exposed during the prespawning period (see Chapter 3). By appropriate temperature manipulation, this stage of readiness can be achieved during the summer as well as during the winter period.[336,337] In The Netherlands, grass carp broodfishes are normally collected from outdoor ponds in the beginning of July, when water temperatures vary between 18 and 21°C. They are transferred into hatchery tanks in which the temperature is gradually increased by 1°C/d, up to 23°C. By that time, depending on the season, fishes are exposed to a total of 1590 to 2128°D. The fishes are injected with pituitary and successfully spawned. The same results were achieved when the fishes were collected at the beginning of January when water temperatures were 3 to 5°C. These fishes were transferred to tanks in which water temperature was increased by 1°C/d, up to 23°C, and were maintained at this level for 20 to 30 d. Induced spawning took place after the fishes had been exposed to the total heat sum of 633 to 968°C. Huisman[336] concluded that the amount of heat needed for fish maturation was not a constant value, but varied depending on starting temperature levels.

Egg development is a long and complex process in which several stages (usually seven) can be distinguished.[139] During the preovulatory stage the process of vitellogenesis is completed, the yolk is synthesized, and accumulated in the egg. The nucleus migrates from the center of the egg toward the micropyle (the opening through which the sperm enters the egg), and the egg absorbs fluids and swells in size (Figure 29). The eggs are approximately 1 mm larger in diameter than immature eggs, and colored in various shades of brown or green. Eggs that are immature or being reabsorbed appear whitish.

When the eggs are in the preovulatory stage, the time is right for hormone injection. Due to the injection, the gonadotropin level in the blood rapidly increases and stimulates egg maturation. During this process, the nuclear membrane disappears, and the first meiotic division takes place; the chromosome number in the egg is reduced to $2n$ ($4n$ in immature eggs) as one half of the chromosomes are discharged with the first polar body. Simultaneously, the follicle is dissolved enzymatically, and the ripe egg falls into the cavity of the ovary and is ready for fertilization. This process in Chinese carp is similar to that in other fish species.[141]

To determine the maturity stage, a small sample of eggs can be taken from the ovary by inserting a tube into the oviduct through the genital opening[319] or by piercing the body wall with the probe.[338] The first method seems less traumatic for the fish and more convenient. However, if the flexible tube is not inserted slowly and carefully, it may injure the oviduct and posterior part of the ovary, making it difficult to obtain ovulated eggs. Piercing the body wall is accomplished very quickly and takes only 2 to 3 s to take a sample. If the operation is performed skillfully, all females will survive. The small wound heals quickly and is visible only as a small pink spot after 2 to 3 d.

B. EGG FERTILIZATION

Two methods are used to fertilize ovulated eggs. The first method consists of placing broodfishes of both sexes in a spawning tank or pond following hormonal injection(s). The fishes are allowed to spawn, and the eggs are fertilized naturally. The second method consists of collecting the spawn by hand stripping and mixing the eggs with milt. Both methods have their advantages and disadvantages which are described below.

1. Natural Fertilization

Tank or pond spawning after injection of Chinese carp allows fishes to spawn when they are physiologically ready, which favorably influences egg quality. Additionally, handling of broodfishes is reduced, thereby minimizing mortality and manpower needs.

Injected broodfishes are released into a spawning tank or pond in appropriately paired sets. Generally, a set is made of one female plus two to four males, and the number of sets used depends on the size of the spawning facility. In Florida (U.S.) successful spawnings were reported when three males were placed with one female in circular fiberglass tanks 1.83 m in diameter.[339] In Japan 15 to 30 males together with 10 to 20 females are spawned simultaneously in a pond 100 to 200 m^2 in area and 1 to 1.5 m deep.[116] Concrete oval ponds constructed with egg collecting chambers are used in China (Figure 30). Mass scale breeding of Chinese carp is performed in Syria[324] and India[329] in cement tanks. In India large-scale breeding is also done in *bundhs,* which are specially constructed ponds in which river-like conditions are simulated.[326]

Ovulation and egg fertilization generally occur within 6 to 12h after hormonal injection. With higher water temperatures, spawning occurs faster. Optimum water temperatures for spawning range from 22 to 26°C. Water flow through the spawning tanks or ponds is crucial. Water flow helps to stimulate the fishes to spawn, and the semibuoyant fertilized eggs are carried through the drain to an egg collector, which is usually made from fine mesh netting.

The main disadvantage of the natural spawning method is that the eggs must be dipped from the collector and transferred to the hatching apparatus. The natural spawning technique also requires more space and water than the hand stripping method and is, therefore, often impractical in large-scale operations. Such operations require that spawning be carried out indoors under temperature-controlled conditions. The efficacy of natural spawning vs. hand stripping is also lower because some females do not release all eggs when spawning. Finally, the natural spawning technique cannot be used for genetic work or the production of sterile fishes.

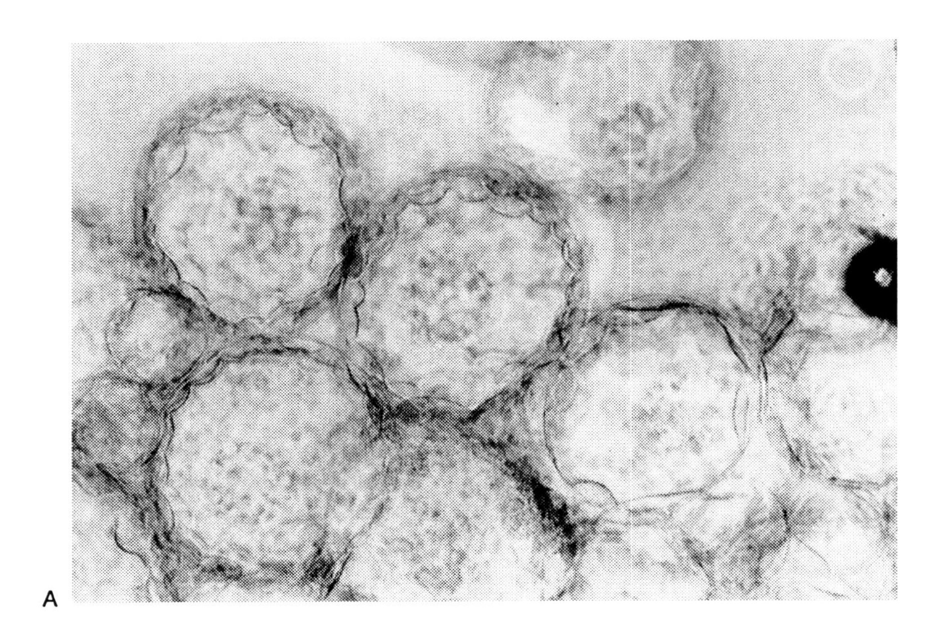

A

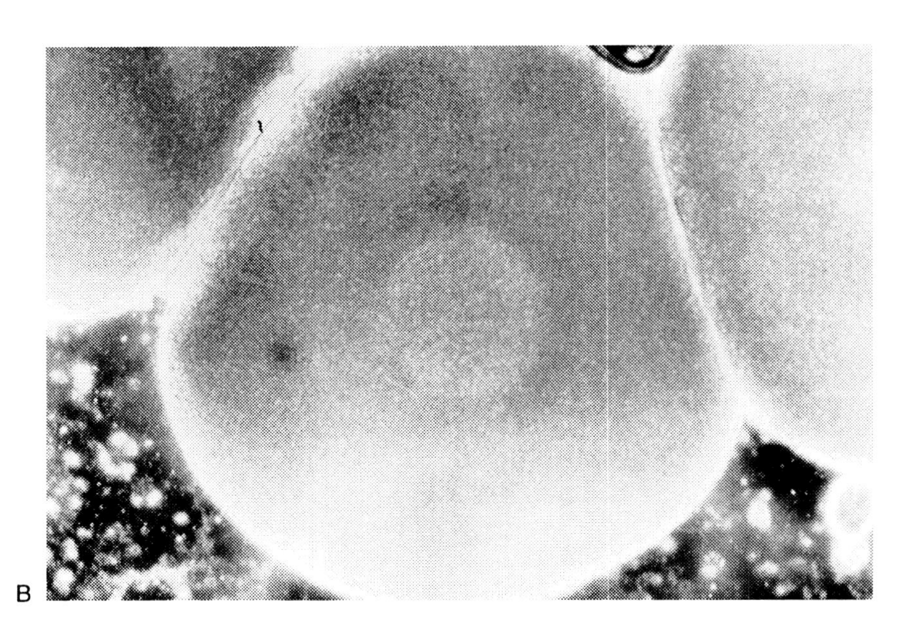

B

Figure 29 Eggs in different developmental stages. (A) Sample of immature eggs from the ovary, (B) eggs prior to nuclear migration, (C) egg with the nucleus moved toward the micropyle, and (D) egg that has begun to break down in the ovary.

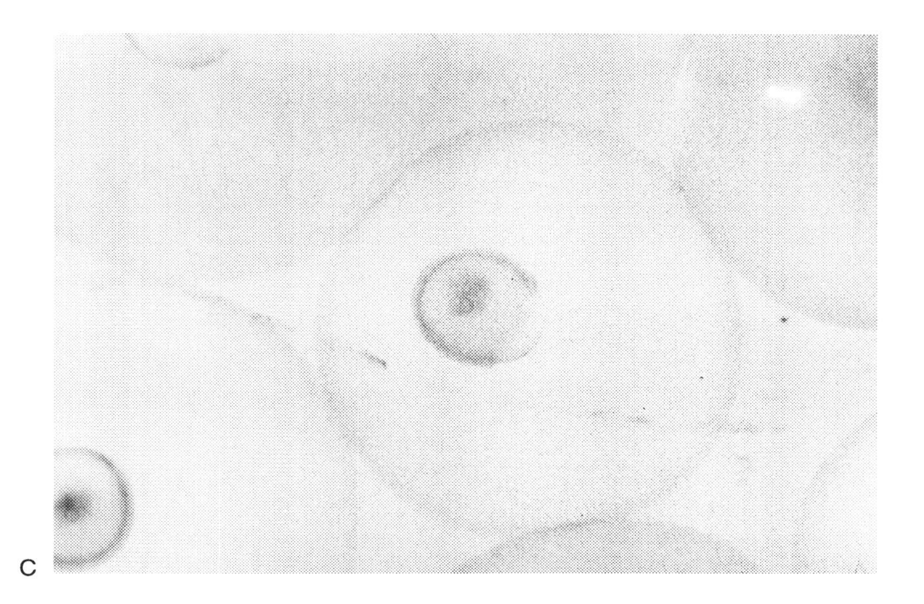

C

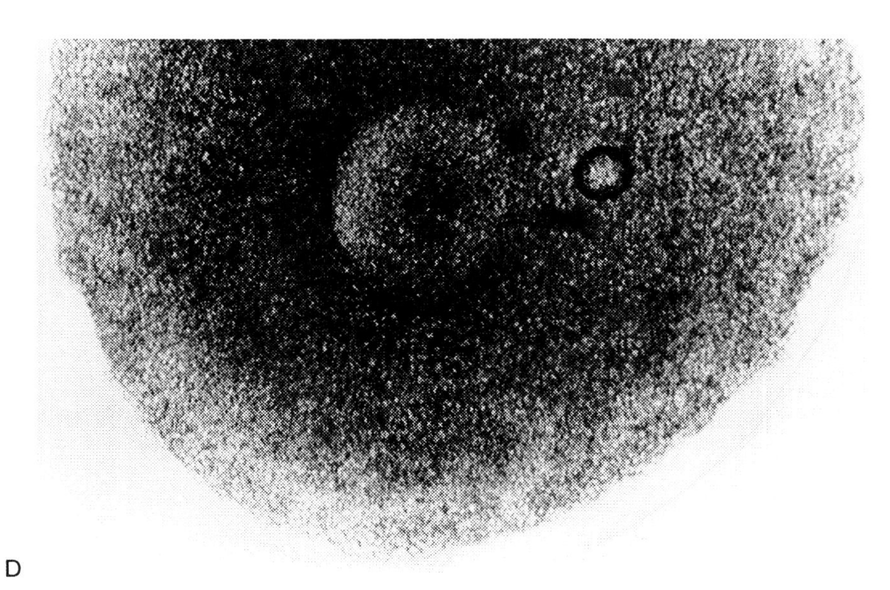

D

Figure 29 continued.

2. Artificial Fertilization

When hand stripping is practiced, it is important to determine the exact time of ovulation. The ovulated eggs quickly overripen; about 50% become overripe within 30 to 80 min. Ovulating eggs detach from the ovarian tissue and are deprived of oxygen supplied by the

Figure 30 Oval spawning pond construction with an egg-collecting chamber (background) and a concrete egg incubator (foreground), used for artificial propagation of Chinese carps in China. (Photographed by K. Opuszynski.)

blood of the female. Therefore, if the eggs remain in the body cavity, their quality deteriorates. The ovulation time after hormonal injection depends mainly on water temperature and ranges from 7 to 24 h at 20 to 28°C. The relation between water temperature and stripping time appears to be linear in the 20 to 30°C temperature range.[340] When the female approaches the expected ovulation time, she should be checked hourly. The abdomen of the female is gently stroked with the fingertips and if eggs are not released, the fish is put back into the holding tank and examined after another hour. When a female is ready for stripping, she is gently lifted from the water and her belly is dried with a towel. The stream of flowing eggs is directed into a dry container (Figure 31). Slight pressure on the belly assists in emptying the ovaries.

If the eggs are kept dry, they can be fertilized within 30 min without affecting fertilization. If water comes in contact with the eggs before the milt is added, the eggs soon begin to swell, the micropyle closes, and fertilization is impossible. Studies have shown that the hatch of larvae is mainly determined by the quality of eggs and not milt, which is usually of high quality. The quality of the eggs can be determined using biochemical methods or morphological analysis under light or electron microscopes;[341,342] these methods have yet to be used in mass-scale artificial propagation practices as they would be too time consuming and costly. Hatchery workers visually inspect the spawn to determine quality. A good spawn contains little ovarian fluid and is grayish-green to brownish-orange in color. The eggs are uniform in size and the variability coefficient of the ovulated egg diameter is only 4 to 8%.[341] Overripe spawns contain a considerable amount of ovarian fluids, and many of the eggs are white in color.

Milt usually can be stripped from males during the entire spawning season, and the same males can be used repeatedly to fertilize eggs. Sperm quality, however, deteriorates as the season progresses and is related to the number of hormonal injections given the

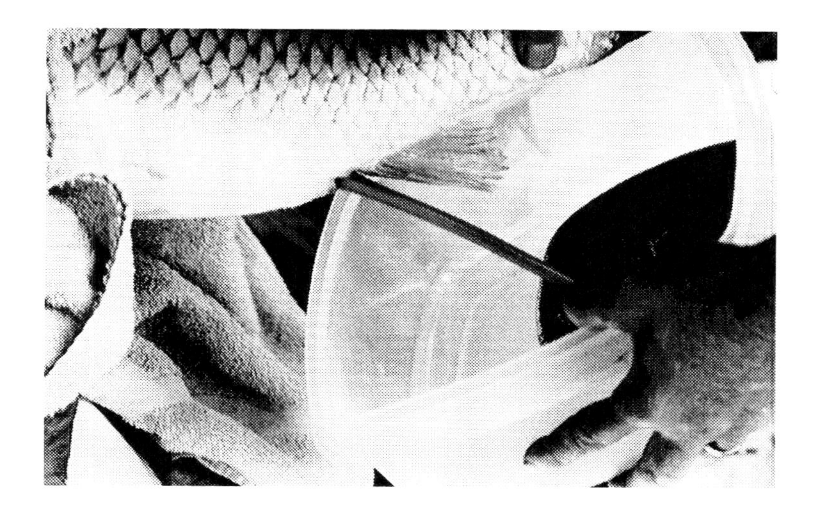

Figure 31 Hand stripping of a grass carp female. The stream of flowing eggs is collected in a dry container. (Photographed by K. Opuszynski.)

male to enhance and continue sperm production. For example, when grass carp sperm quality was compared in the first half (June 9 to July 5) and at the end of the spawning season (July 25), the average time of spermatozoa motility was 92 and 65 s, spermatozoa concentration was 13.6 and 9.6 billion/ml, and the percentages of nonviable spermatozoa were 39 and 48%, respectively. These differences were statistically significant, but if the milt was used in sufficient quantities (2 ml/1000 g of eggs), no visible reduction in fertilization rates was observed.[343]

If the milt is not mixed with water when collected, it can be stored on ice for up to 12 h without lowering the fertilization rate. Trials of milt cryopreservation were also made.[344] Milt of grass carp with various diluents was stored in liquid nitrogen at –196°C. After 1 year, the fertilization rates using this milt were 57 and 99% when fresh milt was used.

Milt from two or three males is usually used to fertilize each spawn to ensure high fertilization rates of the eggs stripped from each female. Only after the milt has been spread over the eggs and thoroughly mixed is water added to activate the sperm. Fertilization takes place immediately. After 1 or 2 min, the fertilized eggs are transferred to the incubating apparatus.

C. EGG INCUBATION

Egg incubation devices must provide continuous water supply to keep the eggs in suspension, to provide oxygen, and to remove metabolic products (carbon dioxide and ammonia). Some incubators allow for removal of the dead eggs, while in others the dead eggs disintegrate, and the remaining portions of the eggs are flushed from the chamber. Good quality water, high in oxygen, and stable temperature are needed for egg incubation. Temperature is the primary factor influencing incubation time. At the optimum temperature of 21° to 25°C hatching occurs after 24 to 28 h. Among other parameters, water hardness is important; 300 to 500 mg/l $CaCO_3$ is recommended for successful hatching of Chinese carp.[345]

D. GENETIC MANIPULATION TO PRODUCE STERILE FISHES

Genetic manipulation of Chinese carp has been performed primarily with grass carp to produce sterile fishes for weed control. Different techniques have been used, including surgical removal of gonads, hybridization, gynogenesis, sex reversal, and triploid induction.

1. Surgical Removal of Gonads

Trials were performed to produce sterile grass carp by surgical removal of the gonads. This method, however, was expensive and time consuming; more importantly, gonad regeneration occurred after surgery. Hence, attention has been focused toward other sterilization methods.

2. Hybridization Between Grass Carp and Bighead Carp

The hybrid resulting from crossing female grass carp ($2n = 48$) and male bighead carp ($2n = 48$) was first reported in Russia[346] and later in Hungary.[347] Hungarian researchers claimed that all their hybrids were triploids ($3n = 72$). Interest in this hybrid occurred later in the U.S., where further studies were conducted. Sutton et al.[348] concluded that both diploids and triploids resulted from the cross of grass carp × bighead carp, and diploids had reduced viability. Allen and Stanley[349] found that diploid, triploid, and tetraploid fishes resulted from this cross. Samples from commercially produced hybrids indicated 82% diploids in 1979 and 76% in 1980, the rest being triploids. In 1981 all the fishes tested were triploids.[350] The high percentage of triploids in 1981 was probably caused by harsh rearing conditions; the weak diploid hybrids died, leaving healthy triploids. The use of hybrid grass carp was abandoned due to problems associated with growth, food habits, and survival.[208,351-354]

3. Gynogenesis and Sex Reversal

Production of monosex fish is another possibility that may be considered to prevent reproduction. If the female is homogametic(XX), as has been found for many fish species, all-female progeny can be produced by gynogenesis. Gynogenesis is the development of sperm-activated eggs without contribution of the male genome. Various radiation and chemical treatments have been used for the genetic inactivation of sperm. Gynogenetic eggs restore diploidy by retention of the second polar body.

Female grass carp and silver carp proved to be homogametic, which made it possible to develop a breeding program for all-female production.[355,356] This technique includes insemination of the eggs with irradiated common carp milt, sex reversal in gynogenetic females, and breeding sex-reversed males (genetic females) with normal females (Figure 32).

These procedures are needed because most gynogenetic embryos are haploids and die before hatching. Less than 1% of eggs yield viable diploids. In contrast, sex-reversed males bred with females produce as many viable offspring as normal males and females. In addition heat shock can increase the yield of gynogenetic diploids. An implant of the androgenic hormone methyltestosterone is effective in many cases to sex reverse genetic females. All-female Chinese carp populations have not found practical application. Stocking these fishes into open waters continues to be a risk due to prior introduction and presence of males.

4. Induction of Triploidy

It was presumed that triploids, having three sets of chromosomes ($3n$), would be sterile because the extra set of chromosomes would lead to the failure of gonad development or the production of aneuploid ($1.5n$) gametes. The union of aneuploid and normal gametes

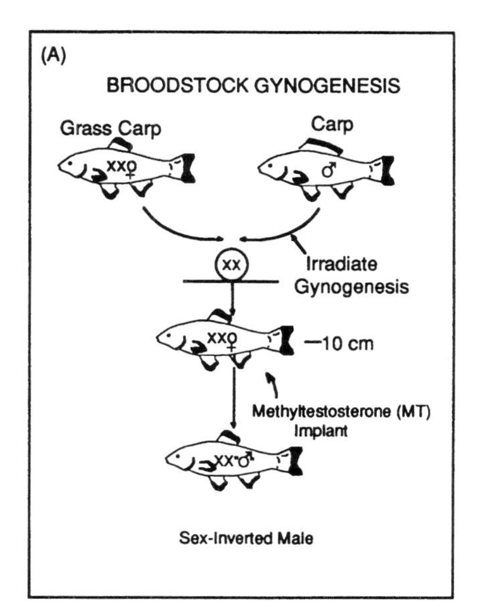

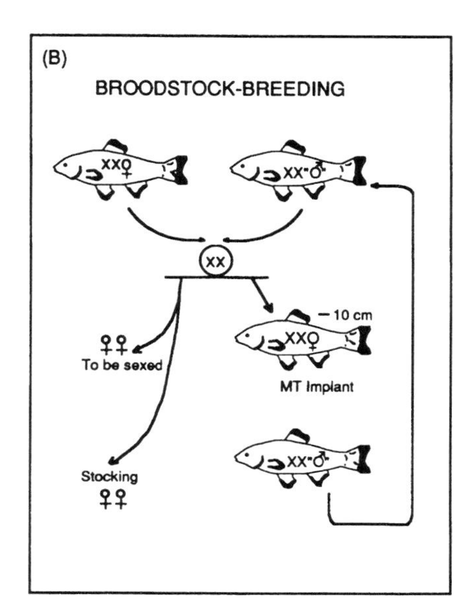

Figure 32 (A) Broodstock production by sex inversion of gynogenetic grass carp female, and (B) progeny of sex-inverted male grass carp and second-generation broodstock production. (From Shelton, W. L., *Aquaculture,* 57, 31, 1986. With permission.)

usually produces inviable offspring. Van Eenennaam et al.[357] reported that some triploid grass carp females were capable of vitellogenesis and incidental ovulation. Triploid males produced sperm in which 60 cells in every billion aneuploid cells contained a true haploid component of chromosomes.[358,359] These haploid cells would differentiate into spermatozoa, which would be capable of fertilizing eggs, resulting in phenotypically normal adults. Although *in vivo* experiments using sperm from triploid grass carp to fertilize eggs of diploid females showed slightly higher survival to 5-month-old juveniles than expected based on the aforementioned theoretical calculations,[357] the probability of triploid grass carp reproduction would be almost nonexistent in natural situations.

The production of grass carp triploids has become the method of choice to obtain sterile grass carp and is widely used by many commercial producers in the U.S. The biological mechanism for triploid induction consists of inhibiting the second meiotic division of the egg. This division takes place when the matured egg is ovulated and the sperm enters the egg. Because one set of chromosomes is ejected from the egg together with the second polar body, the genetic material of the egg is reduced to $1n$. The genetic material of the sperm ($1n$) and the genetic material of the egg ($1n$) then combine to form a diploid ($2n$) embryo. When the second meiotic division is inhibited, two sets of chromosomes are contributed by the egg and one by the sperm which results in a triploid embryo.

Thermal, pressure, and chemical shocks have been used to produce triploid Chinese carp. Thermal shocks consist of exposing the fertilized eggs to an instantaneous temperature increase (heat shock) or decrease (cold shock). Pressure shock involves exposure to high hydrostatic pressure.

Thompson et al.[360] used thermal shock treatments to produce triploid grass carp. Eggs were subjected to a 3.0-min temperature change 18°C below ambient temperature (ambient was 26°C) or a 3.5-min temperature change 12°C above ambient. The best results were obtained when the high temperature treatment was applied 1.0 min after and the low temperature treatment 2.5 min after egg fertilization. The high temperature treatments yielded higher triploidy than the low temperature treatments — up to 87 and 40%, respectively. Other researchers reported somewhat different parameters for heat shock treatments. For example, Cassani and Caton[361] reported studies in which the best shock parameters were 42°C for 1 min duration, starting 4 min after egg fertilization.

Further studies showed the advantage of hydrostatic pressure treatment over heat treatment. Heat treatments generally resulted in somewhat better survival than pressure treatments, but pressure treatments were considerably more consistent with regard to the percentage (>90%) of triploid production. Within the temperature range 22 to 25°C, the optimal parameters for this method are reported to be 7000 to 8000 psi (pounds per square inch) for a duration of 1 to 2 min, starting 4 min after water is added to the eggs and sperm.[361] Increased hydrostatic pressure results in a greater number of triploids, but decreases egg hatch. A pressure of 6000 psi for 1 to 5 min appears to be the minimum level for consistent triploid production of >50% (Figure 33).

Pressure shocks are administered by placing eggs in a steel cylindrical vessel which is closed by a piston fitted with a pressure gauge and relief valve. A hydraulic press is used to apply pressure to the piston (Figure 34). The cost of the equipment needed for pressure treatment is higher than that for heat shock, but the extra cost is offset by the higher efficiency of the pressure treatments. Moreover, because of the higher percentage of triploids, fewer diploid fishes result, which is an important advantage when 100% triploid populations are required.

When the above methods of triploid production are used, every fish must be checked individually to confirm ploidy status. Of the many techniques applied for determining ploidy, use of a Coulter® Counter has become most popular because of the rapid processing time for samples and its ability to immediately separate diploid from triploid fishes. This method consists of determining red blood cell (erythrocyte) nuclear volumes, which are greater in triploid fishes; triploid grass carp have a mean nuclear volume of 14.82 μm^3, whereas the mean diploid nuclear volume is 10.06 μm^3. Fishes must be at least 5 cm long to allow for sufficient blood volume for analysis. Using a Coulter® Counter, a three-person team can evaluate 1600 to 2400 fishes in 8 h.[362]

Flow cytometry is another technique used for determining fish ploidy. This method is based on the evaluation of the amount of fluorescence from cell nuclei stained with a DNA-specific dye. The method is more accurate than the Coulter® Counter method as a result of direct measurement of DNA content within red blood cells. However, it is slower and less convenient because a blood sample cannot be analyzed immediately and the fish must be held individually or tagged until ploidy is determined (the cost of a flow cytometer also is prohibitive). Recent modifications of this technique allow ploidy determination of 250 fishes by a two-person team in a 2-h period.[363] Techniques have been developed to determine ploidy of newly hatched larvae. The entire larval fish is disaggregated, and the resulting single-cell suspension is analyzed using a Coulter® Counter,[364] or cytofluorometry techniques can be used[365] (Figure 35). Using these methods the results of the ploidy treatment can be determined within 24 to 48 h after fertilization. The information gained can be used to modify or repeat treatments as needed. Early ploidy detection is advantageous so producers can stock only high percentage triploid fishes, thus making ultimate use of pond space.

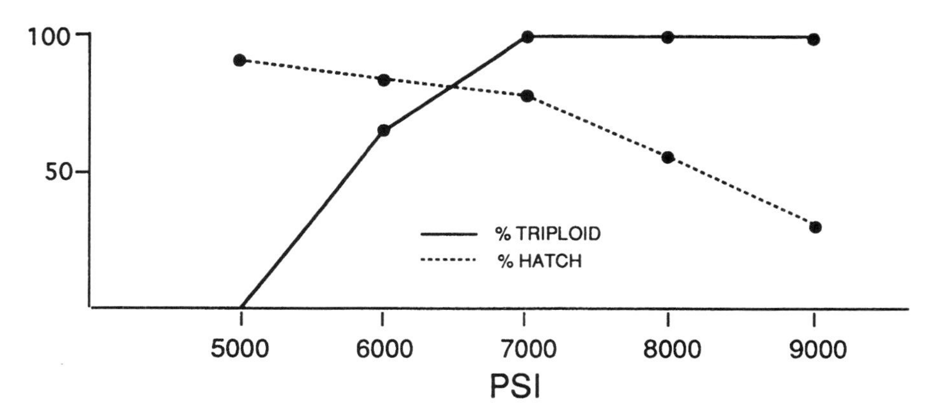

Figure 33 Relative percentage of triploids and percentage hatch from a single hatch of eggs treated 4 min postfertilization for 1-min duration at various pressures. Mean percentage hatch in the control group was 85%. PSI = Pounds per square inch. (From Cassani, J. R. and Caton, W. E., *Aquaculture,* 55, 43, 1986. With permission.)

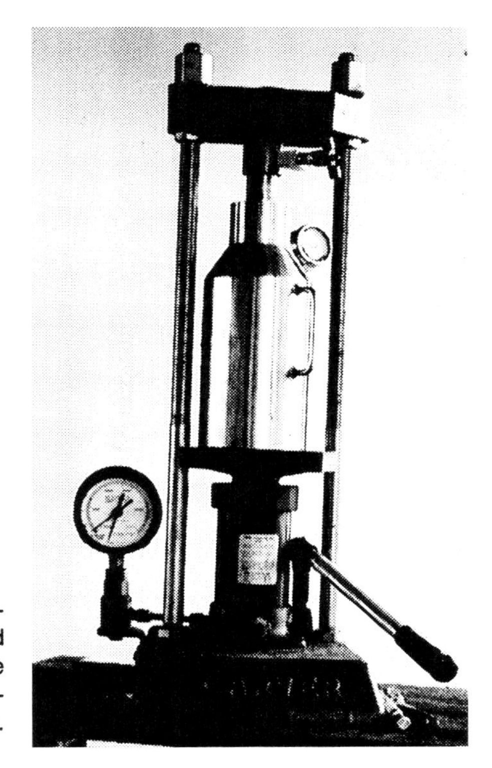

Figure 34 Apparatus for pressure-induced shock treatment. Eggs are placed inside the steel cylinder closed by the piston. The hydraulic press applies pressure to the piston. (Photographed by K. Opuszynski.)

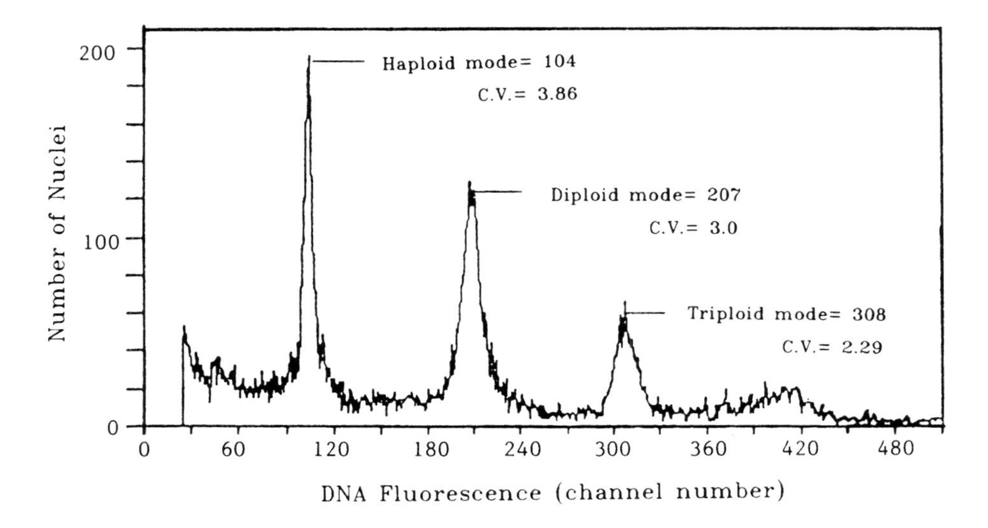

Figure 35 Cytofluorometric analysis of a bighead carp multiembryo homogenate. The histogram displays the distribution of haploid, diploid, and triploid DNA content as measured in relative fluorescence per nucleus. This distribution closely follows the theoretical DNA content distribution for haploids (0.5×), diploids (1.0×), and triploids (1.5×). C.V. = coefficient of variation. (From Aldridge, F. J. et al., *Aquaculture,* 87, 121, 1990. With permission.)

5. Induction of Tetraploidy

An alternative method of producing sterile fishes may result from crossing tetraploid with diploid fishes. Theoretically the offspring produced should be all-triploids, which would circumvent certification of each fish. Although 100% triploidy may not result from this cross (a cross of triploid × diploid rainbow trout resulted in ≥93% triploid progeny[366]), this method may still result in a higher percentage of triploids, with considerably less handling and expense than from the methods currently used.

Tetraploid bighead carp[365] and grass carp[367] larvae were produced using the same methods (heat or hydrostatic pressure shocks). The only difference consisted in employing the shock treatment during a later stage of embryogenesis (Figure 36) in order to produce tetraploids. In this way, instead of inhibition of the second meiotic division, as occurs in induced triploidy, either karyokinesis (chromatid separation) or cytokinesis (cleavage) inhibition takes place. Poor survival of the progeny is a problem with induced tetraploidy. Further trials to refine tetraploid induction techniques are needed.

II. LARVAL CULTURE

Propagation of Chinese carp is possible due to recent progress in artificial propagation techniques. Heavy larval mortality and slow growth rates during initial culture periods, however, are still problems. Therefore, careful planning and preparation before larval culture begins must be completed in order to achieve satisfactory results. The sequence of events that occurs during the culture period is rapid, leaving little time to improvise. During the early larval stages, even a short exposure to adverse conditions can result in high mortality or slow growth. Because larvae are available for only a short period during

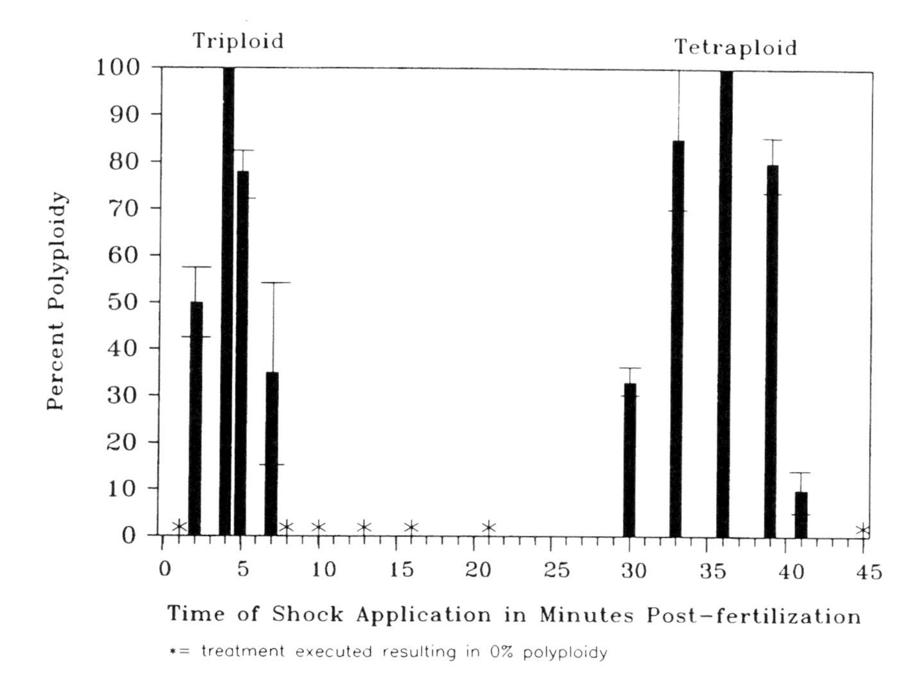

Figure 36 Percentage of bighead carp triploids and tetraploids resulting from hydrostatic pressure shocks of 8000 psi (500 atm) for 90-s duration applied at different times after egg fertilization. (From Aldridge, F. J. et al., *Aquaculture,* 87, 121, 1990. With permission.)

the year, especially in temperate climates, a failure in larval culture may mean that an entire year's effort is lost.

The method of larval culture depends on local climatic, technical, and socioeconomic conditions. The climate determines whether water should be heated and when culture should be started. Different culture strategies can be used, depending on the technology and facilities available. Pond culture is the simplest method from a technological standpoint, whereas tank culture, using recirculating water with fully controlled environmental conditions, is the most complicated. Socioeconomic conditions often determine the type of culture used because of the availability of skilled labor.

Integrated culture methods are worthy of special consideration. In integrated systems the larvae are transferred as they grow from more to less controlled environments, for example, from indoor tanks to cages and finally to ponds. The rationale for doing this is that as they grow, larvae rapidly increase in their resistance to adverse environmental factors. This is especially true as pertains to oxygen deficiency, starvation, live food availability, dry feed utilization, and invertebrate predation control. Even a relatively small increase in fish weight (5 to 6 mg) may substantially improve survival in less controlled environmental conditions.

The management strategy selected depends on the goal and scale of the culture operation. An accessory operation which provides fingerlings for a single fish farm will be smaller than one that specializes solely in fingerling production. The size of the larval culture operation is an important factor in management decisions because not all culture practices can be easily increased when production increases. Currently, the largest larval culture operations raise larvae in ponds.

A. PROBLEMS IN LARVAL CULTURE: PHYSICAL CONDITIONS, FOOD, PREDATION, AND DISEASES

One prerequisite for successful larval culture is a basic knowledge of larval environmental requirements. Important abiotic factors include temperature, oxygen, pH, ammonia, and nitrite; biotic factors include food and predators. Because larval development and growth are rapid during the first weeks after hatching, environmental requirements change during this period. Therefore, the needs of the larvae as a function of growth and development must be precisely specified for all culture periods and then met with the proper management technique.

1. Temperature

Lower (LLT) and upper lethal temperatures (ULT) and optimum temperatures are important concerns for larval culture. Despite detailed studies on the LLTs of larval Chinese carp,[368] it is difficult to determine precisely the lowest temperature that kills fishes because the LLT depends on the age and size of the fish, previous thermal adaptation, and experimental protocol. Additionally, fish responses to low temperatures are complex, and mortality can be delayed.

When temperatures were decreased at a rate of 0.1°C/min, D30 grass carp (D1 is day of hatch) tolerated lower temperatures than did D4 fishes.[368] The increase in tolerance, however, was relatively small (0.01 to 0.07°C/mg body weight). An increase in the acclimation temperature by 3°C caused a 1°C increase in the LLT at both ages. The ultimate lower lethal temperature (ULLT), i.e., the highest LLT possible, was 17°C for Chinese herbivorous carp. The LLT for fishes acclimated to 25°C, which is close to the optimum temperature for reproduction for these species, ranged between 9 and 6°C for D4 and D30 fishes, respectively (Figure 37).

Larval ULTs increased as the acclimation temperature increased up to 34°C. An increase in the acclimation temperature of 3°C resulted in an increase of the lethal temperature of 1°C.[369] The ultimate upper lethal temperature (UULT) is of practical importance for larval culture. The UULT is the lethal temperature that is equal to the acclimation temperature (Figure 37). Knowledge of UULT is important for larval culture because of the relationship that exists between optimum growth temperature (OGT) and UULT:[370] UULT = 0.76 OGT + 13.81.

Once the UULT, which is relatively easy to determine, is calculated, the OGT can be calculated. Ultimate upper lethal temperatures for Chinese carp are similar and range from 43 to 44°C.[369] Consequently, OGTs are 38 to 40°C. Upper lethal temperature changes little from D4 to D50 (0.01° to 0.03°C/d) and can be ignored for all practical purposes.[371]

The OGT is higher than the temperatures that have been used to culture carp larvae; however, few studies have been completed that relate to OGTs. Schlumpberger[372] reported the thermal optimum for growth of silver carp was between 25 and 30°C; Vovk[373] found it was slightly higher, 32°C. Radenko and Alimov[374] grew silver carp larvae at different temperatures and found best growth and survival occurred at 32°C. This temperature was the highest they tested. Opuszynski et al.[371] showed that the weight of silver carp cultured at 35°C tripled in comparison to silver carp that were cultured at 25°C.

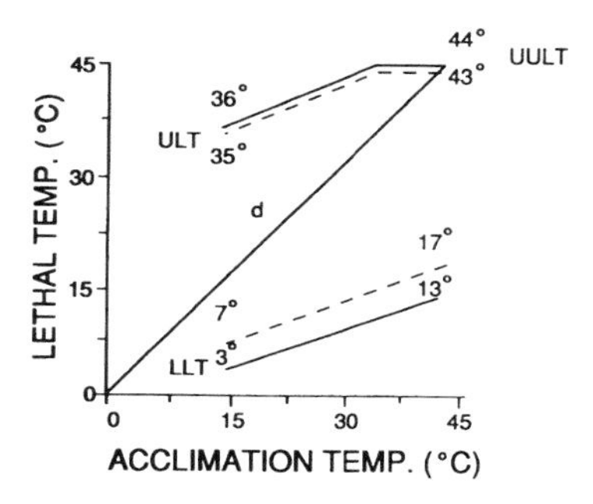

Figure 37 LLT and ULT for 3-d-old (dashed line) and 30-d-old (solid line) grass carp as a function of acclimation temperature. UULT is determined as the point at which the ULT line crosses the diagonal 45° line (d). (From Lirski, A. and Opuszynski, K., *Rocz. Nauk Roln.*, H-101(4), 31, 1988. With permission.)

Raising carp larvae at higher water temperatures than those mentioned above should be tested because the estimated optimum growth temperatures are close to upper lethal temperatures, therefore, accurate temperature control and efficient water aeration would be needed.

2. Oxygen

Lethal oxygen levels (LOLs) for larval Chinese carp are similar to those for common carp larvae.[375] The LOL for D3 silver carp larvae was about 1.8 mg/l, when determined in a 30-min test at 25°C. This value decreased sharply with fish age and leveled off below 0.8 mg/l for fishes that were older than D20 (Figure 38).

No differences in larval carp survival were found when fishes were cultured at 33°C and oxygen saturation was between 23 and 168%.[376] Survival ranged from 89 to 93%; however, significant differences in growth rate were found at different saturation levels. The average weight of fishes after a 14-d culture period was 44 mg at 23% saturation, 64 mg at 43 to 61% saturation, and 80 to 87 mg at 120 to 168% saturation.

3. Nitrogen

Nitrogen gas supersaturation is a real danger in larval carp culture. It occurs when water is heated, especially if heating is combined with changing water pressure caused by pumping. Larvae are extremely sensitive to nitrogen gas supersaturation during the first few days of exogenous feeding; total mortality can occur during a few hours of exposure.[377]

The toxicity of nitrogenous compounds, ammonia and nitrite, may create problems in intensive culture facilities and overfertilized ponds. They are toxic in the nonionized form, and toxicity depends on water temperature and pH. When bighead carp larvae were intensively cultured, no apparent adverse effects were found when ammonia and nitrite ranged from 21 to 45 μg/l nonionized ammonia and 267 μg/l nitrite.[378,307]

4. Natural Food

During the transition period, when larvae weigh between 0.9 and 1.7 mg and transfer from endogenous to exogenous food, they are susceptible to starvation and must quickly find suitable food. This is the critical period in the life cycle. Fishes weighing 5 to 6 mg, however, endure the same starvation period without mortality (Table 22).

Successive changes to larger food items are characteristic for growing larvae (Figure 39). This is important because the fishes optimize energy gains per unit of prey handling time. This follows the optimal foraging model postulated by Krebs.[379] To achieve optimal foraging is to capture and eat the maximum prey size possible.

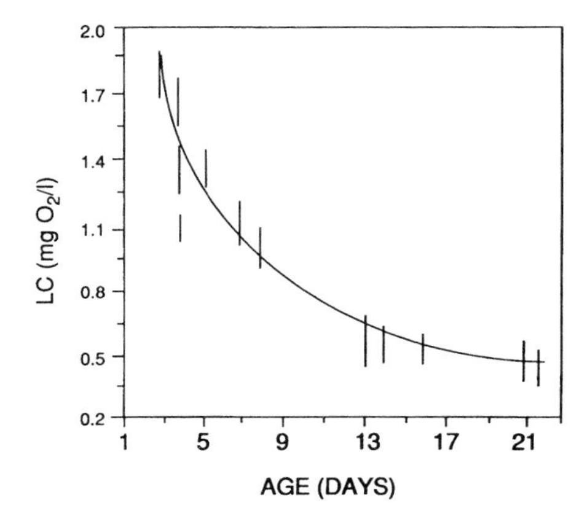

Figure 38 Relationship between age of silver carp and lethal oxygen concentration (LC). LC is an oxygen level at which 50% fish mortality occurred during a 30-min test at 25°C. The vertical bars show standard errors. (From Wozniewski, M. and Opuszynski, K., *Rocz. Nauk Roln.*, H-101(4), 51, 1988. With permission.)

Table 22 **Influence of a 10-d starvation period at 25°C on common, silver, grass, and bighead carp larvae of different weights[495,719]**

	Larvae after yolk resorbtion	
	0.9–1.7 mg	**5–6 mg**
Mortality (%)	50	0
Fish able to commence feeding (%)	50–75	100
Decrease in swimming activity (%)	50–85	0
Loss of weight (%)	35–39	12

During early developmental stages, the maximum size of food consumed is related to mouth size.[254,380] Chinese carp larvae can be rated according to decreasing mouth size in the following order: grass carp, bighead carp, and silver carp. The first foods are rotifers and copepod nauplii.[94,220,381,382] The first foods of grass carp larvae in a pond are rotifers (Figure 11), but they soon switch to copepod nauplii and then copepodites. After 21 d the young fishes feed mainly on cladocera and chironomid larvae.[383] Silver carp larvae also feed on protozoa[171,172] and depend upon smaller food items longer than the other species.[384]

The ratio of larval mouth size to prey size is not the only factor that governs prey size selection. Prey density is also important, as larvae preferentially capture larger prey at higher prey densities. For example, Khadka and Rao[385] found that common carp larvae ate larger prey when zooplankton density increased. At low prey densities, capture of any prey that is encountered is, energetically, the best strategy.

5. Artificial Diets

The lack of proper artificial diets for larval cyprinids is a problem in large-scale culture. Contrary to salmonids, which appear to have a functional stomach before changing from endogenous to external food, cyprinids remain stomachless throughout life and do not have a structurally and functionally differentiated alimentary tract at the time of first feeding. Stomachless fishes have problems digesting and assimilating starter feeds.[386]

Diets based on single cell protein and freeze-dried animal tissues have been successfully used for feeding larval carp.[387,388] Common carp larvae fed exclusively with yeast

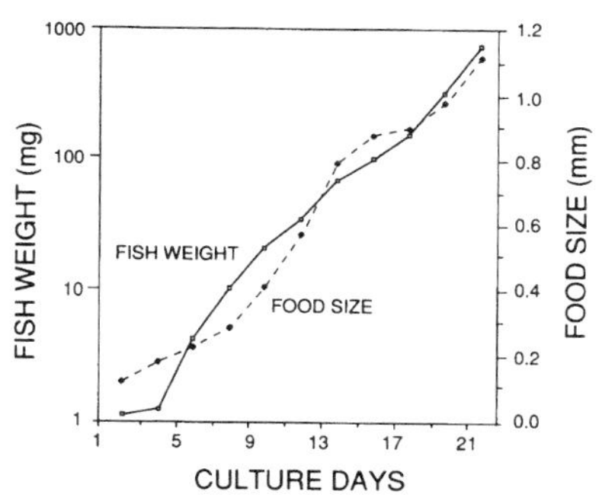

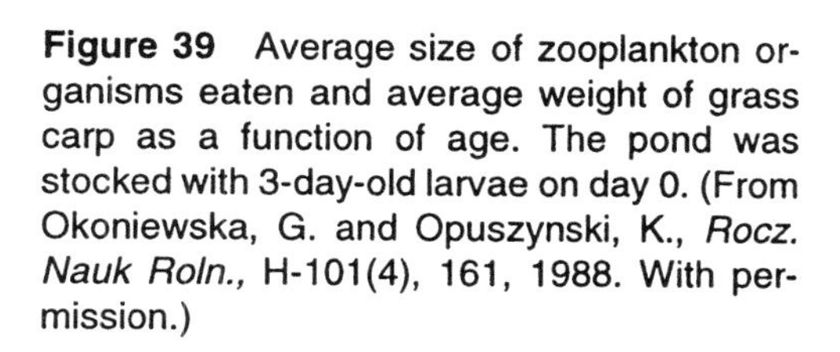

Figure 39 Average size of zooplankton organisms eaten and average weight of grass carp as a function of age. The pond was stocked with 3-day-old larvae on day 0. (From Okoniewska, G. and Opuszynski, K., *Rocz. Nauk Roln.*, H-101(4), 161, 1988. With permission.)

cultivated on petroleum by-products and on beef liver grew to an average weight of >100 mg in 21 d; survival was 87%.[388] The growth rate of common carp larvae fed on live food was still much better.[266] Results reported by Alami-Durante et al.[389] suggest that a breakthrough may be imminent in diet formulations. They found that after 21 d, survival was 95% and mean body weight was 189 mg for common carp fed a yeast-liver diet supplemented with a mineral and vitamin premix. After 10 additional days, when they were fed a commercial trout feed, survival was 95% and mean body weight was 755 mg.

The difficulty of getting cyprinid larvae to accept dry starter diets can be overcome by initially using live food or a mixture of live and dry food and then weaning the fishes onto dry food after 1 to 2 weeks. Bryant and Matty[390] and Dabrowski[295] determined that the lowest weight at which larvae could be transferred from live food to artificial diets was 5 to 6 mg. Opuszynski et al.[270] tested different dry feeds and found that commercial trout starter gave results comparable to EWOS C10 Larvestart when both feeds were administered after an initial 10-d period, when Chinese carp were fed zooplankton.

6. Predation and Diseases

Predation is a cause of high mortality of carp larvae. The most common invertebrate predators are copepods and larval and adult insects.[9] Among vertebrate predators or competitors for food are tadpoles, adult frogs, and other species of fishes.[391] Opuszynski et al.[270] found almost 100% mortality of larval silver carp during the first 6 d of tank culture, when adult cyclopoid copepods were accidentally introduced with zooplankton collected from ponds. Copepods began feeding on the fins of the larvae, causing loss of blood, tissue damage, and microbial infection, which actually induce mortality. The length of the fish is critical. Up to 10 mm they are preyed upon by copepods, but when they grow larger the predator-prey relationship is reversed.[392]

Survival of small silver carp was 2% in a pond heavily infested with threespine stickleback, *Gasterosteus aculeatus*, but ranged from 29 to 36% in ponds in which only a few threespine stickleback were found.[228] Wolny[393] found a negative correlation between water transparency and survival of small fishes in ponds. Low water transparency made it difficult for sight feeders such as birds to prey upon the fry.

Bacteria, fungi, and protozoa can cause high larval mortality and growth retardation. The danger of a disease outbreak is especially high in intensive larval cultures where high fish densities and large amounts of food are used. In larval culture special attention should be given to disease prevention, including disinfection and periodic cleaning of hatchery tanks, nets, and other equipment, and maintaining good water quality. Raising carp larvae at higher water temperatures protects them from *Saprolegnia* and *Ichthyophthirius*, which are among the most serious and frequently occurring diseases.[377] These organisms do not develop above 30°C.[394] Treatment of small fishes is limited to external treatments (dipping, bathing, or flushing) and in most cases is not effective.

7. Interactions of Temperature, Food, and Predation

Factors that influence larval survival and growth do not act separately, and the interaction of these factors has a synergetic effect. If each factor has a sublethal effect, their joint action can result in total fry mortality. Therefore, it is often difficult to determine which factor caused the failure. Several years of experimentation with larval carp in ponds at the Inland Fisheries Institute, Zabieniec, Poland, allowed for the analysis of the influence of temperature, food, and predation on the survival of larvae.[228] All experimental ponds were similar in respect to size (0.2 ha), mean depth (1 m), and water source (river water). All ponds were stocked each year with silver carp larvae from one spawn.

This analysis is based on the following assumptions: (1) if larval survival in a given season is low in all ponds, the influences of temperature, food, and predation on fish survival cannot be separated; (2) if survival is low in some ponds and high in others, temperature can be eliminated; (3) if survival varies among the ponds and at the same time shows a positive correlation with fish growth at the beginning of the season, then a shortage of food may be a contributing factor. It is likely that poor feeding conditions effect fish mortality, mainly through predation; slow growth makes larvae vulnerable to predators for a longer period. For this reason it is difficult to separate these two factors in a pond experiment. In order to separate the influence of feeding conditions and predation, silver carp were also raised in cages. Temperature was increased in some cages with an electric heater and a thermostat to determine the influence of temperature (Figure 40).

In central Europe, low temperature can cause total mortality of Chinese carp larvae in exceptionally cold seasons; in 1976, for example, 100% mortality occurred in cages without temperature control, while over 36% of the larvae survived in temperature-controlled cages situated in the same pond. This substantial difference in survival was caused by a 4°C difference in temperature. When predation was eliminated, survival approached 100%, if temperature and food conditions were appropriate (Table 23).

B. LARVAL CULTURE IN PONDS

The culture of cyprinid larvae in earthen ponds is by far the oldest, most common, and most widely practiced method. It was first employed in China more than 2400 years ago. Management practices in fry ponds are generally similar under temperate[395] and tropical[396] conditions. They include pond preparation, stocking optimization, fertilization, fish feeding, and predator control.

Although natural water temperatures are nearly optimum for the culture of Chinese herbivorous carp in subtropical and tropical climates, low water temperatures are a major concern in temperate climates. Pond culture is widely used in temperate climates, even though high mortality may result. Moderate or satisfactory survival rates can be obtained during seasons in which temperatures are average and above average, but total loss of larvae and young fishes can occur during colder seasons (Table 23). In order to alleviate the problem of cold water temperatures, ponds should be stocked late enough to avoid low and highly variable spring temperatures, but early enough and at a density to allow growth of the fishes to a size that will enable them to survive over the long winter period.

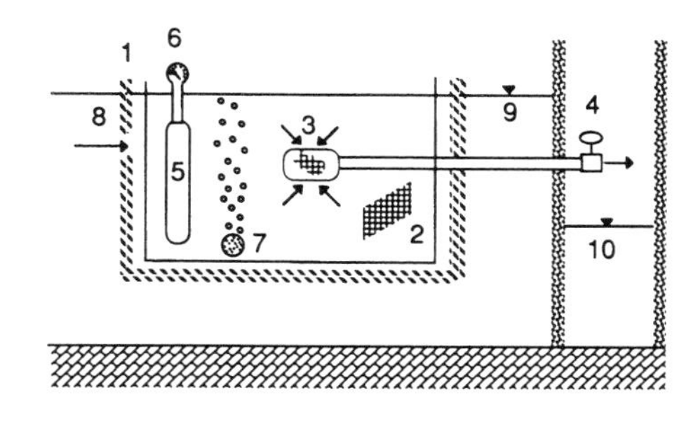

Figure 40 Flow-through cage enabling temperature control in natural temperature drained ponds. A simplified version without thermal insulation and heaters is more commonly used. (1) Thermal insulation (sponge), (2) net cage, (3) outlet screen, (4) outlet pipe with valve, (5) electric heater, (6) thermostat, (7) air stone, (8) inlet, (9) water level in pond, (10) water level in monk.

Table 23 **Causes of silver carp fry mortality in ponds and cages at the Inland Fisheries Institute, Zabieniec, Poland[228]**

	Causes of mortality			No. of	Mean survival
Year	Temp	Food	Predation	trials	and range (%)
Ponds					
1965	+	+	+	3	6 (0–10)
1968		+	+	2	69 (37–100)
1971	+	+	+	5	5 (1–17)
1972		+	+	11	40 (2–100)
Cages					
1973				2	100
1974	+	+		8	1 (0–2)
1975				5	100
1976[a]	+			2	0
1976[b]				3	36 (16–64)[c]

Note: + = cause.

[a] Cages with uncontrolled temperature (lowest recorded temperature = 14°C).

[b] Cages with controlled temperature (lowest recorded temperature = 18°C).

[c] Some fry escaped from the cage.

1. Pond Preparation

Pond preparation includes drying and preparation of the bottom, liming, and fertilization. Pond drying is of great importance. Fry ponds are kept dry in Europe from late summer until the following spring. The pond bottom is plowed and limed after the ponds are dried. Either 1000 to 2500 kg/ha of quicklime (CaO) or 3000+ kg/ha of slaked lime [$Ca(OH)_2$] is used. In order to increase productivity, agricultural crops such as rye, oats, barley, or clover may be cultivated. The vegetation may be harvested and used to feed livestock; more often, the vegetation is plowed under before flooding, improving the soil structure of the pond bottom.

2. Stocking Optimization

The timing of larval stocking after pond filling is an important management decision. After a pond is filled, different aquatic organisms develop and dominate (biological succession). Long-term studies have shown that biological succession in all types of ponds develops in a similar manner and with the same pattern.[397] Even though in adjacent ponds different species may dominate, and community dynamics as well as fish production may differ, the successional pattern is similar.

Immediately after filling, bacteria, heterotrophic protozoa, and heterotrophic algae (such as Euglenoidea) dominate, followed by autotrophic phytoplankton. If larval culture is to be successful, ponds should be stocked when rotifers dominate. Stocking ponds either too early or too late may cause the larvae to starve.[228] If stocking is done too early, the zooplankton community is not well developed, and if stocking is done too late, zooplankton may be too large for the larval fishes. Additionally, populations of predators may have developed. Pond succession rate depends on temperature. A common practice is to stock D4 larvae in ponds that have been filled for 3 to 7 d.

Proper stocking density is also an important management decision. Stocking density must be adjusted to food resources. These resources greatly surpass the fish demands at the time of stocking, but are soon in short supply as fish biomass increases. For this reason, the culture period in larval ponds is short, and the stocking rate is relatively low (Table 24), particularly in temperate climates where the growth period lasts only 3 months (June to August). Grass carp and silver carp fingerlings must grow to over 4 and 10 g, respectively, and accumulate enough fat to successfully survive the winter.[398] For this reason, the proper stocking density is 50,000 to 150,000 larvae per hectare. Yield under such conditions usually does not exceed 300 kg/ha.[399–401]

While polyculture of Chinese carp is a common practice in grow-out ponds, monoculture practices prevail in larval ponds. The feeding habits of the early larvae of these species are similar, and sorting and grading the larvae are burdensome and can cause increased mortality. Two-species polyculture is sometimes used, with good results: common carp and silver carp or common carp and grass carp can be raised together.[228,393]

3. Fertilization

In most cases inorganic fertilization increases fish production; however, declines in fish yield also occur.[402] Fertilization causes increases in fish production in ponds where small algae dominate.[403]

It is generally accepted in Europe that both nitrogen and phosphorus are needed. Recommended N:P ratios range from 4 to 8:1 or 2 to 11:1.[404,405] Many forms of ammonium fertilizers are used. Piotrowska-Opuszynska[406] reported the superiority of urea over other chemical fertilizers in common carp ponds. Release of carbon dioxide during hydrolysis helps lower pH, and gradual decomposition of urea prevents high ammonia concentrations immediately after fertilization. Superphosphate is the primary source of phosphorus in Europe.

Wolny[393] recommended the following rates for fertilization of ponds used to culture larval carp in central Europe. Total nitrogen fertilizer equal to 210 kg nitrogen per hectare should be applied in eight applications. Two applications of 4500 µg nitrogen per liter each should be applied 6 and 3 d before the pond is stocked. The remaining six applications of 2000 µg/l each are applied at 5-d intervals beginning the day after larvae are stocked. Total superphosphate equal to 23.5 kg phosphorus per hectare should be applied in four applications. An application of 900 µg phosphorus per liter is applied 6 d before ponds are stocked. The remaining three applications of 500 µg

Table 24 **Stocking rates and culture systems used to raise carp larvae in ponds**

Country	Stocking density (× 10³/ha)	Type of culture	Final fish size (cm)	Culture period (d)	Ref.
Poland	150	Common carp + silver carp	5–7	30	228, 270
	150	Common carp + grass carp	5–7	30	393
	75	or + bighead carp	4–7		
China	1500–2500	Monoculture	2–3	15–30	426
	1000–1500	Monoculture	3–8	15–30	428
U.S.	250–1250	Monoculture	8–10	21–28[a]	319
Taiwan	2000–3000	Monoculture	2–3	21	412

Note: Monoculture means that either grass, silver, or bighead carp were cultured alone.

[a] Data from K. Opuszynski.

phosphorus per liter each are applied 2, 12, and 22 d after stocking. This recipe was created for a culture period of 31 d.

When Wolny's[393] method is calculated on a weekly basis, post-stocking fertilization rates are 3000 μgN/L/wk and 375μgP/L/wk nitrogen per liter per week. These fertilization rates are high compared to those used in larval fish ponds in the U.S. Culver et al.[407] recommended an N:P fertilization ratio of 20:1, with fertilization rates of 600 μg of nitrogen and 30 μg of phosphorus per liter per week, respectively. Anderson[408] felt that 600 μg of nitrogen per liter per week may be too high for fish larvae which are highly sensitive to nonionized ammonia.

Various locally available plant wastes, animal manures, and night soil (in Asia) are used as organic fertilizers. In the U.S. hay, soybean meal, cottonseed meal, and alfalfa meal are also used. The superiority of either organic or inorganic fertilization is still a controversial issue. Organic fertilization decreases pH and provides a direct source of food for zooplankton. Increased pH is a common problem in larval ponds that are heavily fertilized with inorganic fertilizers. Organic fertilizers, however, can cause oxygen depletion when excessive quantities are used.

4. Fish Feeding

Chinese carp larvae are not fed in ponds in Europe, but is a common practice in Asia and the U.S. In order to make supplemental feeding successful, high stocking densities and high water temperatures (≥22°C) are needed. Fermented and predigested soybean milk is applied several times per day to nursery ponds in Asia. This procedure begins after ponds are stocked with D4 larvae. Soybean milk, however, encourages zooplankton development in ponds rather than nourish the fishes directly. Dry feed, e.g., minnow meal or catfish starter, is used to feed small Chinese herbivorous carp in the U.S. Feeding begins sometime after stocking, when larvae are larger and quickly learn to ingest the dry diet.

5. Predation Control in Ponds

Predation can be a serious problem in ponds, and, if ponds are not prepared properly or if fishes are not stocked at the right time, some predator control may be necessary.

An inexpensive and effective way to control air-breathing predatory insects in ponds is to spray diesel fuel (Europe) or unrefined coconut oil mixed with soap (Asia) on the water surface. These substances should be sprayed at a rate of 30 l/ha. Organophosphate insecticides have been used to kill crustaceans and aquatic insects. However, the crustacean zooplankton recovery period is relatively lengthy after such a treatment,[409] and quantitative and qualitative changes in zooplankton and benthic communities occur, which may be disadvantageous for fish feeding.[397] These chemicals are not approved for use in the U.S.

Although research by Kane and Johnson[410] suggested that TFM (3-trifluoromethyl 1-4-nitrophenol) may be used as a selective tadpole toxicant, a more recent study suggests that these chemicals may not be suitable because doses necessary to kill tadpoles also kill small fishes.[411] Mechanical methods of frog control consist of removal of eggs by hand and killing adult frogs. Larval ponds sometimes are fenced to prevent the entrance of frogs.[412] Wild fishes gain access to larval ponds with incoming water; therefore, different filtering devices are used to reduce this problem.[391] When serious bird predation exists, small ponds can be covered with nets. Different scaring devices are mostly ineffective, as birds quickly learn to ignore them.

C. CAGE AND TANK LARVAL REARING

Cage and tank rearing is an intensive type of culture that makes it possible to raise a large number of small fishes in a limited space. The advantages of intensive culture are reduced land area; improved survival due to predator control; convenient feeding and better utilization of commercial dry diets; possibilities of water quality control, including temperature; and ease in harvesting of fishes. Intensive culture, however, requires complex facilities, greater supervision, skilled labor and management, and more time.

1. Rearing Facilities

Flow-through cages have been designed to raise larvae in zooplankton-rich ponds.[354,413] Different cage arrangements are used to provide water flow to cages situated in drained (Figure 40) and undrained ponds (Figure 41). Cages can be insulated and provided with electric heaters to control temperature in cold ponds, but such facilities are not economically feasible if electricity is expensive. Flow-through cages not only protect larvae against predation and enable continual delivery of live food, but also make it easy to train larvae to accept dry feed. Nearly 100% survival can be attained in cages, provided that favorable water quality and food exist in the pond. A major disadvantage is that water quality in the cages is largely dependent on conditions in the heavily fertilized ponds.

Where aquaria or tanks are used, water quality is maintained either by a flow-through or by a recirculating system. Surface water or well water can be used as water sources. In cold climates, when indoor tanks are used, water heating is cost effective because high stock densities are used, and good survival and early fingerling production occur.[414] Rottmann et al.[271] reported that larval culture can be done without water exchange when a simple and inexpensive airlift sponge is employed as a biological filter. This method, if proved feasible in large-scale culture, is advantageous, as the cost of heating water is low and food is not washed from the tanks. The results of intensive larval culture in different facilities are summarized in Table 25.

2. Larval Feeding with Natural Food

Zooplankton can be produced in ponds, harvested, and subsequently delivered to the fishes, or intensive culture of live food can be used to feed the fishes.

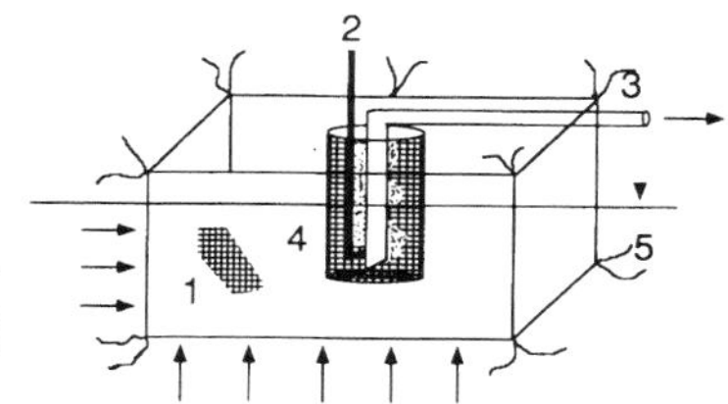

Figure 41 Flow-through cage designed for undrained ponds. (1) Cage net, (2) compressed air, (3) air lift, (4) air lift screen, (5) water level.

Table 25 **Results of intensive culture of Chinese carp larvae using different methods of culture in different climates**

Method and place/climate	Stocking density (individuals/l)	Culture period (d)	Lowest temp recorded (°C)	Survival (%)	Mean weight (mg)	Ref.
Flow-through cages in drained ponds; natural water temperature; Poland/temperate	30–40 30–40	15 21–28	13[a] 18[b]	0–1 100	1 15–340	413
Flow-through cages in undrained ponds; natural water temperature; Florida, (U.S.)/subtropical	20	10	26	98–100	18–56	354
Tanks; natural water temperature; Florida/subtropical	20	10	26	86–88	18–56	354
Indoor aquaria; sponge-filter system; temperature control	13–57	21	25[c]	55–99	66–503	271

[a] Exceptionally cold season.

[b] Average temperature season.

[c] Data from K. Opuszynski.

In zooplankton ponds heavy fertilization is used to promote zooplankton development. Such heavy fertilization cannot be used in ponds stocked with larval fishes because good water quality standards must be preserved. Zooplankters are resistant to low oxygen concentrations and high ammonia levels. Lincoln et al.[415] reported that the rotifer *Brachionus rubens* became abundant in ponds when nonionized ammonia concentrations declined below 20 mg/l. Sewage purification ponds can be used to produce live fish food because mass development of zooplankton often occurs there.[416] In most cases, however, special zooplankton culture ponds are situated close to the larval culture facilities.

Rotifers must occur in peak numbers in zooplankton culture ponds when larvae commence exogenous feeding. It is not easy to synchronize these events. The use of several zooplankton culture ponds that are filled at 3-d intervals is the best way to ensure that adequate food exists when it is needed. Chemicals must be used to kill existing organisms in ponds that cannot be completely drained. In Europe these chemicals include copper sulfate, liquid ammonia, and organophosphate insecticides.[397] The latter group is of special interest because of their selective action and short degradation time.[417] The organophosphate insecticides Dylox, Flibol, Foschlor, and Neguvon kill crustaceans and encourage rotifer development. These chemicals are used in concentrations necessary to kill crustaceans (1000 μg of active ingredient per liter), are safe to fish larvae, and can be used in the ponds that are stocked. Opuszynski and Shireman[409] found up to 15,000 rotifers per liter in Dylox-treated ponds as compared to <300 rotifers per liter in untreated ponds. The first rotifer peak occurred 9 d after the Dylox treatment at 24°C. These chemicals, however, are not approved for use in the U.S.

Collecting zooplankton from zooplankton ponds is labor intensive in large-scale larval culture operations. Different zooplankton collectors can be used, but all have the same shortcoming. If the net mesh size is small enough to collect rotifers, it clogs rapidly with algae and debris and must be cleaned often. This problem is less serious when larger crustaceans are collected because the mesh size is larger. Feeding fry in tanks with zooplankton-laden water delivered directly from zooplankton culture ponds proved to be very efficient.[354] If the water quality in the zooplankton culture pond is not satisfactory for larvae, the water in the fish tanks can be aerated or the tanks can be supplied with clean water or both can occur.

Different live foods can be intensively cultured for feeding small Chinese carp. Rotifers *(B. rubens),* cladocerans (*Daphnia* sp.) and *Moina* sp., brine shrimp, and nematodes (*Panagrellus* sp.) are among the most commonly cultured food organisms.[319] Zooplankton culture procedures are more complicated and more labor intensive than the management of zooplankton culture ponds. High population densities can be obtained, however, in culture facilities (e.g., 500 rotifers per milliliter), which makes zooplankton collection easy. Another advantage of intensive culture is that the nutritional quality of the culture organisms can be improved by prefeeding them selected diets. For example, considerable progress has been made recently with the enrichment of rotifers and brine shrimp with polyunsaturated fatty acids (HUFA).[418,419] Low growth rates and high mortality of fish larvae are often related to deficiencies in polyunsaturated fatty acids in the live food.

Brine shrimp are widely used where dried eggs are available.[420] They are advantageous as they are simple to hatch, and egg decapsulating is simple.[421] Rottmann et al.[271] compared *B. rubens,* brine shrimp, and *Panagrellus* sp. as a food for intensively cultured grass carp and bighead carp larvae. Although all the live foods tested gave satisfactory results, fishes that were fed rotifers were significantly larger at the end of the 3-week feeding trials.

3. Larval Feeding with Dry Diets

Feeding larvae with live food in culture facilities having controlled environmental conditions is cost effective for only a short period of time, when total fish biomass is low. The transition period from natural to artificial diets can be accelerated when the natural diet is gradually replaced by dry feed and when both fishes and feed are kept at high densities. The Dutch procedure, which is used in large-scale integrated tank-and-cage culture, is a good example.[422] Grass carp larvae are stocked in 120-l aquaria at densities of about 330 larvae per liter, and are fed with brine shrimp during the first 5 to 7 d. The brine shrimp are steadily replaced by pelleted trout starter. The hatchery-raised fingerlings are then transferred to cages and are fed exclusively with pellets using demand feeders.

Larval feeding methods with dry foods still need refinement regarding protein requirements, intake levels, and feeding frequencies. When bighead carp larvae were fed isocaloric diets (290 kcal digestible energy per 100 g) having protein levels ranging from

20 to 50% in 5% increments, weight gain and increase in total length were highest for larvae fed the 30% protein diet.[307] This protein level, however, is relatively low compared to the 41 to 43% protein requirement for grass carp larvae.[267] Feeding rates of 10, 20, and 30% of body weight per day and feeding frequencies of one, three, and five times daily were tested in another study on bighead carp.[378] Highest growth and survival were achieved in the 30% ration fed once daily. This finding regarding feeding frequency is in contrast to present practices that favor frequent feeding.

4. Predation Control in Intensive Culture

Predation is relatively simple to prevent in intensive culture. If larvae are fed in cages or tanks with zooplankton collected from ponds, the zooplankton-laden water can be filtered through properly sized netting. Flow-through cages have been designed to raise larvae in zooplankton-rich ponds. Rotifers and copepod nauplii flow into the cage, while the predatory zooplankton is retained by the fine net.

D. UTILIZATION OF HEATED EFFLUENTS FOR LARVAL CULTURE

The use of heated effluents from power stations for larval carp culture in earthen ponds, tanks, and cages is a widespread practice in Hungary, Germany, The Netherlands, Poland, and the Commonwealth of Independent States, where there are numerous power stations with water cooling systems. Effluent temperature is usually high enough to increase temperatures in flow-through ponds when the exchange rate is about 2 d (Figure 42). This increase in water temperature can result in substantial production increases. After 21 d of culture, grass carp survival and weights in static and heated-effluent ponds were 2% and 206 mg and 63% and 363 mg, respectively.[423]

The integration of indoor tanks and outdoor cages or raceway culture using heated effluents is an attractive possibility in temperate climates because the growing season can be extended from April until October. This combined and integrated culture system is used in Germany[424] and The Netherlands.[422] In The Netherlands fingerlings are raised in temperature-controlled tanks in a greenhouse, and then transferred to cages situated in the cooling water discharge canal of an electric power station. The temperature of the discharge water is about 7°C higher than the ambient water temperature.

III. CULTURE OF MARKETABLE FISHES

Chinese carp are grown primarily as a food fish in most regions where they are native or introduced. In the U.S., however, they are raised mainly for aquatic weed management or environmental improvement or both. In Europe and Asia grass carp are almost always grown in polyculture with other species. Polyculture practices may differ depending on climate and socioeconomic and cultural conditions, but are generally based on the concept that one or two species provide the main crop (usually the species for which there is the greatest market demand) and subsidiary species utilize additional food resources.

A. PRINCIPLES OF POLYCULTURE

The basic objective of any fish culture is to produce the greatest quantity and quality of the desired size fish at minimum cost. Fish yield is directly proportional to the number of fishes stocked into the pond. In single-species culture (monoculture), fish production increases to a maximum stock density, then declines if fish density continues to increase. Therefore, it must be assumed that the stocking densities used are those that result in the maximum yield (Figure 43, M). The problem, however, is more complicated, because as

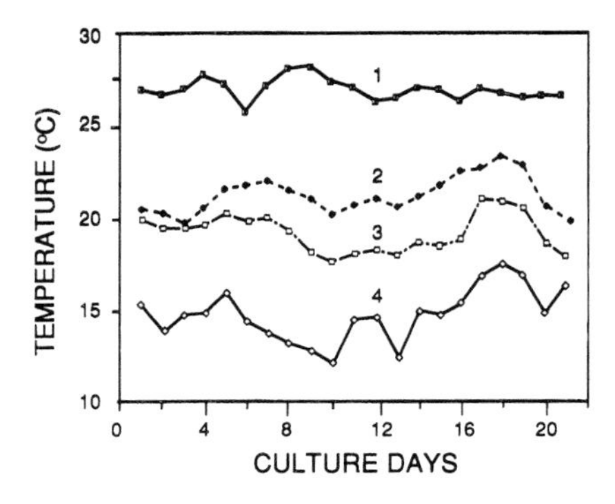

Figure 42 Mean water and air temperatures during larval grass carp culture, Goslawice, Poland. Ponds were stocked on 21 June 1980. (1) Heated-effluent inflow water, (2) flow-through ponds, (3) static water ponds, (4) air temperature. (Data are from Okoniewska, G. et al., *Rocz. Nauk Roln.*, H-101(4), 111, 1988. With permission.)

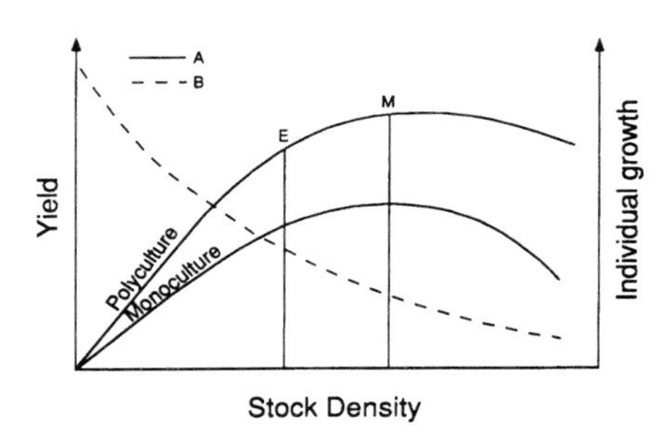

Figure 43 Relationship of stocking density, yield, and individual growth rate in mono- and polyculture systems. (A) Yield, (B) individual growth rate, (E) economic yield, (M) maximum yield.

the stocking density increases the growth of individual fishes declines. Hence, a stocking density designed to produce a desired final fish size and yield to optimize economic benefits must be used (Figure 43, E).

The relationship of stocking density, fish size, and yield remains essentially the same even though several fish species are grown in polyculture. The influence of the subsidiary fish species on yield of the primary fish can differ depending on species composition, numerical relationships, food availability, and culture procedures. Subsidiary fish species (1) can decrease the yield of the primary species (may be acceptable if the yield of subsidiary fishes offsets this decrease), (2) cannot alter the yield of the primary species, or (3) can increase yield if a synergistic effect exists between the primary species and subsidiary species. The results of a polyculture system are unacceptable if lower total profit is achieved when compared to monoculture of the primary species.

It must be stressed, however, that the net gain achieved through polyculture is not easily measured. Increased production from a polyculture system may be apparent when the primary species is understocked. In this case the possibility exists that by increasing the stocking density of the primary species a higher yield may be obtained by adding additional numbers of the subsidiary species. Many studies designed to determine the effect of polyculture on pond production are inconclusive because these studies lacked a control (monoculture) group. The control group should contain the same number and size of primary fishes as in the stock of primary fishes in polyculture, and should produce the maximum yield under the given set of conditions (Figure 44). Maximum common carp production was obtained at a stock density of 800 individuals per hectare, and increasing the density to 900 individuals per hectare actually decreased production. The addition of

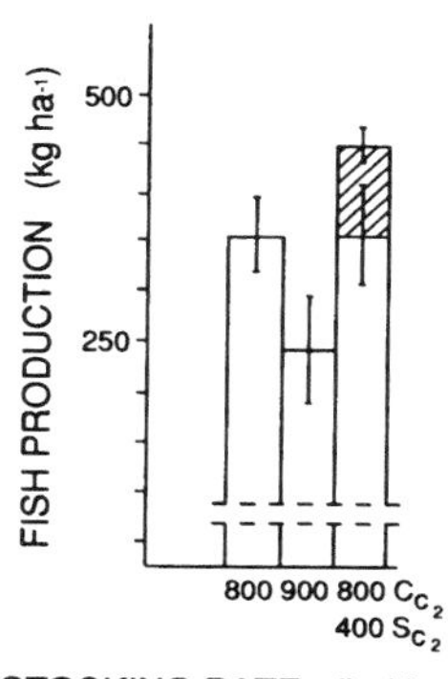

Figure 44 Increase in the maximum production of carp ponds due to additional stock of silver carp in Poland. Fishes were not fed, and ponds were not fertilized. Empty bars = common carp, hatched bar = silver carp. Stock density per hectare of 2-year-old common carp = C_{c_2}, and 2-year-old silver carp = S_{c_2}. Vertical lines = standard deviations. (From Opuszynski, K., in *Managed Aquatic Ecosystems,* Michael, R. G., Ed., Elsevier, Amsterdam, 1987, 63. With permission.)

silver carp caused an increase in production, which was impossible to achieve by manipulation of common carp stock densities.

Several factors must be considered when searching for the optimal polyculture system. The selection and the numerical combination of fish species should depend on food habits, suitability of environmental conditions, availability of natural and supplementary food, and market demand. Chinese carp are a suitable species for polyculture systems in many regions of the world. Polyculture systems including Chinese carp directly utilize primary production for fish production. This is doubly advantageous: (1) direct food competition with other cultured fish species that are either not herbivorous or feed on different plant food is avoided, and (2) fish flesh is converted from phytoplankton and macrophytes, which are the most abundant food resources in any aquatic ecosystem (Figure 45).

B. REGIONAL REVIEW OF CULTURE PRACTICES

1. Chinese Methods of Polyculture

Although fish culture became a common practice in China around 1000 B.C., only common carp were cultured. Polyculture was not introduced until the 6th century A.D. It is said that polyculture in China was caused by the coincidence that the name of the emperor of the Tang Dynasty was Li; his name sounds like the Chinese name for the common carp. Because the name of Emperor Li was considered sacred, it was inadmissible that common carp could be killed and eaten, so the Chinese looked for alternative species. Several species were selected, each with different feeding habits, and that is how Chinese polyculture began.

The contemporary polyculture system in China may involve some or all of the following species: grass carp, silver carp, bighead carp, black carp *(Mylopharyngodon piceus),* common carp, Chinese bream *(Megalobrama amblycephala),* crucian carp *(Carassius auratus),* and, in warmer areas, mud carp *(Cirrhina molitorella)* and different species of tilapia. Of the above species, grass carp and Chinese bream consume macrophytes; silver carp, phytoplankton; bighead carp, phytoplankton and zooplankton; black carp, mainly molluscs; and common carp, mud carp, crucian carp, and tilapias consume invertebrates, plant materials, and detritus.

In ponds the proportion of the above fishes varies depending on environmental conditions and food availability. Generally, grass carp, silver carp, bighead carp, and common carp comprise up to 90% of the total fish biomass in the ponds. The other species

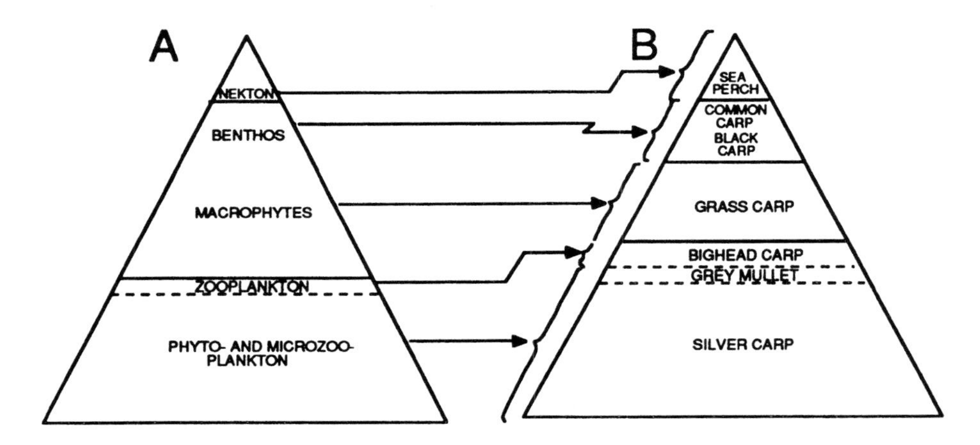

Figure 45 Weight percentage interrelationships between the standing crop of different fish food in ponds (A) and the yield of different fishes (B) in polyculture in Taiwan. (From Tang, Y. A., *Trans. Am. Fish. Soc.*, No. 4, 708, 1970. With permission.)

utilize the remaining food resources that would be wasted. When the supply of aquatic vegetation or land plants or both is abundant, grass carp is the major species. When animal manure is abundant, silver and bighead carp are the dominant species. Black carp, which have high market value, may be the major species if molluscs can be easily collected in quantity from adjacent waters. The various combinations of species used in polyculture ponds are given in Table 26.

The preferred market size of fishes ranges from 0.5 to 1.5 kg. It usually takes from 2 to 3 years to grow them, except in the warmest and most fertile southern regions of China, where it takes 1 year. Two types of polyculture practices are used to produce marketable-sized fishes in China, namely mixed-age and multigrade culture.[425-427] Both practices assure fast growth rates by manipulating stocking density and avoiding overcrowding caused by increased fish biomass during the growing season.

Mixed-age or stocking and harvesting in rotation culture is primarily used in ponds that can be drained. In this system ponds are stocked with three age groups of fingerlings of each chosen species. Ponds are seined approximately every 2 months and the fishes that have reached marketable size are harvested and are replaced by fingerlings of the same species.

The multigrade method, which is also locally termed multigrade conveyer culture, is mainly used in ponds that cannot be drained. A combination of species is grown in a series of ponds from fingerling to marketable size. As the fishes grow, they are graded, and the larger ones are transferred to the ponds containing larger fishes. The last pond in the series contains the harvestable fishes of market size. This method assures maximum growth of fishes as the stocking density is regulated along with growth, as well as adjusted for the productive capacities of the ponds. This method, however, requires a large number of ponds for the production of marketable fishes and a continuous supply of fry and fingerlings to avoid any disruption in the production cycle.

The production capacities of Chinese ponds are greatly enhanced by fertilization and fish feeding. The use of organic fertilizers has been a common practice for centuries. Organic fertilizers consist mainly of green waste, compost, animal manure, and human

Table 26 **Fish combinations used in polyculture in China and Taiwan in percentage of total number (%TN) or total weight of fishes (%TW)**

	Location					
	Pearl River[425]		Guangdong[425,426]		Taiwan[412,426]	
Fish species	%TN	%TW	%TN	%TW	%TN	%TW
Grass carp	**55**	24	12	19	9	3
Silver carp	16	12	**65**	8	3	—
Bighead carp	10	7	10	**47**	6	3
Black carp	—	**42**	—	2	—	—
Mud carp	—	—	—	19	**82**	—
Tilapia	—	—	—	3	—	**86**
Common carp	—	3	5	2	—	6
Miscellaneous	19	12	8	—	—	2

Note: The primary fish is printed in bold type.

waste. The use of the latter is a traditional practice and it is estimated that "night soil" constitutes as much as one third of the total organic fertilizer resource of China.[425] The use of human waste, however, can cause the spread of disease. To counteract this, human wastes undergo anaerobic fermentation in a closed chamber prior to application.

Until recently, fertilization rates were a matter of tradition and experience rather than scientific consideration. It was believed that the feces from 45 pigs, 30 cattle, or 3000 ducks could sustain fish production in a one hectare pond.[428] The total amount of organic manure applied is high and can reach >300 ton/ha/year.[426] Based on Chinese empirical data, 100 kg of manure should produce 1 to 1.5 kg of silver or bighead carp.[429] In actuality the application rate of organic fertilizer is governed primarily by water transparency and oxygen levels in ponds.

Recent Chinese studies provide more accurate data on the effect of manure application upon fish yield in a polyculture system consisting of silver, bighead, common, and crucian carp.[430] Net fish yield was directly proportional to the amount of manure applied over the range 0 to 48 kg dry weight manure per hectare per day (the experiments were carried out for 115 to 201 d in different years). Each 10 kg/ha/d increase in the manuring rate resulted in 1.2 kg/ha/d increase in net fish yield. Silver and bighead carp accounted for about 75% of this increase. The conversion ratio of manure to fish biomass was 8.3 kg dry manure to 1 kg fish wet weight. Fish yield averaged 10.2 kg/ha/d in manured ponds and only 4.3 kg/ha/d in control ponds, where inorganic nutrients were added at a rate equivalent to the nutrient concentration in the manure. These data clearly show the importance of heterotrophies in the food chain using the Chinese method of organic fertilization.

Inorganic fertilization of ponds is a relatively new practice, and not enough data have been collected to allow for evaluation. It was reported, however, that the use of inorganic fertilizers, mostly superphosphate, in Taiwanese ponds resulted in higher yields compared to that obtained with traditional organic fertilization.[431] Superphosphate proved also to be more economical than organic fertilizers; 1 kg of P_2O_5 produced nearly 10 kg of marketable fishes. This increase in yield was due mainly to improved growth of the planktivorous fishes. Silver carp growth rate increased almost three times after fertilization, and bighead carp growth increased by about one third or more. It must be mentioned, however, that these results were obtained in a polyculture system in which grass carp constituted only 10% of the total stock. If more grass carp had been used, the beneficial effects of inorganic fertilization might not have been conspicuous, because grass carp feeding activity enhances pond fertility.

Grass carp and Chinese bream are fed available plant material. Not only are the grasses and vegetables grown on the pond dikes used, but residues from vegetable crops and by-products and waste materials from the food industry are also used (bean cake, grain residues, rice bran, wheat husk, cabbage seed cake, cottonseed cake, etc.). Among terrestrial plants, stems and leaves of legumes, leaves of a variety of green vegetables, and even leaves of many kinds of trees are used. Promising results have been obtained in the replacement of cereals with leaves of *Zizania latifolia* in grass carp diets.[432] Many kinds of aquatic plants are also used to feed grass carp including *Wolffia arrhiza,* duckweed, *Spirodella polyrhiza, Vallisneria spiralis, Potamogeton malaianus, P. maackianus, P. crispus,* hydrilla, and *Najas minor.*[131]

Where the number of grass carp in a polyculture system is high, the input of plant material may exceed 150 ton/ha.[426] About 30 to 50 kg of land plants produce 1 kg of grass carp. With aquatic plants this ratio ranges widely from 37 kg for *Wolffia* to 101 kg for *Vallisneria.* Because the grass carp is not a very efficient food converter a great deal of the ingested food returns to the pond as feces. These feces, containing partially digested plant matter, serve as food for other fishes and act as an excellent organic fertilizer, stimulating plankton growth. It is impractical, therefore, to separate the effects of fertilization and supplementary feeding when grass carp are present. A common saying by Chinese fishermen is "feed one grass carp well and you will feed three other fishes".

The Chinese polyculture system is highly productive (Table 27), ecologically efficient, environmentally sound, cost effective, and is based on using herbivorous fishes to shorten the food chain. Fish production is incorporated into the AAA system, which consists of agriculture, animal husbandry, and aquaculture (Figure 46). The agriculture subsystem produces feeds for animal husbandry, grass for feeding fishes, and plant components for the preparation of pelletized fish food. The grass fields are fertilized with human manure obtained from cities. Animal husbandry includes piggeries, dairy farms, chicken, and duck or goose farms. These systems produce manure, which along with the human manure, is used to fertilize fish ponds. In addition, chicken manure may be used as a component of pelletized fish food. Some of the organic matter along with fish feces sink to the bottom and enrich the soil of fish ponds. The concentration of nitrogen, phosphorus, and potassium in the pond mud may exceed almost 200 times that in the water.[429] Pond mud is collected several times a year and applied to adjoining vegetable and field areas. Based on Chinese data, 50 kg of fishes produces enough bottom sediments to fertilize over 0.6 ha of crop land.[425]

Although the greatest production of herbivorous fishes occurs in managed ponds, these fishes are also grown on a mass scale in lakes, rivers, channel enclosures, and more recently, in cages.[433] Because Chinese carp do not reproduce in stagnant water, millions are stocked in lakes every year. It is recommended that the size of fishes at stocking be 16.5 to 23 cm.[434]

Recently, tilapia, specifically Mozambique and Nile tilapia, have been added to Chinese polyculture systems. Because the winter water temperatures are too low for these fishes, they must be wintered under greenhouses or in ponds receiving industrially heated water effluents. Tilapia have not been an important component in Chinese polyculture, but they have become the dominant fishes in Taiwanese polyculture (Tables 26 and 27).

Table 27 **Fish production in Chinese polyculture systems**

Production (kg/ha)	Remarks	Ref.
3,200	Guangdong, average for 2,400 ha	426
4,500	Pearl River delta, av. net yield for 28,000 ha	427
9,000	Yangtze River delta, highest net yield for 857 ha	427
9,750	Pearl River delta, highest net yield	427
11,430	Kiangsu Province, intensive culture, av. for 170 ha	425
32,500	Kiangsu, Wusi District, Tai Hu Commune, a record yield	131
39,600	Taiwan, tilapia primary species (see Table 26), pelleted feed, automatic feeders, paddlewheel aerators	412

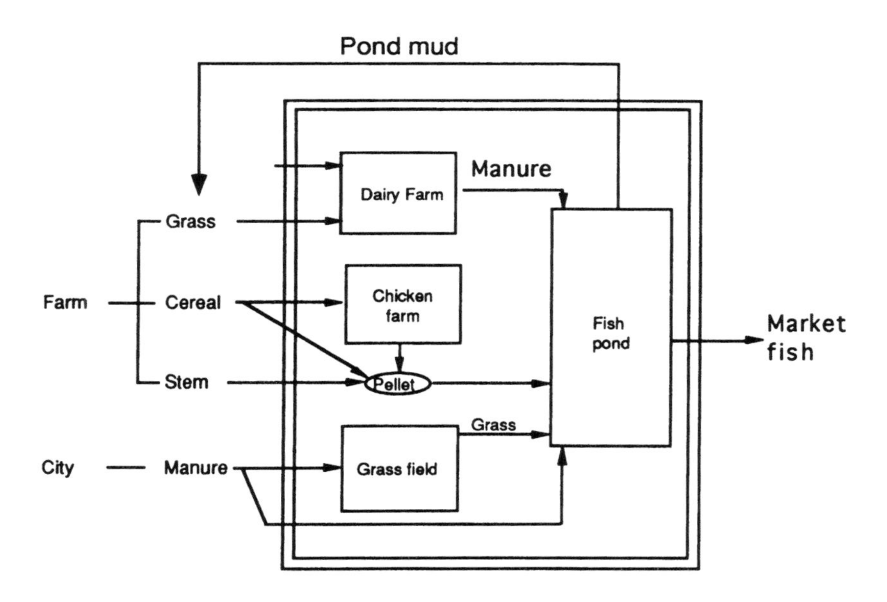

Figure 46 Structure of an integrated system consisting of agriculture, animal husbandry, and aquaculture (AAA system); Nanhui Fish Farm near Shanghai, China. (From Li, Sifa, *Aquaculture,* 65, 105, 1987. With permission.)

The Mozambique tilapia was first introduced into Taiwan in 1946 and since then four other species have been added, including *Tilapia zillii,* nile tilapia, blue tilapia *(T. aurea),* and *Sarotherodon hornorum.* Most of the tilapia are produced in polyculture with Chinese carp, but production practices are very diverse, including duck-cum-fish, chick-cum-fish, hog-cum-fish culture, or intensive monoculture production using pelletized food.[412] In polyculture, the yield of tilapia alone may reach 36 ton/ha/year where all male hybrids (female blue tilapia × male Nile tilapia) are used. At these high production levels mechanical aerators must be operated to increase pond oxygen.

In Taiwan striped mullet are sometimes grown in polyculture with Chinese carp or with milkfish. The striped mullet is a euryhaline fish that can be cultured in brackish or freshwater ponds. It feeds mainly on benthic detritus. In polyculture with Chinese carp, striped mullet at a stocking density of about 1000 to 3000/ha reach 300 to 400 g in 9 to 10 months. In intensive polyculture yields of 1800 kg/ha of mullet and 7200 kg/ha of milkfish can be expected in 7 months.[412]

2. Polyculture in Europe

Warm water pond culture in Europe has been dominated for centuries by common carp monoculture. Carp culture started very early and was already well developed by the late Middle Ages (14th to 16th centuries) in central and western Europe. Common carp have become traditional fasting and holy day diets for Christian and Jewish populations. Trials to incorporate native fish species into common carp ponds were unsuccessful because of the wide food range of the common carp, which consume basically all species of invertebrates eaten by other fishes. Herbivorous fishes of commercial importance did not exist in Europe prior to the introduction of Chinese carp.

Chinese carp utilize food resources that are not utilized by common carp (Figure 47), and effectively control undesirable vascular plants in ponds. Although this fish, especially where water plants are scarce, also consumes supplementary food given to common carp, the total food consumption in polyculture ponds does not exceed that in common carp monoculture ponds. This is due to the increased fertility of ponds caused by grass carp feces.[399] Production of grass carp is limited, however, by the amount of aquatic weeds in the carp ponds. The Chinese method of feeding grass carp with plants is impractical in Europe because labor costs are too high. Attention has instead been focused on silver and bighead carp. The algae and zooplankton on which they feed can be easily stimulated by fertilization.

An important question arose as to whether silver or bighead carp were more suitable for polyculture in Europe where subsidiary fishes should not decrease common carp yield. Although bighead carp, due to more rapid growth, give higher yields than silver carp at the same stocking density, they cause considerable decreases in common carp yield (Table 28). Although both silver and bighead carp do not consume the supplementary food given to common carp, the proportion of invertebrates in the bighead carp diet is greater and the species consumed are similar to those eaten by common carp. Bighead carp consume large cladocerans and copepods and also chironomid larvae, which constitute the basic food of common carp.[229] Silver carp consume smaller zooplankton, mainly *Bosmina longirostris,* and rotifers, which are not important to common carp. Hence, silver carp is more suitable than bighead carp for polyculture ponds in Europe.

Assuming that some food competition exists between silver carp and common carp, and that the food of the silver carp can be increased by inorganic fertilization, a practical problem arose as to how much the stocking density of silver carp could be increased without adversely affecting common carp production. To answer this question similar ponds were stocked with the same density of common carp, which results in maximum economic yield under local conditions (Poland), and with an increasing stocking density (0 to 12,000/ha) of silver carp. All ponds were fertilized (290 kg N + 47 kg P_2O_5 per hectare) and common carp were fed barley. Silver carp stocking caused a decline in common carp production. This decline, however, was low (260 kg/ha) and occurred at the low silver carp stocking density. An increase in silver carp stocking density from 8000 to 12,000 individuals per hectare did not cause an additional decrease in common carp

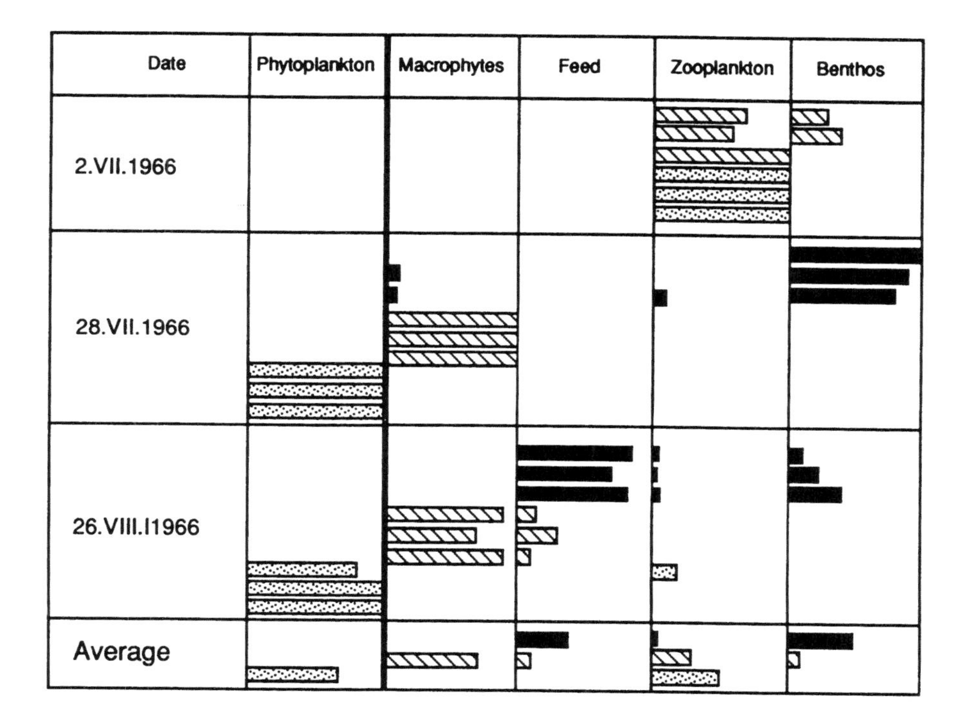

Figure 47 Percent composition of fingerling food of common, grass, and silver carp in polyculture, Zabieniec, Poland. Common carp were not stocked as of 2 July 1966. Three horizontal bars in each column depict three replications. (From Opuszynski, K., *Ekol. Pol.*, 27, 93, 1979. With permission.)

production (Figure 48). Total production was increased by almost 800 kg/ha at the highest silver carp density. Silver carp growth rates decreased considerably with increased stocking density, indicating that in polyculture with common carp maximum density is determined by intra- rather than interspecific relationships. Causes for the decrease in common carp production were not clear. The decrease could not be attributed solely to direct food competition between the two fish species as the standing crop of phytoplankton and chironomid larvae were higher in the polyculture ponds.[228,400]

A considerable increase in pond production has been obtained in Hungary with the introduction of Chinese carp.[9] The highest relative increase in total yield (common carp + Chinese carp) occurred in less fertile ponds where common carp yield was low. With increased fertility, common carp yield increases and becomes higher than that of silver, bighead, and grass carp (Table 29). Increases in silver carp yields and decreases in the grass carp yield, along with an increase in common carp production, constitute a typical phenomenon in fertile ponds.

In Hungary, when common carp and silver carp are integrated with pig farming (fish-cum-pig farming), high fish production of 2300 kg/ha is attained.[435] Inorganic fertilizers

Table 28 **Influence of silver and bighead carp on common carp production in ponds, Zabieniec, Poland**

Stocking (individuals/ha) and species	No. of replicants	Individual growth (g)	Survival (%)	Production (kg/ha)	In relation to control (kg/ha): Decrease of common carp production	In relation to control (kg/ha): Increase of total production
2000 C	3	671	83	1101	—	—
2000 C	3	608	91	1103	0	132
1500 Sc		108	90	130		
2000 C	3	578	94	1092	9	258
3000 Sc		98	93	267		
2000 C	2	546	82	873	228	33
1500 Bh		182	96	261		

Note: C = common carp, Sc = silver carp, Bh = bighead carp. All ponds were fertilized with urea at 200 kg N/ha and superphosphate at 40 kg P_2O_5/ha. Common carp in all ponds were fed sorghum *ad libitum*.

From Opuszynski, K., *Aquaculture,* 25, 223, 1981. With permission.

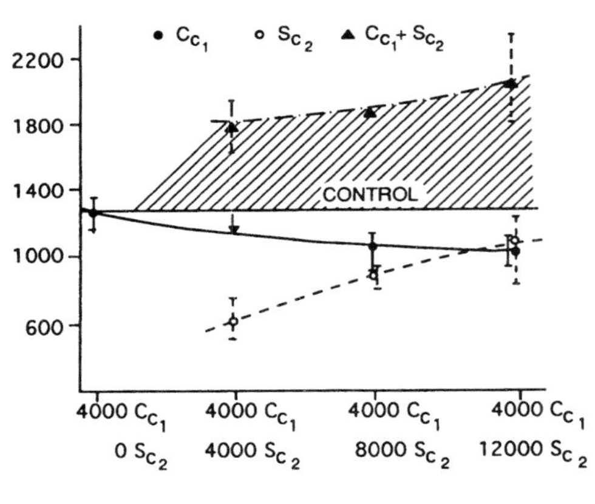

Figure 48 Influence of various stocking rates of silver carp on the production of common carp in ponds, Zabieniec, Poland. C_{c_1} = 1-year-old common carp, S_{c_2} = 2-year-old silver carp. Control line = carp production in monoculture at a stocking rate of 4000/ha. Hatched area = increase in total production due to silver carp. Vertical lines = standard deviations. (From Opuszynski, K., in *Managed Aquatic Ecosystems,* Michael, R. G., Ed., Elsevier, Amsterdam, 1987. With permission.)

and supplementary feeding were not used. Liquid pig manure was applied daily through sprinklers over the pond surface. Thirty-six to 60 pigs produced enough excreta to fertilize a 1-ha pond. The daily fish production in this system (18.0 kg/ha) was comparable to fish polyculture-cum-pig farming in India; however, the growing period in Hungary lasted 120 to 130 d, which was much shorter than in India.

Since the late 1950s, polyculture of Chinese and common carp has been widely practiced in the Commonwealth of Independent States.[436-438] The best results have been obtained in the warmer southern regions. For example, in the Krasnodar region (the Ukraine), Chinese carp comprised approximately 50% of the total yield in carp ponds, ranging from 1175 to 2360 kg/ha.[439]

Table 29 **Increase in total yield (without feeding and fertilization) due to introduction of Chinese carp into Hungarian ponds**

Monoculture: common carp	Yield (kg/ha) Polyculture: Common carp	Grass carp	Silver carp	Bighead carp	Total	Percentage increase compared to monoculture (%)
100	200	300	100	100	700	700
200	300	300	200	200	1000	500
300	450	150	300	300	1200	400
400	500	100	330	300	1230	300
500	550	70	350	300	1270	250
600	600	50	370	350	1370	230
800	800	50	390	350	1590	200
1000	1000	50	400	350	1800	180

From Antalfi, A. and Tölg, I., *The Herbivorous Fishes*, Panstwowe Wydawnictwo Rolnicze i Lesne, Warsaw, 1975, 270 pp.

In Bulgaria intensive polyculture of common, silver, and grass carp was tested using high-protein pelletized food and organic and inorganic fertilization. Under such conditions, common carp contributed 76%, silver carp 21%, and grass carp 3% to the total yield of 6292 kg/ha.[306] In other experiments in Bulgaria common carp were grown with silver carp and black buffalo *(Ictiobus niger)*, imported from the U.S. These species contributed 60, 17, and 23%, respectively, to the total yield of 5022 kg/ha.[307] In the former Czechoslovakia, 1547 kg/ha of 3-year-old marketable fishes was obtained (1177 kg common carp, 278 kg silver carp, and 92 kg grass carp) in ponds fertilized with superphosphate and urea; common carp were fed with wheat grain.[440] In the former East Germany pond production under intensive culture conditions, with grass carp as the primary species, amounted to 1709 kg/ha, including 1098 kg of grass carp, 352 kg of silver carp, and 259 kg of common carp.[441] Since 1971 more than 50 highly eutrophic lakes have been stocked with silver and bighead carp. Stocking rates ranged from 30 to 820 fishes per hectare, with an average individual weight of 350 g at stocking. Silver carp reached an average weight of 5.5 kg and bighead carp reached 11 kg after 7 years. Harvesting began 2 to 3 years after stocking.[299]

3. Polyculture in Israel

Fish culture in Israel was begun in 1939, but polyculture became a common practice only a few decades ago. Nevertheless, high production levels have been achieved, along with a great volume of practical experience. In the beginning common carp were the only fish grown. Fish farmers noticed, however, that due to the introduction of food and the climatic conditions the ponds became highly productive. The algae produced was not being utilized. To utilize this resource they began to use tilapia (*T. zilli,* Galilee cichlids, and blue tilapia) along with the common carp. These species are indigenous to Israel. At some farms, mainly those located along the seashore, two species of mullet (striped mullet and *Mugil capito*) were also cultured. Chinese carp have been used since 1965. At present the yield in polyculture ponds averages approximately 3000 kg/ha and in intensive culture ranges from 8000 to 10,000 kg/ha. In exceptional ponds production ranges from 20,000 to 25,000 kg/ha.[276,277]

Polyculture consisting of common carp, silver carp, and a species of tilapia was found to be the most successful combination. A thorough analysis of the interrelationships among these species was conducted by Yashouv[442] and Reich.[443] In combined culture with silver carp each of the species gave much higher yields than in monoculture. The addition of silver carp increased common carp yield by 22% and tilapia by 11%. The relationship between common carp and blue tilapia was more complex and can be defined as asymmetric interspecific competition. Blue tilapia did not affect the growth of common carp unless tilapia densities were very high, and, even so, the decrease in common carp growth was slight. The presence of common carp, however, caused tilapia to grow more slowly in all cases. In regards to the effect of common carp and tilapia on silver carp growth, it was always beneficial. These bottom-feeding species enhance silver carp growth rates by stirring up bottom sediments, thereby transporting nutrients and detritus to the water column.

The interactions among different fish species in polyculture were confirmed by Milstein.[444] The synergistic effect between common carp and tilapia and between common carp and silver carp occurred only at lower stock densities of either tilapia or silver carp. Above a density of about 1000 individuals per hectare of silver carp, a decline in the growth rate of common carp and tilapia occurred. This growth reduction was small at 1300 silver carp per hectare, but substantial at 2600/ha. In the presence of common carp and tilapia the growth rate of silver carp increased at both densities.[445]

Because of the increasing cost of supplemental feeding in intensive polyculture systems, studies have been conducted to determine if fish feed can be substituted by fertilizers. In the polyculture of common carp, silver carp, grass carp, and tilapia a yield of 4121 kg/ha was reached without supplementary feeding, but liquid cow manure was added to the ponds daily throughout the season. To produce 1 kg of fishes, 22.6 l of liquid manure (12% dry weight) was required. This equals a conversion ratio of 2.7 kg dry manure to 1 kg of fishes. Although 6282 kg/ha were obtained when fish were fed high-protein pellets, the food conversion ratio using pellets was 2.46, making production more expensive.[304]

The superiority of organic over inorganic fertilizers was discussed in older Israeli papers. It was argued that organic fertilizers were more efficient because they are based on heterotrophic organisms (bacteria and protozoa) feeding directly on the decomposed organic matter. Under these conditions, the production of natural food is dependent only on the amount of manure supplied and is not limited by light penetration as is the case when inorganic fertilizers are used to stimulate phytoplankton production. Recent studies, however, indicate that reevaluation of these assumptions is in order and that some controversy still exists concerning fertilization practices. In ponds receiving inorganic fertilizers and manure the organic matter in the manure contributes only slightly to fish growth, as the food chain originating with algae accounted for over 90% of the fish yield.[446] Algae were harvested either as live algae (by silver carp and possibly by tilapia) or as microbial processed detritus (by common carp, tilapia, and grass carp). The fish yield in ponds receiving inorganic fertilizers was not significantly different from that in the ponds receiving inorganic fertilizers plus manure.

4. Grow-Out of Chinese Carp in the U.S.

In the U.S. grass carp are grown primarily for vegetation control. The fishes must be large enough to avoid predation when stocked in open water bodies (see Chapter 5). Interest has been increasing, however, in using Chinese carp in polyculture with channel catfish as the primary species. The use of Chinese carp for animal waste management, resulting in additional fish production at the same time, has also been tested experimentally.

The first trials of intensive culture of grass carp in circular tanks were made using duckweed as food.[209] High survival (over 90%) and satisfactory growth rates (from 3 to 73 g in 88 d) were obtained. Feeding fishes with wet aquatic vegetation proved to be impractical for intensive culture because a great quantity of vegetation had to be fed daily, which was labor intensive and caused water quality problems at the high stocking density. Further experiments showed that in the intensive tank culture of larger fingerlings (about 50 g initial weight) fresh plant material could be successfully replaced by a complete catfish cage-culture diet.[99]

The most common current practice for growing grass carp fingerlings is to stock them in earthen ponds with macrophytes. At high stocking rates, the aquatic plants are quickly consumed by the fishes, and production must rely on supplementary feeding. In Alabama, at the stocking rate of 8000 fingerlings per hectare, only 5% of the food ration consisted of natural food, including ostracods, cladocerans, and filamentous algae.[447] The grass carp depended primarily on supplemental diets. They were fed either catfish pellets or a food containing either 19 or 38% dried alfalfa plant *(Medicago sativa).* The alfalfa diets were vitamin and mineral enriched and formulated to be equal to the catfish diet in estimated digestible energy (2.9 kcal/g) and digestible protein (28%). The fishes fed the diet containing 19% alfalfa grew faster (from 61 to 216 g in 84 d) and had a lower food conversion rate (1.4) than the fishes fed the other diets.

The beneficial effect of the inclusion of high-quality roughage such as alfalfa in the grass carp diet was not confirmed by further studies.[448] Fingerling grass carp averaging 5 g were stocked in earthen ponds at a rate of 11,875 fishes per hectare and were initially fed a commercial trout feed for 41 d, followed by a 23-d feeding period with floating catfish starter feed. Subsequently, the fishes were fed for 108 d with either a commercial catfish food or a grass carp floating food containing 15% dehydrated alfalfa meal. The catfish and grass carp feeds were similar in crude protein (35%) and estimated digestible energy (3.1 kcal/g). No significant differences occurred in growth rates between the foods used. The average yield was 1766 kg/ha. This study indicates that grass carp feed containing alfalfa meal gives results as good as commercial catfish feed. Also, grass carp may reach the size needed for weed control in one season in the southern U.S., where water temperatures range from 26 to 34°C during most of the growing period.

Promising results were attained in polyculture of the channel catfish as the basic species together with silver and bighead carp. The latter two species apparently do not compete for food with catfishes, which are grown using commercially produced pelletized feeds. They take advantage of the plankton that is not consumed by the catfishes. A more accurate evaluation of the impact of Chinese carp on channel catfish production in ponds has yet to be made. The authors dealing with this problem mentioned "little difference,"[449] "slight increase,"[450] or "not an appreciable increase"[451] in catfish yield. Whatever the influence, silver or bighead carp or both yielded from 465 to 1965 kg/ha,[313,449] and the net income was $1080/ha higher with polyculture when compared to the catfishes alone.[450] Polyculture of channel catfishes, freshwater prawns *(Macrobrachium rosenbergii),* Chinese carp, and tilapia was also tested.[452] The average total net production was 3149 kg/ha: 1554 kg catfishes, 351 kg prawns, 467 kg hybrid female bighead carp × male silver carp, 190 kg grass carp, and 587 kg tilapia.

High fish production was reached in the first attempt in North America to utilize swine manure in the polyculture of Chinese carp and American fishes.[305] Silver, bighead, and grass carp were stocked with common carp, hybrid buffalo (female *Ictiobus cyprinella* × male *I. niger*), channel catfish, and largemouth bass *(Micropterus salmoides)*. Ponds receiving wastes from 39 and 66 pigs per hectare were tested. The fishes were not fed. Total fish production after 170 d was 2971 and 3834 kg/ha at the lower and higher manuring levels, respectively. Silver carp accounted for 61 and 65% of the total production.

5. Culture Methods in Other Regions

In India polyculture of Indian major carp has been traditionally practiced for thousands of years using stocking ratios of 3:3:4 or 3:6:1 of catla, rohu, and mrigala, or 3:5:1:1 of catla, rohu, mrigala, and calbasu *(Labeo calbasu)*.[453] The catla is a surface or water column feeder; its food consists of planktonic algae, rotifers, crustaceans, insects, and decaying plant material. It grows fast under favorable conditions — to over 1.5 kg in the first year and to over 5 kg by the end of the second year. The meat is considered a delicacy by the local population. The rohu is predominantly a water column feeder, consuming phytoplankton, vegetable debris, zooplankton, detritus, and mud. It grows rapidly and has good eating quality. The mrigala is a bottom feeder, subsisting on blue-green and filamentous algae, diatoms, small pieces of higher plants, detritus, and mud. It grows fast and is highly esteemed as a food fish. The calbasu has a diet similar to the rohu.[454,455]

Only after 1961 were Indian major carp used in polyculture with Chinese carp. Fish production in polyculture, with improved management practices, has increased from 600 kg/ha to the highest recorded production, 10,500 kg/ha.[453] This technology, by analogy to the so-called Green Revolution in Agriculture, is referred to as the "Fish Revolution" or "Aquaplosion" in India. The proportions of the fish species used in polyculture are different (Table 30), and different fertilization and feeding practices are used to take full advantage of locally available raw materials and to make optimum use of the food resources existing in local bodies of water. Cow manure is a traditional organic fertilizer in India, but inorganic fertilizers are also used. Supplementary feeds are empirically formulated mixtures of brans (wheat, corn, rice, etc.) and oil cakes (groundnut, mustard, coconut, etc.). They are not prepared for any particular species, but serve all the species in the pond.

An example of polyculture resulting in production of 9389 kg/ha (net production was 9088 kg/ha) is given in Table 30. Rohu and silver carp contributed most to the stock (24% each), but rohu accounted for only 12% of the total production. Silver carp had the highest share in the yield (31%), followed by common carp (22%) and grass carp (19%). The culture period lasted 1 year. Water temperatures varied from 22° to 39°C. Cow manure was used at a rate of 10 t/ha/year and superphosphate, urea, and ammonium sulfate were also applied at 425, 188, and 125 kg/ha/year, respectively. Supplemental feed, including groundnut oil cake and rice bran, was given at 18 ton/ha. Aquatic weeds at 40 tons/ha were supplied for the grass carp.

Under intensive polyculture conditions, feed constitutes about 67% of the total production costs.[456] Maximal growth of Indian and Chinese carp occurs when the dietary protein content is 45%.[457] Therefore, attempts were made to increase grass carp numbers to replace the supplemental food with aquatic and terrestrial vegetation.[458] Grass carp accounted for 50% of the total stock density, with catla, rohu, mrigala, silver carp, and common carp at 10% each. Fertilizers and prepared feeds were not supplied, but grass carp were fed with a total quantity of 119 ton/ha of aquatic weeds and terrestrial grasses. Relatively high yields ranging from 12.3 to 15.1 kg/ha/d (the culture lasted from March to December) was obtained at remarkably low input costs of $0.08 to 0.12/kg of fishes.

Table 30 **Fish combinations used in polyculture in India as percentages of total number of fishes (%TN) and total yield (%TY)**

	Percentage in polyculture					
Fish species	%TN[721]	%TY[721]	%TN[721]	%TY[721]	%TN[710]	%TY[710]
Catla	10	19	25	31	9	6
Rohu	30	27	30	33	24	12
Mrigala	12	14	12	14	9	9
Silver carp	25	30	8	12	24	31
Grass carp	10	5	2	1	9	19
Common carp	13	5	22	9	19	22
Others					6	1
Total yield (kg/ha)	5253		4506		9389	

The method was cost effective but labor intensive, and could be recommended only in those areas where labor is inexpensive and submerged aquatic vegetation such as hydrilla, *Potamogeton,* and *Ceratophyllum* or terrestrial grasses and vegetable wastes are abundant.

In search for new food resources for grass carp, suitability of waste mulberry leaves was tested and was found to be on a par with lucerne leaves and superior to hybrid napier.[459] Grass carp culture was also conducted in bamboo cages where fishes were fed vegetation. Lucerne gave the best fish growth, followed by paragrass and hybrid napier.[460] The importance of grass carp in Indian polyculture is confirmed by observations that their excreta may enhance the growth of indigenous carp to about the same extent as fertilization with cow manure.[461] Although cow manure is widely used, integrated fish-cum-pig farming was tested. The excreta of 35 to 40 pigs was adequate to fertilize a 1-ha pond and to support fish production of 18 kg/ha/d.[435] Polyculture consisted of grass and silver carp with Indian carp. The growing period lasted 270 to 365 d, with water temperatures ranging from 19° to 34°C.

In Southeast Asia aquaculture developed gradually over hundreds of years by first employing Chinese culture methods that were extended and modified to suit local conditions. Grass, bighead, silver, and common carp are the principal species used in varying combinations in Malaysia, Thailand, and Singapore.[462] In Malaysia yields per hectare in polyculture range from 1200 to 2700 kg, depending on stocking densities and management practices. Grass carp are fed with duckweed and *Wolffia,* water kangkong *(Ipomas reptans),* and hydrilla grown in small ponds adjacent to fish ponds. Land grasses such as napier or guinea grass *(Panicum maximum)* are also used with leaves of tapioca, sweet potato, yam, papaya, and banana. Other supplementary feeds include rice bran, coconut cake, peanut meal, peanut cake, sago, boiled crushed maize, and soyabean.[463]

The digestibility of locally available grasses by grass carp appeared to be very low (<20% dry matter). Once the cell wall of the grass was ruptured, however, grass carp were able to efficiently absorb the released soluble proteins. This information was used in Malaysia to formulate a nutrient-balanced pelletized food containing 30% napier grass for grass carp. When these grasses were dried, ground to 2 mm diameter, and included in the feed, their digestion coefficients were higher than those of maize, copra cake, or rice bran.[464]

Besides traditional polyculture, pig-cum-fish farming, including Chinese carp and Mozambique tilapia, was tried with successful production and financial results in

Malaysia.[465] Cage culture is also widely used and has a long history in Southeast Asia. Cages are made of wood, bamboo, rattan (an Asiatic palm), and canvas. In Indonesia the cages are set in the streams that pass through villages and carry heavy loads of animal and human wastes, which stimulate rich plankton development. Bighead carp cage culture was also tested in the Philippines.[151] No differences in the growth rate were found among the fishes fed 40 and 20% protein diets, and those not fed at all. The fishes in all three treatments attained sexual maturity and were spawned after hormonal injection. Successful results of cage culture of bighead and silver carp were also reported in Nepal.[466] In Lake Begnas the harvest of both species from cages contributed from 26 to 48% to the total annual harvest from the lake. The caged fishes were not fed and utilized the plankton resources of the lake with little competition from the indigenous species. Little is known about growing Chinese carp in Africa. Experimental polyculture of silver and bighead carp together with Indian carp resulted in excellent yields on Mauritius.[467] The silver carp is considered a suitable fish for polyculture with common carp in ponds receiving cattle manure and inorganic fertilizers in South Africa.[468]

Milkfish farming is the most important brackish water monoculture in Southeast Asia, including the Philippines, Indonesia, Taiwan, Thailand, Vietnam, and Malaysia.[455] The milkfish inhabits the tropical waters of the Indian and Pacific Oceans, but has not been an important sea fishery. Milkfish are successfully cultured because they are able to live in a wide range of temperatures (8.5 to 42.7°C) and salinities (0 to 158‰).[412]

The milkfish is herbivorous, feeding primarily on benthic algae and occasionally on phytoplankton and small invertebrates. The green filamentous algae *Enteromorpha* and *Chaetomorpha* form a bottom mat in extensively managed ponds. The fishes graze upon this mat, but the food conversion rate is relatively low (20 kg of algae for 1 kg of fishes produced). In intensive culture situations where organic and inorganic fertilizers are used, the algal mat consists mainly of blue-green algae *(Anabaena, Anthrospira, Lyngbya, Microcoleus, Oscillatoria,* and *Spirulina),* which is inhabited by numerous bacteria and protozoa. The conversion rate when milkfishes feed on this mat, locally called "lab-lab", is greater (15:1); therefore, higher fish yields are obtained. Different management methods and culture systems give variable yields averaging between 600 and 2500 kg/ha.[455]

The most important management practice in these systems is to keep a balance between the algal mat standing crop and fish biomass. The mat should not be overgrazed by the fishes, but if the feeding pressure is too low, the mat continues to grow until it detaches from the bottom and floats to the surface. This is undesirable because the floating mats cause the productive capacity of the pond to decrease.

The milkfish fishery in the large brackish lagoons of Laguna Bay (90,000 ha) near Manila, the Philippines, are impressive.[469] These lagoons are extremely fertile due to the high concentration of sewage from surrounding villages and cities. Besides traditional pond culture, a new system using fish pens has been developed. The term "fish pen" has a broad meaning in Asia. It can include natural depressions in the ground along the shoreline, which are temporarily filled with water during the rainy season, as well as enclosures made by damming or netting off parts of a body of water. In Laguna Bay milkfish culture pens covered 4800 ha in 1973 and 7000 ha in 1974, and by 1983 the total pen area reached 10,000 ha, providing a yield between 4000 and 7000 kg/ha/year. Total milkfish production in the Philippines is between 200,000 to 250,000 ton/year, and it is predicted that it will eventually reach 500,000 tons.

IV. DISEASES

Chinese carp are susceptible to a variety of infectious diseases and have a rich parasite fauna (Table 31). Most of the diseases and parasites of Chinese carp have been reported from culture conditions because high stocking densities, enriched waters, and stress stimulate pathogens. Also, more studies have been conducted on cultured fishes than on wild populations. Worldwide introduction of Chinese carp enhanced the possibility of the spread of fish diseases and parasites.

After the first few years of introduction of Chinese carp into the former U.S.S.R., parasites such as *Crytobia branchialis* (Protozoa) and *Capillaria amurensis* (Nematoda) were found. Despite strong regulations and quarantines, 27 additional parasite species were introduced.[470] In order for a quarantine to be effective it must be in place for a long period. For example, when fishes were introduced into Germany, only 25% of the total parasites they harbored were immediately identified. After a 2- to 4-year quarantine period, an additional 50% were identified, and the remaining 25% were identified after 5 to 10 years of quarantine.[471] Two species of cestodes *[Bothriocephalus gowkongensis (= acheilognathi)* and *Khawia sinensis]* and a nematode *(Philometra lusiana)* were among the most dangerous parasites introduced to Germany. *B. gowkongensis* heavily infested common carp in eastern Germany. This parasite was found on 220 fish farms, where the average rate of infestation of common carp stocks was 14%.[126]

A number of parasite species of Far East origin were detected in Chinese carp in Hungary.[472] Among them were the protozoans, *Balantidium ctenopharyngodontis, C. branchialis, Eimeria sinensis, Entamoeba ctenopharyngodontis, Myxidium* sp., and *Spironucleus* sp., and the helminths, *Amurotrema dombrowskajae, B. gowkongensis, Dactylogyrus ctenopharyngodontis, D. lamellatus,* and *D. nobilis.* Only a few of these parasites, including *C. branchialis, Spironucleus* sp., and *B. gowkongensis,* have established themselves on native fishes. Of these, only *B. gowkongensis* became widely distributed and is now one of the most significant pathogens in carp ponds. It is believed that *Dactylogyrus suchengtaii,* a gill parasite of silver carp, arrived in Hungary through rivers from Romania, where it was previously introduced with fishes imported from China or the former Soviet Union.

No precautionary measures were taken when grass carp were introduced into Florida (U.S.); however, no exotic parasites were identified.[473] Both *B. gowkongensis* and *D. ctenopharyngodontis* were introduced into other areas of the U.S.

In order for tapeworms or trematodes to become established in a new area, proper intermediate hosts must be present. For example, *B. gowkongensis* and *D. ctenopharyngodontis* became established in the U.S. because a suitable intermediate host was available. Other parasites did not become established because a suitable intermediate host was not available. If mature fishes or fishes exposed to parasites in their native range are introduced it is likely that some parasites will survive in the new habitat, even though the fishes are quarantined. In order to counteract this spread of parasites, parasite-free fish larvae that have not begun to feed should be introduced.

Chinese carp, when introduced to a new location, may not be immune to indigenous parasites and infectious diseases; therefore, they can be easily infected by local pathogens and may suffer heavy losses. Interested readers are directed to the extensive literature available on this subject.[11,22,126,474–478]

Table 31 **Infectious disease agents and parasites of Chinese carp**

	Ref.		Ref.
Viruses		***Protozoa** (continued)*	
Rhabdovirus sp.	11	*Chilodonella* sp.	11
R. carpio	11	*C. cucullulus*	22
R. GRV	722	*C. cyprini*	11
		C. hexasticha	22
Bacteria		*Chloromyxum* sp.	11
Achromobacter sp.	11	*C. cyprini*	11
A. curydice	722	*C. nanum*	11
A. pestifer	722	*Costia necatrix*	11
Aeromonas sp.	11	(= *Ichthyobodo necator*)	
A. punctata	11	*Cryptobia* sp.	11
A. salmonicida	11	*C. branchialis*	11
var. *achromogenes*		*C. cyprini*	11
Bacillus cereus	722	*Dexiostoma campylum*	22
B. megaterium	722	*Eimeria carpelli*	11
Carp erythrodermatitis	722	*E. cheni*	22
bacteria		*E. mylopharyngodonis*	11
Citrobacter sp.	724	*E. sinensis*	11
Flavobacterium aquatile	722	*Enamoeba ctenopharyngodontis*	11
Flexibacter columnaris	11	*Epistylis* sp.	11
Micrococcus luteus	722	*E. lwoffi*	11
M. flavus	722	*Euglenosoma caudata*	11
Myxococcus piscicola	11	*Frontonia acuminata*	22
Paracolobactrum aerogenoides	722	*F. leucas*	22
Pseudomonas sp.	11	*Glaucoma pyriformis*	11
P. dermoalba	22	*G. scintillans*	22
P. fluorescens	722	*Glugea* sp.	722
P. fragi	722	*Hemiophrys macrostoma*	11
P. putida	722	*Hexamita* sp.	11
Staphylococcus aureus	727	*Icthyophthyrius* sp.	11
		I. multifiliis	11
Fungi		*Myxidium* sp.	11
Branchiomyces sanguinis	11	*M. ctenopharyngodonis*	11
Saprolegnia sp.	11	*Myxobolus dispar*	11
Ichthyophonus hoferi	722	*M. ellipsoides*	11
		M. drjagini	722
Protozoa		*M. pavlovskii*	722
Apiosoma sp.	22	*Sessilia* sp.	22
A. cylindriformis	11	*Sphaerospora carassii*	11
A. magna	11	*Sphaerosporidae lieni*	725
A. minimicro nucleata	11	*Spironucleus* sp.	11
A. piscicola	11	*Tetrahymena pyriformis*	11
Balantidium	11	*Thelohanellus oculi-leucisci*	11
ctenopharyngodontis		*Trichodina* sp.	11

Table 31 (continued) **Infectious disease agents and parasites of Chinese carp**

	Ref.		Ref.
***Protozoa** (continued)*		***Trematoda** (continued)*	
T. bulbosa	11	*D. spathaceum*	11
T. carasii	11	*Diplozoon* sp.	722
T. domerguei	11	*D. paradoxum*	11
T. meridionalis	11	*Fasciolata* sp.	722
T. nigra	11	*Gyrodactylus* sp.	11
T. nobilis	11	*G. ctenopharyngodontis*	11
T. ovaliformis	11	*G. elegans*	726
T. pediculus	11	*G. kathariner*	11
T. reticulata	11	*G. medius*	722
Trichodinella sp.	722	*G. wageneri*	722
T. subtilis	722	*Metagonimus yokogawai*	11
Trichophrya sp.	11	*Opisthorchis* (= *Chlonorchis*) *sinensis*	11
T. piscium	722		
T. sinensis	11	*Posthodiplostomum* sp.	22
T. variformis	723	*P. cuticola*	11
Tripartiella sp.	11	*Sphaerostoma bramae*	722
T. bulbosa	11	*Tetracotyle* sp.	11
T. lata	11	*T. percae fluviatilis*	11
Trypanoplasma sp.	722	*T. variegata*	11
Zschokkella nova	11		
		Cestoda	
Trematoda		*Biacetabulum appendiculatum*	11
Amurotrema dombrowskajae	11	*Bothriocephalus gowkongensis* (= *acheilognathi*)	11
Ancyrocephalus subaequalis	11		
Apharyngostrigea curnu	11	*B. opsarichthydis*	722
Aspidogaster amurensis	11	*Diagramma interrupta*	22
Cotylurus communis	11	*Khawia sinensis*	11
C. pileatus	11	*Ligula intestinalis*	11
Dactylogyrus sp.	11	*Triaenophorus lucii*	722
D. aristichthys	22	*T. nodulosus*	11
D. ctenopharyngodontis	11		
D. hypophthalmichthys	726	***Nematoda***	
D. inexpectatus	726	*Capillaria amurensis*	722
D. lamellatus	11	*C. pretrushewskii*	722
D. magnihamatus	11	*Capillaria* sp.	11
D. nobilis	22	*Philometra* sp.	11
D. scrjabini	726	*P. lusiana*	11
Diplostomum sp.	11	*Philometroides lusii*	722
D. indistinctum	11	*Rhabdochona denudata*	11
D. macrostomum	11	*Skrjabillanus amuri*	722
D. mergi	11	*Spiroxys* sp.	11
D. paraspathaceum	11		

Table 31 (continued) **Infectious disease agents and parasites of Chinese carp**

	Ref.
Hirudinea	
Hemiclepsis marginata	722
Piscicola geometra	722
Arthropoda	
Argulus sp.	11
A. foliaceus	722
Ergasilus sp.	722
Lernaea sp.	11
L. ctenopharyngodontis	11
L. cyprinacea	11
L. elegans	11
L. piscinae	22
L. quadrinucifera	11
Neoergasilus longispinosus	11
Paraergasilus medius	11
Sebekia oxycephala	11
Sinergasilus lieni	11
S. major	11

Chapter 5

Utilization of Grass Carp for Aquatic Weed Control

I. NUISANCE PLANTS

Nuisance aquatic plants are those that in some way cause serious problems in bodies of water. These problems can be either real or perceived. In different bodies of water the nuisance rating of a particular plant may change according to the intended use of the body of water. A plant may be a nuisance in one place because of coverage or volume infestation, but may be advantageous in another system because it provides cover and food for fishes and wildlife.

The negative effects aquatic plants have on the environment were listed by Pieterse:[479]

1. They impede the movement of water in irrigation and drainage ditches and canals
2. They hinder navigation
3. They interfere with hydroelectric generation
4. They trap silt particles, thus increasing sedimentation
5. They decrease fisheries when in dense situations
6. They adversely affect recreation (swimming, boating, angling)
7. They increase water loss by evapotranspiration
8. They increase health hazards by creating habitats favorable to vectors of human diseases

Aquatic plants are classified into floating, submerged, and emergent categories. Floating plants are probably the worst pests because they are noticeable, have a high growth rate, block light, which causes oxygen depletion below the mat, and when dense can block waterways and can become floating islands when the mat is invaded by other species. The water hyacinth is probably one of the most noxious of the floating plants (Figure 49), causing innumerable problems outside its native range. Spencer and Bowes[480] list two additional floating plants that are problematic on a global scale: *Pistia stratiotes* and *Salvinia molesta.* Other floating plants can cause problems in localized areas.

Submerged plants are usually not as noticeable as floating plants, but also cause serious problems in bodies of water. Most submerged plants have wide ecological tolerances, and can spread by vegetative growth. Six species were identified by Spencer and Bowes[480] as being problematic globally due to their prolific growth and reproduction: *Ceratophyllum demersum, Potamogeton crispus, P. pectinatus* hydrilla, *Myriophyllum spicatum,* and *Elodea canadensis.*

Emergent species are usually restricted to shallow water, but can become dense along shorelines. Five species are listed as globally important by Spencer and Bowes:[480] alligator weed *(Alternanthera philoxerides), Echinochloa crus-galli, Phragmites australis,* torpedo grass *(Panicum repens),* and cattail (*Typha* sp.). Emergent plants can reproduce vegetatively from nodes, stolans, or rhizomes[480] and by sexual means. Cattails, for example, produce seeds that are distributed long distances by the wind.

Figure 49 Water hyacinth mat. (Photographed by K. Langland.)

II. MECHANICAL, CHEMICAL, AND BIOLOGICAL METHODS FOR AQUATIC PLANT MANAGEMENT

Noxious aquatic weeds are controlled primarily by mechanical, water level manipulation, chemical, and biological methods.

A. MECHANICAL CONTROL

Hand removal methods were the only methods available until mechanical harvesters were developed. Hand removal of aquatic plants with various hand-held tools (rakes, hoes, forks, etc.) is still used today, especially in areas where inexpensive labor is available.[481] Hand methods are restricted and are of little use in large bodies of water, but may be useful in small, selected areas where a selective approach to management is needed.[482]

Mechanical harvesters come in many varieties, sizes, and shapes. Some are designed only to cut and remove aquatic plants, whereas other machines used to remove aquatic plants are not designed solely for that purpose, but have other functions. For example, dredges, backhoes, and draglines, designed for construction purposes, are also used to remove aquatic weeds from irrigation ditches and canals. A number of reviews are available that discuss the history of their development and describe the various types of machines.[482–485]

Floating aquatic weed harvesters have been developed to cut and remove floating and submerged vegetation. These machines usually employ reciprocating cutter bars, which are lowered to a required depth to cut the plants (Figure 50). Early cutters did not remove

Figure 50 Weed harvester used to harvest submerged aquatic weeds. (Photographed by K. Langland.)

the cut vegetation from the water. Harvesters are now available which not only cut the plants, but pick them up on a conveyer belt and dump them on a deck where they can be off-loaded onshore or to barges that transport the cut material to shore.

Plant harvesting has several disadvantages, including costs of harvester maintenance, transportation costs, ineffectiveness in large bodies of water, or fast-growing plants that require repeated treatments (plants grow faster than they can be harvested). Advantages include the removal of nutrients from the water; plants that can be removed quickly from selected areas; selective removal of plants, leaving desired vegetation; and the possible production of useful materials from the harvested plants.

B. CHEMICAL CONTROL

The use of chemical control methods is more widespread than mechanical methods and they are more economical. Large areas can be treated more effectively because chemicals can be applied more effectively over a wide area, and have longer lasting effects. However, further knowledge of their proper application is needed for their use. If improperly used they can have negative side effects not only for aquatic organisms and wildlife, but for humans as well.[486]

The objective of any herbicide treatment is to get the active ingredient to the plants to be treated. Floating and emergent plants are more easily treated as the chemical can be sprayed onto them directly, whereas submerged plants are treated by applying the chemical directly to the water. Therefore, the amount of herbicide used is based on the volume of water to be treated rather than on the plant surface area treated.

An herbicide formulation consists of an organic or inorganic active ingredient, an inert carrier, and in some cases adjuvants (wetting agents, emulsifiers, or antifoaming agents). In the U.S. herbicides must be registered by the U.S. Environmental Protection Agency (USEPA). Of the almost 200 herbicides currently registered only nine are labeled for aquatic use, and two of these are used only for irrigation ditches in 17 western states. The seven active ingredients that remain available for aquatic weed control are copper, dichlobenil, diquat, endothall, fluridone, glyphosate, and 2,4-D.

These herbicides are classified as either contact or systemic herbicides. Contact herbicides act quickly and are generally lethal to all plant cells they contact. Because the active ingredient does not move within the plant, they are effective only where they contact the plant and are more effective on annual plants. Perennial plants (woody plants) are defoliated by contact herbicides, but they quickly resprout from the unaffected part of the plant. Aquatic plants that are in contact with sufficient concentrations of the herbicide are affected, but regrowth occurs from the protected part of the plant beneath the hydrosoil. Because the entire plant is not killed, retreatment is necessary, sometimes several times a year. Endothall, diquat, and copper are contact herbicides.

Systemic herbicides are absorbed into and move within the living plant. Those herbicides absorbed by plant roots are referred to as soil-active herbicides and those absorbed by the leaves are referred to as foliar-active herbicides. Dichlobenil, fluridone, glyphosate, and 2,4-D are systemic aquatic plant herbicides. Systemic herbicides are more selective and act slowly in comparison to contact herbicides. They are more effective for controlling perennial and woody plants.

Aquatic herbicides must be taken up by the target plants quickly in sufficient amounts to be toxic, but remain nontoxic to other organisms. Because herbicides are applied in very low concentrations, if applied properly, they are not persistent in the water treated and disappear faster when applied to large bodies of water because of dilution.

Herbicides also can be classified according to the number of plants they control. Broad-spectrum herbicides allow control of most vegetation within the treatment area, whereas selective herbicides control only certain plants. Broad-spectrum herbicides are used for total vegetation control in areas where bare ground is preferred. Glyphosate is an example of a broad-spectrum aquatic herbicide. Others, including endothall, diquat, and fluridone, are used as broad-spectrum herbicides, but also can be used as selective herbicides under certain conditions. An example of a selective herbicide is 2,4-D, which can be used to control broadleaf plants with little impact on grasses. Selectivity also can be related to the method of application. An herbicide can be selective by carefully placing the herbicide on the target plant and avoiding nontarget plants. Selectivity also can be accomplished by the rate of application. For example, water hyacinth can be controlled selectively among spatterdock when used at recommended rates. Spatterdock is controlled when higher rates are used or if granular formulations of the same chemical are used. Selectivity also depends upon the ability of a plant to alter or metabolize a herbicide so it no longer has herbicidal activity. Some herbicides affect only very specific plant biochemical pathways; therefore, they control only those plants that have that particular pathway. Selectivity may also depend upon the growth phase of the plant. For example, during early growth upward transport of food reserves occurs so that soil-active herbicides are most rapidly absorbed and moved upward to the actively growing portion of the plant and points of herbicide activity. Herbicides that act through the foliage (glyphosate) are least active during this stage of plant growth, but are most active when perennial plants are completing their annual growth cycle. At this time materials are translocated downward toward the roots where the herbicide is active. It has been noted that some plant parts are more susceptible to herbicides. For this reason some plant parts can be killed without harming the entire plant. For example, it is possible to use diquat, a broad-spectrum herbicide, to selectively manage water hyacinth among bulrush stems. The exposed (emergent) stem of the bulrush is affected and the extensive rhizome and root system of the plant is unaffected, so the plant can regrow quickly.

C. WATER LEVEL MANIPULATION

Water levels can be raised or lowered to manage aquatic plants. When lowered, aquatic plants are exposed to the elements, and other mechanical means of control can be used more easily. When raised, water covers the plant, inhibiting light penetration, causing them to die. This control method is limited primarily to reservoirs where water can be drained through a control structure.

Drawdown is the most common method and has been used for many years not only to manage aquatic plants, but to manage fish populations and to oxidize and consolidate bottom sediments. When this method is employed, water usage must be considered as drawdown will transmute recreation, esthetics, and agricultural uses of the body of water. It is also necessary to have a source of water for refilling.

Drawdowns are usually conducted during the winter as most of the adverse impacts of drawdown are minimized, and in the northern latitudes the exposed plants are subjected to freezing temperatures. Summer drawdowns can be used, but during these events a greater chance exists for oxygen depletion, causing fish kills and greater impacts to other recreational water uses. There is also greater likelihood for the spread of emergent plants such as cattails and willows.

There is no doubt that drawdowns will alter the vegetation community, however, the results obtained after the drawdown may not be always desirable. In general, submerged plants have variable responses to drawdown, whereas emergent plants are likely to increase.

Drawdown effects were reviewed by Cooke,[487] who concluded that few aquatic plants were consistently controlled in a range of lakes in the U.S. A number of reasons were given for the limited success of the drawdown programs:

1. Plants are protected by formation of a protective barrier to desiccation.
2. Soil type influences rate of desiccation; sandy soils will dry faster than clay soils.
3. Nuisance plants may be more resistant to drying than other plants or parts of the plant (seeds or tubers) may be resistant to drawdown.

In addition to these direct effects he listed several other negative factors:

1. Water restrictions are imposed on water usage.
2. The release of nutrients from decaying vegetation can cause algae blooms after reflooding.
3. Additional plants grow in the remaining drawn down portion of the reservoir.
4. Dissolved oxygen reductions occur in shallow water, increasing fish kills.
5. Fish spawning and migratory fish movements may be changed.

The advantages of drawdowns include the low cost of the management procedure, consolidation of bottom sediments, and possible benefits to the fishery. In addition other control methods can be used more efficiently and inexpensively while the reservoir is down.

III. SURVEY OF GRASS CARP USAGE FOR AQUATIC WEED CONTROL

The potential of grass carp for weed control has probably provided more reason for its spread than its suitability as a food fish. The grass carp did not become a species for weed control until methods for its propagation were developed in the 1950s and 1960s. Once propagation methods were developed, broodfishes could be moved to different countries, negating the need to import vast numbers of small fishes.

A. FORMER U.S.S.R. AND EASTERN EUROPE

Herbivorous fishes were first introduced into the former U.S.S.R. in 1937, but these introductions were unsuccessful. The first successful introduction occurred in 1949, when fishes from the Amur River basin were introduced into the Moscow area.[488] Between 1958 and 1963 approximately 8 million fry were imported from China. In 1961, after the first successful breeding of the grass carp, they were introduced throughout the U.S.S.R. to increase fish production and to control aquatic weeds. A number of experiments were carried out to determine the usefulness of grass carp for weed control.[183,185] The most successful introduction of grass carp occurred in the Kara Kum Canal, where natural reproduction occurred. Complete weed control and increased fishery production occurred in this water system.[109,157,184,258]

Nickolsky and Aliev[258] reported that in some of the reservoirs where the grass carp controlled weeds and where silver and bighead carp were not present, phytoplankton blooms occurred. Other systems were stocked with grass carp, but either the introductions were unsuccessful or recruitment was lacking. In the Volga River where intensive stockings were made each year, only a few fishes were produced (1 fish per hectare). After 20 years, similar results were obtained in the Kuban, Syr Darya, and Ili Rivers. Yefimova and Nikanorov[489] discussed the prospects of weed control in Ivan'kovskoyo Reservoir. Other reports of weed control in the U.S.S.R. include a 66% reduction in an irrigation canal within 2 months[490] and utilization in ponds.[491,492] Negonovskaya[310] reviewed the results of stocking grass carp in the former U.S.S.R. and concluded that stocking was unsuccessful in the central regions if underyearling fishes were stocked. More success was obtained in southern latitudes, where the major predator (pike, *Esox lucius*) of the grass carp was limited. He concluded that faster growth rates in the southern region allowed for shorter predation times.

Grass carp were introduced into most eastern European countries in the early 1960s primarily as a food fish. Therefore, research was directed toward reproduction and the use of these fishes for food production.

Grass carp were successfully spawned in Poland in 1970 and used in polyculture with common carp and the other Chinese carp,[399,493] and investigated for weed control.[191,205] Because climatic conditions caused difficulties in spawning grass carp and caused slow growth and low larval survival, the use of thermal effluents was investigated.[494,495] Considerable work has also been done in Poland relating to diets and feeding of herbivorous fishes.[256,270]

Grass carp were first introduced into Hungary in 1963, and this has positively influenced the development of Hungarian fish farming, sport fishing, and aquatic weed management.[496] Investigations on the genetics and productions of sterile fishes,[347,497–499] the importance of exotic fishes in Hungary for fish farming,[435,496] and diet formulation[267,268] have been conducted since their introduction.

Grass carp were introduced into the former Czechoslovakia and Bulgaria, where they controlled a variety of weed species,[488,500] but again they were used primarily as a food species.[488] In the former East Germany grass carp were not only used for food production, but also were used successfully to remove aquatic plants from canals and irrigation ditches more inexpensively than with other methods.[291]

B. WESTERN EUROPE AND UNITED KINGDOM

Many western European countries have investigated grass carp for weed control and as a species to raise in combination with common carp. Researchers in The Netherlands have reported on the use of grass carp for weed control.[291,501-503] Grass carp were first imported to The Netherlands from Hungary in 1966 and from Taiwan in 1968.[503] Results from a long-term field study begun in 1973 indicated that grass carp were a promising means of weed control; therefore, the individual managers of the bodies of water were given permission to experiment with grass carp under certain restricted conditions.[504] This research program was conducted by a working group made up of individuals from the various organizations having weed control responsibilities in The Netherlands. This group concluded that the grass carp was an efficient tool to manage aquatic plants under field conditions; that there would be no natural reproduction in The Netherlands, so the grass carp would not be spread; and that the grass carp should be released under certain defined conditions. These conditions were that grass carp would not be stocked unless a weed problem arose, the body of water was suitable for grass carp to live in, the interests of nature conservation were taken into account, and that the maximum stocking rate was 250 kg/ha. Permission to stock grass carp was granted by the Ministry of Agriculture and Fisheries on the advice of local civil servants. The following conditions of permission were to be met: the grass carp must be delivered from disease-free stock, all stocking areas had to be enclosed by a barrier and permission to stock covered a period of 5 years. This procedure has not changed since its inception.[503] The area stocked by grass carp has increased steadily since 1977. It was estimated that about 1500 ha were stocked in 1987. An evaluation of stocked sites indicated that in normal years, when winters were not severe, stockings were successful >50% of the time and insufficient control occurred in 15 to 29% of the sites.[503]

Preliminary experiments in the U.K. indicated that the grass carp could control aquatic plants, and stocking densities were determined to remove specific amounts of vegetation from ponds.[505] If understocked in systems containing a variety of vegetation, grass carp consume the most palatable species first, which allows uneaten species to increase, resulting in no net decrease in vegetation. In such situations supplementary removal of the unpalatable species may be required using herbicides.[506] The use of grass carp is tightly controlled in British waters as a separate license is required for each stocking, and this requires an investigation of each site. Prior to each stocking the Nature Conservancy Council, which is the government agency having statutory responsibility for environmental conservation in the U.K., is consulted by the licensing authority.[486]

The use of grass carp in western Germany was described by von Menzel.[507] In the northern areas of Germany weed control usually requires dense grass carp stocks, which may not be economical.[508,509] Riechert and Trede[510] reported that 2- to 3-year-old grass carp showed promise for the control of water hyacinth. Bohl[471] discussed the history of grass carp introduction to Germany and researched the possibility of diseases being introduced with the carp. He also discussed the treatment of diseases and parasites carried by introduced grass carp.

C. UNITED STATES

Grass carp were first imported into Alabama and Arkansas in 1963. Arkansas researchers immediately began studying grass carp in lakes. Successful weed control was attained with grass carp in Lake Greenlee in less than 1 year, and by 1975 aquatic macrophytes were being controlled in over 100 large lakes.[511,512] Research in Alabama was related more to its plant-feeding selectivity in small pools and weed control potential in small ponds.[188,513,514]

Since these initial studies, a tremendous amount of research was devoted to the grass carp and its potential for weed control, and to the possible adverse impact the grass carp may have on endemic fish populations, water quality, and other aquatic organisms.

Most of the initial work in the U.S. was accomplished in aquaria, tanks, and ponds to determine the efficacy of the grass carp for weed control.[515–517] In these studies researchers found that grass carp ate a wide variety of aquatic plants in quantities great enough to be a potential biological control in natural systems.

Research then proceeded to small and mid-sized lakes, where additional evaluation occurred.[518–522] Due to the success in these systems, grass carp were more recently stocked into larger systems in Texas (Lake Conroe, 8100 ha), Alabama (Guntersville Reservoir, 27,935 ha), South Carolina (Lake Marion, 45,000 ha), and Florida (Lake Istokpoga, 8097 ha).

Even though grass carp were stocked in many areas, the primary concern was the possibility that the grass carp would spawn in U.S. waters, causing declines in sport fishing, which is a multibillion dollar industry. As discussed earlier, the grass carp has spawned naturally in the U.S. In order to alleviate the reproduction problem in areas in which natural reproduction has not occurred, work began to develop a sterile grass carp for weed control. The first attempt was a cross between the grass carp and the bighead carp. Although this hybrid was reported to be sterile, it did not consume enough weeds to be used for weed control.[208,523] The production of gynogenetic and androgenetic fishes was also investigated. Morphological, biochemical, and cytological analyses indicated that androgenetic and gynogenetic grass carp have pure inheritance from a single parent.[524,525] All the fishes produced proved to be females, indicating sex determination through a system of female homogametry[526–528] (see Chapter 4, Section I.D). More recently, a sterile triploid grass carp has been developed.[361,529] This fish has proven capabilities as a biological control for aquatic plants and is currently used for that purpose.[530,531]

Presently 37 states allow use of either grass carp diploids or triploids for weed control. Five states have no restrictions on use, three states allow use of diploids after obtaining a permit, one state (Colorado) allows use of diploids in the east and triploids in the west, and 27 states allow only triploid stocking. Eight states allow importation of triploid grass carp only for research purposes, and Colorado allows both diploids and triploids only for research purposes (Table 32).

D. OTHER COUNTRIES

Some African nations have had limited experience with the grass carp. The species has been recommended for weed control in Kenya,[532] Sudan,[533] Ethiopia,[534] and Egypt.[535] Two countries, South Africa and Egypt, have experimented with grass carp. The fish was used in South Africa for incorporation into polyculture systems,[294] and in Egypt for weed control purposes. In 1976 an experiment was conducted in a small drainage canal near Cairo. The positive results of this experiment prompted the Egyptian government to continue grass carp research.[503] In cooperation with Dutch researchers hatcheries were built to produce grass carp, and a large-scale field experiment was conducted in a large irrigation canal. Initial stocking rates were based on data from The Netherlands (200 to 300 kg/ha), and were not sufficient to control vegetation in the deeper, larger canal where the experiment was conducted. It was found that small fishes (1.5 to 30 g) could be used because large predators were absent. The cost of weed control with grass carp was less

Table 32 **State importation of grass carp (U.S. Fish & Wildlife Service)**

State	Accept	Permit required	Ploidy inspection required	Visual disease inspection	Research purposes only	Authorized dealers only
Alabama	Diploid			*		
Arizona	Triploid	*	*	*		*
Arkansas	Diploid					
California	Triploid	*	*	*		
Colorado	Trip.–west Dip.–east	*	*		*	*
Connecticut	Triploid	*	*			*
Delaware	Triploid	*	*		*	*
Florida	Triploid	*	*	*		*
Georgia	Triploid	*	*			*
Hawaii	Diploid	*		*		
Idaho	Triploid	*	*	*		
Illinois	Triploid	*	*			*
Indiana	Triploid	*	*		*	
Iowa	Diploid	*				
Kansas	Diploid					
Kentucky	Triploid		*			*
Louisiana	Triploid	*	*			*
Mississippi	Diploid	*				
Missouri	Diploid					
Nebraska	Triploid	*	*			*
Nevada	Triploid	*	*	*		*
New Jersey	Triploid	*	*		*	
New Mexico	Triploid	*	*			
New York	Triploid	*	*	*	*	
North Carolina	Triploid	*	*			*
Ohio	Triploid	*	*			*
Oklahoma	Diploid					
Oregon	Triploid	*	*	*	*	
Pennsylvania	Triploid	*	*		*	
South Carolina	Triploid	*				*
South Dakota	Triploid	*	*			
Tennessee	Triploid	*				
Texas	Triploid	*	*		*	
Virginia	Triploid	*	*			*
Washington	Triploid	*	*	*	*	
West Virginia	Triploid	*	*			
Wyoming	Triploid	*	*	*		

Note: States that allow diploid grass carp also allow triploid; all states not listed ban grass carp. Many states conduct random ploidy inspections on distributors (revised October 27, 1993).

than with conventional methods in the main canal, where water was maintained year-round; however, in smaller canals, where water was drained during January, grass carp offered little savings. Although grass carp were an effective and economical means of controlling weeds in Egypt, their popularity as a food fish could cause overfishing in populated areas.[536]

Herbivorous fishes, including grass carp, were introduced into Israel in the 1960s to increase production in culture ponds.[44,537] Since 1968 grass carp have been used to control vegetation in an ichthyofauna reconstruction project in the Israel National Water System.[300] The results obtained from this project have been positive and the ichthyofauna system has been in continued use.

Grass carp have been introduced into many Asian countries for control of submerged weeds and water hyacinths. They were first introduced into India in 1959 from Hong Kong[214] and have been used in polyculture systems with other species of carp. Growth rates were found to be superior to those of Indian carp. They were also used to control submerged plants and duckweed in culture ponds[538,539] and weeds in irrigation canals.[540] Grass carp were used in Haliji Lake, Pakistan, but the stocking rate was not dense enough to effectively control the aquatic plants.[541,542] Limited success has been achieved in these countries, not only because the fishes do not consume floating plants in appreciable amounts, but also because the fishes are removed by local people before they are large enough to be effective.[543]

Grass carp have also been introduced and evaluated for weed control in a number of other countries (Table 33). Although the fishes were introduced for research and evaluation it does not necessarily mean that they are used in that country. For example, grass carp are not permitted for use in Canada and Australia. New Zealand recently reviewed their grass carp stocking policies. At present only sterile triploid fishes are allowed, and between 1987 and 1993 they were released into 13 locations. Because it is unlikely the grass carp will reproduce naturally and that any adverse environmental impact will result from its use, it is felt that the precaution of using only sterile fish in New Zealand should be reviewed.[544]

IV. IMPACT OF CONTROL METHODS

The impact of the weed control method used depends primarily upon the amount of weeds removed. Nonselective control methods such as the grass carp will affect the environment according to the amount of vegetation removed. If plants are removed by any control methods the impact to the system will be similar.

A. CHEMICAL CONTROL

Most data collected on the toxic effects of herbicides in natural systems indicate that they are relatively nontoxic to aquatic organisms and their residues in water and soil are short-lived due to uptake by target plants, nontarget plants, dilution, and sediment losses. A number of authors have reported on the possible chronic effects that herbicides may have on fishes. Studies were conducted that document the sublethal effects of aquatic herbicides, including histological changes,[546,547] stress response,[548] gill damage,[547,549] behavioral changes,[550,551] and reproduction.[552] It is difficult to evaluate these results because of the differences in environmental conditions. These studies do indicate that fishes suffer pathological changes when exposed to herbicides. These changes are either severe enough to cause death or to cause physiological changes that last for a certain period of time and then recovery occurs.

Table 33 **Introductions of grass carp**[11]

Country	Date	Source	Purpose	Ref.
Afghanistan	1966–67	China	Culture	728
Argentina	1970	Japan	Experimental weed control	729
Austria	1960	Romania	Experimental	730
Bangladesh	1976	?	Culture	731
Bulgaria	1964	U.S.S.R.	Polyculture	488
Burma	1969	India	Culture	732
Cambodia	?	?	Culture	455
Canada	?	?	Experimental	217
Cuba	1966	U.S.S.R.	Experimental	733
Czechoslovakia	1961–65	U.S.S.R.	Polyculture	488, 500
Denmark	?	?	Experimental	734
Egypt	1976	U.S.	Experimental culture and weed control	535
England	1964	Hungary	Experimental weed control	702
Ethiopia	1975	Japan	Weed control	534
Fiji	1968	Malaysia	Experimental weed control and culture	735, 736
Germany (eastern)	1965	U.S.S.R.	Experimental weed control	737
Germany (western)	1964	Hungary	Weed control	471
Hong Kong	?	China	Culture and weed control	279
Hungary	1963–66	China and U.S.S.R.	Polyculture	488
India	1959	Hong Kong and Japan	Culture and weed control	738, 739
Iran	1966	U.S.S.R.	Experimental	740
Iraq	1968	Japan	Culture	741
Israel	1952	?	Polyculture	713
	1965	Japan	Polyculture	276, 277
Italy	1972	Yugoslavia	Experimental culture	742

Table 33 (continued) **Introductions of grass carp**[11]

Country	Date	Source	Purpose	Ref.
Japan	1878	China	Culture	102
	1943–45	China	Culture	116
Java	1949	China	Culture	743
Kenya	1970	?	Culture	744
Korea	1967	Taiwan	Experimental culture	745
Loa, PDR of	1968	Japan	Culture	746
Malaysia	1930	China	Culture	747
Mexico[a]	1960	Taiwan and China	Weed control and culture	119, 120, 748
Nepal	1966–67	India and Japan	Culture	749
	1972	Hungary	Culture	750
Netherlands	1968	Taiwan	Experimental weed control	736
New Guinea	1965	Hong Kong	Culture	751
New Zealand	1966	Malaysia	Experimental weed control	203
Nigeria	1972	?	Culture	752
Pakistan	1964	China	Weed control and culture	542
Panama	1977	?	Culture and weed control	753
	1978	U.S.	Weed control	754
Philippines[b]	1966–69	?	Culture	755
Poland	1964–66	U.S.S.R.	Culture	493

Romania	1959	China	Polyculture and weed control	488
Sarawak	?	Hong Kong and Taiwan	Polyculture	463
Singapore	?	?	Culture	455
South Africa	1967	Malaysia	Experimental	715
Sri Lanka	1949	China	Culture	743
Sudan	1973	?	Culture and weed control	533
Sumatra	1915	China	Culture	743
Sweden	1970	Poland	Experimental weed control	756
Taiwan[c]	?	China	Polyculture	117, 118, 757
Thailand	?	China	Culture	743
United Arab Emirates	1968	Hong Kong	Experimental culture and weed control	732
Uruguay	?	?	Experimental	176
U.S.[d]	1963	Malaysia and Taiwan	Experimental weed culture	104
U.S.S.R.	1937,	?	Culture and weed control	103, 758
(European and central Asian)[a]	1950s (1954–59)	China	Culture	111
Vietnam	1969	Taiwan	Culture	732
Yugoslavia	?	?	Culture	759

[a] Have established populations; [b] Reportedly breeding in Pampanga River; [c] Has reportedly bred in reservoirs; [d] Reportedly breeding in the Mississippi River.[121]

From Shireman, J. V. and Smith, C. R., *FAO Fisheries Synopsis*, 135, 86, 1983.

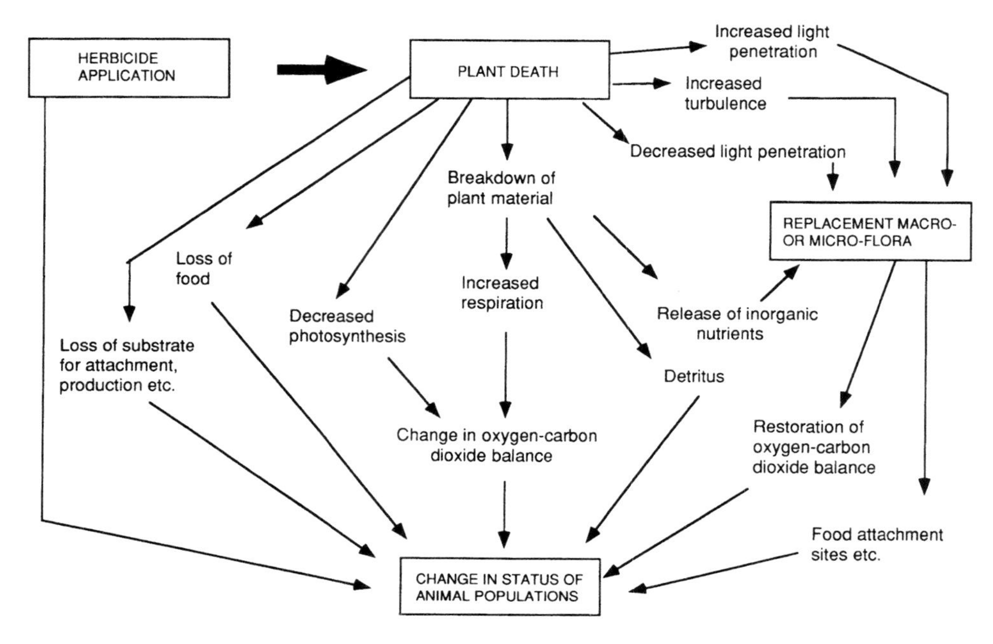

Figure 51 Effects of aquatic weed control and faunal changes likely to occur. (From Brooker, M. P. and Edwards, R. W., *Water Research,* Vol. 9, 1975, pg. 1. With permission.)

Water quantity parameters usually are not affected by herbicides, and if changes occur they are usually short lived. The most detrimental effect of plant control can be the establishment of algae blooms from the release of nutrients from decaying plants. The degree of severity depends upon the nutrients in the system and the amount of weeds controlled. Major problems can occur in small systems where very little water circulation and soil deposition occurs. Brooker and Edwards[553] discussed the changes that can occur due to the elimination of weeds and their subsequent decay, which are summarized in Figure 51.

B. MECHANICAL CONTROL

Mechanical harvesting of aquatic plants is often considered the most environmentally sound method of removing nuisance plants and has been touted as a method of removing nutrients from the system.[554] Several authors[555–557] have shown that rooted aquatic macrophytes pump nutrients from the sediments and that cutting the stems may actually release nutrients to the overlying water.[556–558] The release of nutrients from the injured plants and the decay of plants left in the water after cutting can cause algal blooms or growth of filamentous algae. Differences have also been found in the amount of algal biomass in shallow and deep water after macrophytes are harvested. Significantly greater algae biomass occurred in shallow water postharvest, and in deeper water, algal biomass decreased as macrophyte biomass decreased.[559] The relationship between nutrients and mechanical harvesting, according to Wade,[482] is poorly understood.

Mechanical removal of macrophytes removes fishes and invertebrates along with the harvested weeds. Dredging has greater effects on bottom macroinvertebrates, as the bottom sediments are removed along with the plants. Invertebrate recovery time can range from 8 to 9 months[560,561] to over 2 years[562] in dredged areas. If the substrate is not removed and the weeds are cut, recovery time is not as long, even though many invertebrates are removed along with the weeds.

Removal of aquatic plants by harvesting can influence fishes in several ways. Fishes that utilize vegetation for spawning can be affected if weeds are removed near the spawning season. The absence of food organisms found in the harvested weeds may affect food habits and growth. Small fishes that use macrophytes for shelter are susceptible to predation by larger fishes. This can be either beneficial or detrimental depending upon the fish populations present in the system. Fishes can be removed directly by the harvesting process.[563,564] With mechanical harvesting these effects can be minimized by timing the harvest and harvesting in areas that minimize the impact on the fishery. Another advantage is that the impact caused by mechanical harvesting is not long term.

C. GRASS CARP

As stated previously all control methods can have effects on the ecosystem. The grass carp has been studied extensively in this regard, especially in the U.S., the U.K., and The Netherlands. The possible effects of grass carp on the ecosystem are complex and depend upon the stocking rate, size of fishes stocked, macrophyte abundance, size of the system, and complexity of the ecosystem. The potential effects of stocking grass carp into systems with high and low macrophyte densities were summarized by Shireman and Smith[11] (Table 34).

Considerable research has been directed toward the food habits of the grass carp and their possible competition with other species for the food resource. Under polyculture conditions they have been implicated as competitors for zooplankton with other carp species.[191,399] There is some disagreement as to the length of time grass carp fingerlings consume zooplankton and benthic organisms, which influences the degree of competition. Sobolev[174] stated that grass carp switched completely to macrophytes at 50 mm. Opuszynski[399] found that grass carp fed exclusively on animal food (chironomids and planktonic crustacean) until they transferred to plant food at 36 to 43 mm. Watkins et al.[179] reported that 32% of invertebrate organisms were consumed by fishes from 50 to 100 mm. Once grass carp switch to macrophytes the amount of animal food in the diet is minimal. The possibility exists, however, that larger grass carp may compete with other species for benthic food once the macrophytes are removed.[111,565] However, Kilgen and Smitherman[196,566] found little competition with sport fishes in sparsely vegetated ponds. In ponds in Georgia, animal remains constituted only 0.02% of the stomach volume of 286 to 588 mm grass carp.[567] Colle et al.[194] reported that although benthic invertebrates were abundant in Florida ponds, only trace amounts were observed in the stomach contents of grass carp 63 to 220 mm total length. Zooplankton were also numerous in the pond, but not eaten.

The changes that occur in zooplankton abundance are not caused by grass carp grazing, but are due to the removal of vegetation and possible grazing by other fish species that are enhanced by vegetation removal. For example, the decline in zooplankton in Lake Conroe, Texas, after aquatic plants were removed, was attributed to increased predation by shad,[568] which increased fourfold after vegetation removal. Vegetated habitats support diverse zooplankton populations,[569] represented primarily by species inhabiting the littoral zone. Species composition, however, may depend upon trophic state.[570,571] When the vegetation is removed the zooplankton community shifts to more pelagic forms.[569]

Aquatic vegetation harbors many macroinvertebrates.[282,568,572–575] It has also been reported that macroinvertebrates are site specific, inhabiting only particular species of

Table 34 **Potential effects of stocking grass carp in an ecosystem**[11]

Grass carp			
Moderate stocking		**Intensive stocking**	
Low macrophyte density	**High macrophyte density**	**Low macrophyte density**	**High macrophyte density**
Macrophyte control and moderate nutrient increase in sediments, increase in emergent plants, possible reduction in recruitment of photophilous spawners, potential plankton and benthos increases, exposure of plant-inhabiting animals to predation, possible production increases of predators (gamefish, etc.)	Partial and/or temporary control and moderate nutrient increase in sediments, increase in emergent plants; changes the same, but not as extreme as with complete macrophyte control	Macrophyte elimination (overcontrol), initial nutrient increase in water and sediments Phytoplankton bloom, reduction in recruitment of photophilous spawners, possible changes in benthos population, exposure of plant-inhabiting animals to predation and elimination, shift from littoral to pelagic species Possible increase in detritivores, decreased oxygen levels and pulses, slight reduction in pH	Macrophyte control, nutrient release to water and sediments, and temporary increase in emergent plants Possible reduction in recruitment of photophilous spawners, probable increased predation on plant-inhabiting animals, probable production increases of predators (gamefish, etc.) Possible increase in detritivores Decreased oxygen levels and pulses and a reduction in pH Increased alkalinity

From Shireman, J. V. and Smith, C. R., *FAO Fisheries Synopsis*, 135, 86, 1983.

aquatic plants.[576] Watkins et al.[575] found that in Orange Lake, Florida, the vegetated area of the lake harbored a significantly greater number of macroinvertebrates than the nonvegetated area. Hydrilla contained the greatest number of organisms of the four vegetation types sampled. Removal of macrophytes, therefore, could cause a reduction in the number of macroinvertebrates, as reported by Klussmann et al.[568] Removal of vegetation, however, does not necessarily forecast reduced macroinvertebrate numbers. For example, Leslie and Kobylinski[577] reported significantly greater numbers of total benthic invertebrates in Deer Point Lake, Florida, after vegetation removal. Species composition, however, changes from those that crawl and live on substrates to those that burrow in the sediments.

D. WATER QUALITY

It is difficult to evaluate the effects of grass carp on the water quality of different bodies of water. This difficulty arises because of the different evaluation methods used, size of the body of water stocked, hydrology (static or flow through), natural nutrient relationships caused by regional geology, amount of vegetation removed, and climate (regional and local).

Prior to grass carp stocking the amount of vegetation is estimated in the body of water to be stocked. After the fishes are stocked the vegetation is monitored along with water chemistry parameters. The next step is to relate the changes that occur to the grass carp stocking. If changes occur they are caused by nutrient release from broken stems, decay of uneaten plants, and decaying feces. The nutrients made available through this process can cause increased phytoplankton,[192,578,579] however, phytoplankton increases do not occur in all cases,[515,580,581] and in some cases phytoplankton decreases.[582] Whether phytoplankton blooms occur is probably related to the amount of vegetation controlled and the trophic state of the lake. Canfield et al.[579] reported that the percentage of the total lake volume infested (PVI) with macrophytes significantly influences chlorophyll concentrations in Florida lakes. Major changes in lake chlorophyll did not occur until the PVI values were >30%, levels which are far in excess of what the general public will accept. In nutrient-poor lakes algae abundance changes very little, even if large amounts of vegetation are removed. In nutrient-rich lakes algae is present even though macrophytes are abundant. When macrophytes are removed from these systems it is likely that algae blooms will occur and persist. In most cases when macrophytes have been removed, phytoplankton increase, but they decline as nutrients are used. The phytoplankton respond as if fertilizer has been added to the system.

Studies conducted to determine water quality changes after the introduction of grass carp indicated that temperature and oxygen values are relatively unaffected. Oxygen levels after macrophyte removal are usually maintained by an increase in phytoplankton. After reviewing water quality data from a number of multiyear field studies it is evident that most of the parameters measured remain unchanged, and changes in a single parameter are not consistent among studies (Table 35). Water quality parameters are therefore difficult to predict and changes are related to the amount of vegetation controlled, external nutrient loading, water level fluctuation, and sedimentation.[580,583] In order to evaluate these changes realistically the regional geology, hydrology, and limnology of the body of water should be considered. In addition water quality parameters may increase over the short term (first year), but will return to prestocking levels in subsequent years.[584]

Aquatic plant fish relationships have been studied extensively and were reviewed in 1993 by Kilgore et al.[585] Early researchers, working in small recreational ponds, often

Table 35 **Water quality parameters and percent vegetation at end and start of field studies after grass carp stocking**

Study length (years)	Location system	Parameter												
		Ph	Cl*a*	K	Ca	TN	TP	TA	NO_3	COD	Turbidity	% Veg (start)	% Veg. (end)	Ref.
4	Lake Baldwin, FL	0	+	+	+	0	+	–		–		79%	0	584
3	Illinois ponds (annex ponds)	–	+				0	+	0	+		Variable		760
1 (1975)	Indiana ponds	0	0	+			0	0	0		+	Variable	0	515
1 (1976)		0	0	0			0	0	0		0	Variable	0	
3	Suwannee Pond, FL	0	0				0			0	0	51%	3%	761
	Pasco Pond, FL	0	0				–	0		+	+	49%	2%	
	Madison Pond, FL	0	0				0	0		0	0	97%	32%	
	Broward Pond, FL	–	0				0	0			0	13%	2%	
3.5	Lake Bell, FL	–	0			0	+	0		0	+	88%	4.2%	580
9	Deer Point Lake, FL	–				+	0	0	0	0	+	61%	26.3%	
	Clear Lake, FL	0	0			0	+	–	0	–	+	89.2%	1.1%	
	Lake Holden, FL	0	0			0	0	0	0	–	+	59.2%	3.2%	
4	Martha, Burkette, FL		+			+	+				+	90%	6%	530
7	Lake Conroe, TX	–	+	0			0	+		+		45%	0	568
4	Red Haw Lake, IA	–	–				0	+	–		–	25%	2.3%	582
4	Lake Susannah, FL	0	+			0	+	0				60%		569

Definitions: + = increase, - = decrease, 0 = no change. cl*a*, Chlorphyll*a*; TN, total nitrogen; TP, total phosphorus; TA, total alkalinity; and COD, chemical oxygen demand.

stated that aquatic plants were detrimental as they contributed to the imbalance of fish populations.[586] Recent research in larger natural systems seems to indicate that aquatic plants are essential in the system if a healthy fish population is to be maintained. Researchers, however, cannot agree as to the level of aquatic plants that are needed. Compounding the problem is that certain species of plants or combinations of plants seem to be better fish habitats than others.[587] Plant density also has been shown to be important in determining the species composition, abundance, and condition of fishes.[587–589] Hoyer et al.[590] state that the relationship between vegetation and fish abundance is more complicated and that trophic state must be considered. In lakes in which large amounts of aquatic macrophytes exist the trophic status of the lake may be underestimated. A positive relationship has been shown between trophic status as measured by chlorophyll-*a* or total phosphorus with fish standing crops in reservoirs and natural lakes.[591,592] Therefore, removing the macrophytes in lakes with higher trophic states may actually cause fish standing crops to increase. It should be stated, however, that species composition may change, shifting to species that rely more heavily on a phytoplankton-based food chain. In Lake Baldwin, Florida, after the removal of vegetation by grass carp, the harvestable game fish biomass increased fourfold and largemouth bass recruitment was as high or higher than during years when macrophytes were abundant.[584] Dense weed beds can provide too much structure, interrupting the predator and prey relationship, and thus causing fish production to decline.[593] Others have reported that largemouth bass production decreases with reduced vegetation.[594] These studies indicate that the relationship among fish abundance, growth, and production are complex. It appears that an intermediate level of plant density may be optimal.[589,595,596] Canfield and Hoyer[597] sampled fish populations in 60 Florida lakes having different amounts of vegetation. Their data suggested that greater numbers and biomass of fishes existed in lakes with vegetation levels between 15 and 80%. In lakes near the extremes the probability of having a lower standing crop of fishes was increased. They found that certain species were associated with vegetation, and when vegetation increased these species also increased, whereas other species responded in the opposite manner (Table 36).

The most comprehensive report of the effects of grass carp introduction on fish populations was accomplished in Arkansas. According to Bailey[512] the introduction was effective in controlling vegetation in more than 100 lakes (+20,000 ha). Although there were some significant fluctuations in some aspects of the fish populations in some lakes, there was a lack of any general trend, indicating to Bailey that other causal relationships existed. In eight heavily vegetated lakes (submersed vegetation 40 to 80%) total standing crops of fishes after 5 years were within the levels prior to stocking. Identical findings were observed in moderately vegetated lakes (10 to 40%), nonvegetated lakes containing grass carp, and control lakes where grass carp were not stocked. Shad populations fluctuated more in lakes from which greater vegetation was removed. The total standing crop of catchable bass, bluegill, redear sunfish, crappie, and young of the year of these species fluctuated with both increases and decreases in total yield. No trend was evident in either direction. Condition factors of largemouth bass appeared to be improved. Colle and Shireman[589] reported changes in the condition of largemouth bass, bluegill, and redear sunfish with changing levels of hydrilla in two Florida lakes. Bluegill and redear sunfish condition factors were reduced when most of the water column was filled with hydrilla. Harvestable largemouth bass had reduced condition factors once hydrilla coverage was >30% and smaller bass had reduced condition factors once hydrilla coverage exceeded 50%.

Table 36 **Mean fish species percent composition (by weight) of the total fish biomass estimated with blocknets**

			Area covered with aquatic macrophytes (%)			
Fish species	*n*	*r*	**0–25**	**26–50**	**51–75**	**75–100**
Species with decreasing percent composition with increasing macrophyte coverage						
Gizzard shad	21	–0.44[a]	28.12	11.21	0.25	0.93
Threadfin shad	28	–0.32[a]	15.48	19.06	6.98	1.99
Species with increasing percent composition with increasing macrophyte coverage						
Bluespotted sunfish	27	0.34[a]	0.93	0.85	3.60	2.13
Bowfin	17	0.59[a]	2.48	3.69		7.77
Dollar sunfish	22	0.47[a]	0.11	1.17	0.81	3.72
Golden topminnow	32	0.33[a]	0.11	0.51	0.61	0.61
Least killifish	17	0.45[a]	0.00	0.01		0.02
Mosquitofish	52	0.32[a]	0.06	0.10	0.07	0.57
Sailfin molly	11	0.74[a]	0.12	0.10		0.45
Tadpole madtom	13	0.46[a]	0.08	0.27	0.78	0.49
Warmouth	65	0.67[a]	2.89	16.89	10.33	20.25
White catfish	11	0.55[a]	0.77	0.02		10.56
Yellow bullhead	35	0.41[a]	0.69	1.75	3.18	1.75
Species with no trend in percent composition with increasing macrophyte coverage						
Black crappie	40	–0.13	5.20	4.19	2.72	3.13
Bluegill	65	–0.15	30.34	31.96	31.43	22.23
Blue tilapia	14	0.05	9.47	2.80	15.41	
Bluefin killifish	28	0.11	0.61	0.08	0.91	0.77
Brook silverside	44	–0.21	0.38	0.05	0.10	0.07
Brown bullhead	32	0.01	2.65	0.27	2.61	3.58
Chain pickere	9	0.14		4.97		6.41
Everglades pygmy sunfish	11	0.08	0.02	0.10		0.03
Flagfish	5	0.24		0.79	0.04	0.91
Florida gar	20	0.29	2.39	1.34	1.00	4.59
Golden shiner	47	–0.10	4.18	4.26	5.94	1.57
Lake chubsucker	39	0.01	11.85	8.75	6.65	14.13
Largemouth bass	65	0.12	14.71	18.16	18.64	18.60
Lined topminnow	19	0.07	0.18	0.39		0.35
Longnose gar	3	0.31	0.01			0.02
Pirate perch	4	–0.60	0.10	0.07		0.05
Redbreast sunfish	8	0.38	0.84	0.13		
Redear sunfish	51	0.04	10.68	15.70	7.86	11.29
Redfin pickerel	11	0.16	1.19	0.57		2.40

Table 36 (continued) **Mean fish species percent composition (by weight) of the total fish biomass estimated with blocknets**

Fish species	*n*	*r*	Area covered with aquatic macrophytes (%) 0–25	26–50	51–75	75–100
Seminole killifish	33	0.03	0.56	2.10	0.02	0.37
Spotted sunfish	23	0.35	0.30	1.77	3.01	0.87
Swamp darter	43	0.12	0.03	0.06	0.08	0.05
Taillight shiner	11	–0.05	0.27	1.01		0.05

Note: The values are listed by species for the percent area covered with aquatic macrophyte groups. *n* represents the number of lake samples in which a species was found and *r* is the correlation coefficient for the relation between percent area covered with aquatic macrophytes and percent composition for each species in those lakes.

[a] = Significant at $p \leq 0.10$.

Data from Reference 597.

In Lake Conroe, Texas, after macrophyte removal by grass carp, the abundance and biomass of forage fishes changed.[568] Forage fish standing crop was reduced 60% in the littoral zone. Both gizzard and threadfin shad increased due to the increase in phytoplankton as did several minnow species that do well in open water. The density and biomass of most sunfish species were significantly reduced. Removal of vegetation exposed these fishes to increased predation, which is probably responsible for their reduction. The number of age 1 bass was reduced; however, bass >240 mm total length did not decline. The number of intermediate-sized bass declined after vegetation removal, but biomass differences did not occur. The growth of first-year bass increased, allowing them to enter the fishery earlier. Sampling indicated that both black and white crappie declined over the course of the study. A strong year class occurred in 1979 when vegetation coverage was 20%. In subsequent years a strong year class was not observed. Growth of crappie increased, which may have been due to increased forage availability. Maceina and Shireman[598] reported increased growth of black crappie in Lake Baldwin, Florida, after hydrilla reduction. They attributed this increased growth rate to the establishment of the threadfin shad population. In Lake Conroe white and yellow bass abundance also increased significantly, probably a result of increased forage or increased spawning success. In Gunterville Reservoir, Alabama, an experiment was conducted to eliminate weeds in a large embayment with grass carp. No adverse effects occurred on the total fish community, there was a shift from sunfish to shad, largemouth bass growth increased, and total fish standing stock increased.[599]

The relationship between vegetation and fish populations is very complex. Removal of vegetation can cause changes in water quality and fish populations. However, the trophic status and regional abiotic factors of the lake must be considered before a complete evaluation can be made. Most of the studies reviewed agreed that small fishes may be decreased as vegetation is removed, but total fish production may not be affected. Fishes using vegetation for spawning substrate will be reduced.[191,488,493] Although eliminating vegetation by any means in a body of water will cause similar changes, it should be emphasized that utilization of grass carp can cause complete elimination of vegetation for a considerable number of years.

V. MANAGEMENT CONSIDERATIONS USING GRASS CARP

As stated in previous chapters the grass carp is an efficient aquatic weed eating fish. If overstocked it will remove all palatable vegetation in a body of water. If understocked it is unlikely that the effects produced by the fish will be noticed.

A number of factors should be considered before a body of water is stocked. It is virtually impossible to answer the question of how many fishes are required for each body of water with a blanket statement. Due to possible predation small fishes (<25 cm) should not be stocked.[600] The growth rate of the fishes and consumption rates of different sizes of stocked fishes should be known in order to predict the amount of vegetation that will be eaten over the course of the initial stocking. Plant selection also varies with size, as larger fishes will eat plants that are hard tissue plants or plants too large for small fishes to consume. Knowing these parameters the stocking rate can be adjusted during the control period. Water temperature must be considered as feeding rates are dependent upon water temperature. Depending upon the latitude, the number of active feeding days will change. Stocking rates, therefore, should be adjusted for local climate conditions. Zonneveld and van Zon[502] stated that effective weed control could not be expected when water temperatures were below 18°C. Below these temperatures consumption is not great enough to offset plant growth. The species of aquatic plants in the system must be identified as the grass carp consumes aquatic plants at different rates, depending upon their palatability. Highly preferred plants are eaten quickly. Stocking rates must be higher if the body of water contains plants that are eaten reluctantly. It is also significant that grass carp plant palatability changes as temperature increases. At higher temperatures they will consume plants that were not consumed at lower temperatures.

As with any management plan the user must decide prior to stocking what stocking will achieve. It is easiest to use a stocking rate that will eliminate all vegetation. In bodies of water in which boating, swimming and other surface water sports are envisioned, complete macrophyte removal may be the best plan. In systems in which fishing and wildlife management are included in the goals, a stocking rate should be used that removes only a prescribed portion of the aquatic vegetation. It is important to understand that using grass carp for vegetation control is not an immediate response. The plants in the system that are not consumed continue to grow; therefore, it could take several seasons to obtain desired results, and in some cases it is necessary to periodically add additional fishes. Such fishes, however, should not be added too quickly.

A number of models have been developed to determine stocking rates that estimate the number of fishes needed to achieve various levels of vegetation control.[601–604] It should be emphasized, however, that these models are only as good as the information entered into them, and they should not be considered as being applicable to systems over a wide geographical range unless data are available to refine the model to fit area conditions.

Although grass carp have been used worldwide for aquatic weed control, the authors present a few case studies that illustrate different stocking strategies.

In Illinois two strategies (serial and batch stockings) have been developed for stocking grass carp after considerable research and data were used to develop models that were applicable to Illinois conditions.[605,606] When serial stockings are used fishes are stocked at predetermined intervals to control vegetation. These regularly scheduled stockings, according to Wiley et al.,[607] usually minimize the total number of fishes required to achieve stocking objectives. When batch stockings are used fishes are stocked in a single batch, and stocking additional fishes is spread over a longer time period. The authors usually recommend serial stocking as it is more efficient, cost effective, and allows for greater control over the amount of plants removed. Stocking rates are based on region, plant species (including palatability and abundance), consumption rate, and plant biomass. Stocking rate tables have been developed according to plant groups and the geographic area in which the body of water is located. These tables are used to determine the number of 10-in. (25 cm) fishes per vegetated acre (0.4 ha) that should be stocked initially and the interval and number that should be stocked at the second stocking. The Illinois system is one of the most refined systems and appears to be suitable for use by pond owners. Many other agencies often stock fishes according to a set range of fishes per surface hectare of water. When this system is used the number of fishes stocked is based upon the experience of the person making the stocking recommendation.

Another approach to vegetation management with grass carp is to integrate control methods. Plant biomass can be reduced by incorporating a second or third control method. Shireman and Maceina[522] discussed the possibility of using grass carp and either chemical or mechanical control to reduce plant biomass during the first year, and then stocking enough grass carp to consume hydrilla regrowth and a portion of the old growth each year. It is necessary to estimate the yearly growth rates of the target plant, including the decrease, during winter senescence. This method is feasible in monoculture systems, but in systems in which grass carp have a number of plants to choose from it may be more difficult to maintain desired levels of vegetation. For example, Fowler and Robson[506] report that grass carp changed a diverse plant community into a community containing only a few unpalatable plants which increased in coverage. They stated that understocking should be avoided in systems in which both palatable and unpalatable plants prevail. In these situations another treatment could be used in conjunction with grass carp to reduce the biomass of unpalatable plants.

In Florida lakes, if the objective is to leave vegetation in the system, a herbicide treatment is made prior to stocking. Five to eight fishes are stocked per hectare of the vegetation that was previously controlled by the herbicide treatment. Mortality is estimated at 20%. In southern Florida a higher stocking rate is needed. Although this stocking strategy works well and achieves the stocking objectives, restocking is needed in 5 years. In small ponds in which complete vegetation removal is required, grass carp are stocked alone at 49 to 247/ha. The high stocking rates are needed where less palatable plants exist. Grass carp at the rate of 124/ha are used for hydrilla control in these systems.[768]

Grass carp are used in waterways of The Netherlands to remove vegetation. In these systems a stocking rate of up to 250 kg/ha is used. If higher stocking rates are used, vegetation is eliminated completely. Complete vegetation removal is considered undesirable as it will cause drift of bottom materials and bank erosion when the fishes graze the zone immediately above the water level.[503]

In Imperial Valley, California, the Imperial Irrigation District uses grass carp to remove vegetation from irrigation canals and ditches. The grass carp are an ideal control because they cannot cause damage to land crops as herbicides might and they are less expensive to use than mechanical methods. Two different stocking rates are used. The maximum stocking rate is determined by California Fish and Game. In the western irrigation area, where hydrilla was a problem in the large canals and reservoirs, the goal was to eradicate hydrilla. In these systems up to 247 grass carp per hectare are permitted, or 160 grass carp per kilometer. On the east side, where no hydrilla is found, fishes can be stocked to as high as 49 fishes per hectare or 32 fishes per kilometer. At the beginning of the stocking project 960 km of hydrilla infestation existed. Currently, <1.6 km is infested. The grass carp consume any regrowth that occurs. The objective is to ensure that no hydrilla remains in the system; therefore, the canals are routinely checked for hydrilla. In lateral ditches in which the grass carp leave plants, some mechanical control is used.[769]

From the above discussion it is obvious that the grass carp is a viable aquatic weed control method, and if used carefully, some vegetation can be maintained in the system. In most cases it is more cost effective than mechanical or chemical methods.[502,503,608] The stocking rate used should be based upon a management objective. To achieve this objective, local conditions, including climate, amount of vegetation, size of the fishes, and the palatability of the vegetation in the system should be considered before they are stocked.

A. CONTROL OF STOCKED GRASS CARP POPULATIONS

If grass carp are overstocked in a body of water, the management agency responsible may want to remove excess fishes from the system. They are difficult to remove unless a fishery has been developed for them. In the Amur River basin fishery regulations were imposed as a result of overfishing concerns.[132,147] Gorbach[142] recommended that fishing should be suspended for 10 years. Generally the Russians consider that their introductions of grass carp provided positive weed control and enhancement of the existing fishery; therefore, fishing is not allowed for several years after stocking.[157] It was also reported by Dubbers et al.[609] that grass carp have not been successful in Egypt because overfishing by the local population could not be kept under control. Similar reports are available for Japan, where most of the grass carp have been fished out of Lake Kasumi and Lake Kita.[116] Van der Zweerde[503] remarks that poaching is a serious problem, especially immediately after the fishes are stocked, as they are accustomed to being fed in the hatchery. After they are free for a period of time they become hook-shy and are difficult to catch.

In several countries the grass carp have been touted as a sport fish. A research project was conducted in England to determine the utilization of grass carp as a sport fish in small ponds.[610] Grass carp proved to be a very popular fish, as it is strong and fights well. The fish takes a variety of baits including maggots, bread paste, and flake. The fish has also been reported to have angling value in The Netherlands[502] and Germany.[611] It is doubtful that grass carp will become a prized sport in the U.S.; however, the fish is caught on hook and line and success rates apparently depend on bait selection and the amount of vegetation present.[198,567,612–614]

Standard fish collecting methods such as electrofishing, gill nets, haul seines, and fish toxicants have been used to collect grass carp. Although these methods have been successful to a certain degree, none have been developed to remove large numbers of grass carp. Under some conditions rotenone may exhibit marked selectivity to grass carp as opposed to other species.[615–617] It is doubtful that any of these conventional methods will be successful for removing large numbers of grass carp. A pellet consisting of Bermuda grass and rotenone was recently developed which is selective for grass carp; however, it is still being field tested.[768]

Chapter 6

Utilization of Phytoplanktivorous Fishes for Counteracting Eutrophication

I. INTRODUCTION

Eutrophication is a global problem caused by nutrient enrichment, notably phosphorus and nitrogen. The process is enhanced by human activities (anthropogenic eutrophication) and is of increasing concern because of its deteriorating effect on water quality. The most visible effects of eutrophication are excessive algae blooms, which may make water unsuitable for other purposes. In extreme cases of eutrophication algae decomposition may cause oxygen depletion, resulting in the death of most organisms living in the body of water.

The established methods used to counteract eutrophication reflect classic or traditional limnological approaches to lake ecosystem structure and function. This approach, often described as the "bottom up" hypothesis,[618] views freshwater ecosystems as operating through nutrient availability, which influences each consecutive trophic level of the food web from primary producers (macrophytes, phytoplankton) to secondary producers (benthos, zooplankton) and to tertiary producers (fishes). Methods of counteracting eutrophication which are based on this traditional limnological approach are physical in nature, e.g., water aeration, chemical precipitation of phosphates, or bottom sediment removal. These methods have been used successfully, but they are very costly (e.g., in Lake Erie costs for phosphorus reduction shared by the U.S. and Canadian governments exceeded $10 billion).

Biological or biomanipulation methods to control eutrophication recently have received considerable attention and have been intensively tested. Biomanipulation[619] methods are designed to manipulate food-web organisms to control nutrients and improve water quality.[620] Contrary to "bottom up" techniques, biomanipulation methods are referred to as "top down" or "trophic cascade" methods.[618,621] These methods consist mainly of implementing changes in community structure to increase phytoplankton grazing or to limit nutrient availability, or both. As the interest in biomanipulation increased, fishes became the organism preferable for manipulation. This is well illustrated by the increased number of papers dealing with the environmental effects of fishes (Figure 52).

Two reasons account for the popularity of fishes for biomanipulation. First, fish populations can be manipulated as they can be easily removed, stocked, or the species composition changed using standard fishery management techniques. Second, fishes have important commercial and recreational values. Generally, two biomanipulation procedures with fishes have been tested. The first consists of using phytoplanktivorous fishes to reduce phytoplankton biomass by direct grazing. The second procedure is achieved through indirect depression of phytoplankton through fish-derived alternations in zooplankton communities and nutrient cycling. Additional comments are needed to make the second procedure clear.

The role of zooplankton to keep phytoplankton in check has been recognized for a long time. For example, Wright[622] found that zooplankton consumed 85% of the daily net primary production in Canyon Ferry Reservoir (Montana). Similarly, in a Canadian lake

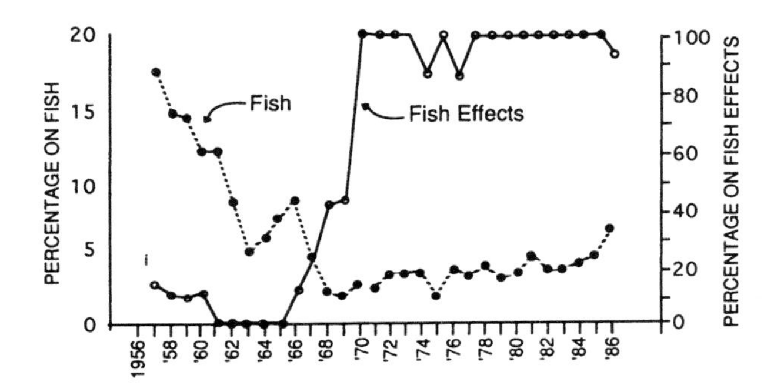

Figure 52 Percentage of papers published in *Limnology and Oceanography* pertaining to fishes, and the percentage of those pertaining to fish effects. (From Northcote, T. G., *Can. J. Fish. Aquat. Sci.*, 45, 361, 1988. With permission.)

66% of the phytoplankton was consumed, and in another lake, the total daily primary production was consumed by zooplankton.[623] In the 1980s mathematical models were used to quantify the amount of zooplankton grazing needed for a significant reduction in phytoplankton biomass.[624] The efficiency of zooplankton grazing, however, depends not only on zooplankton biomass, but also on community structure. Small zooplankton, such as rotifers or small crustaceans, are ineffective phytoplankton grazers. Not only do they consume less algae, but for their body size they excrete greater amounts of phosphorus. Large zooplankton consume more phytoplankton and excrete relatively smaller amounts of phosphorus into the water. For example, it was calculated that if the small zooplankton species in Lake Wingra were replaced with a larger species that was formerly abundant in the lake, a two orders of magnitude reduction in phosphorus would occur (Figure 53).

Since the classic works of Hrbácek and colleagues[625,626] and Brooks and Dodson,[627] the impact of planktivorous fishes on zooplankton communities has been carefully studied (for reviews, see Lazzaro[34] or Northcote[618]). They found that severe fish predation could cause dramatic decreases in the abundance and biomass of large zooplankton and shift the structure of the zooplankton community toward dominance by small species (Figure 54). Thus, the feeding activity of planktivorous fishes may indirectly stimulate eutrophication by removing the large zooplankton. This is also true regarding bottom-feeding fishes. They may increase the amount of phosphorus and nitrogen available for the phytoplankton by disturbing the bottom sediments and also by releasing nutrients from sedimented fish feces.[628–630] Hence, by decreasing the biomass of planktivorous and benthic fishes, the biomass of large zooplankton will increase, reducing the nutrient concentration in the water, and consequently decrease the biomass of phytoplankton.

II. EUROPEAN EXPERIENCE

In the wake of Chinese carp introduction into European waters great expectations arose regarding the use of silver carp to control eutrophication. Scientists and managers were interested for two reasons. First, because there were no indigenous planktivorous filter feeding fishes in Europe, their potential in controlling algae was unknown. Second, carp are demanded as food fishes in Europe; therefore, the introduction of silver carp might solve two problems: improve water quality and obtain additional fishes for food.

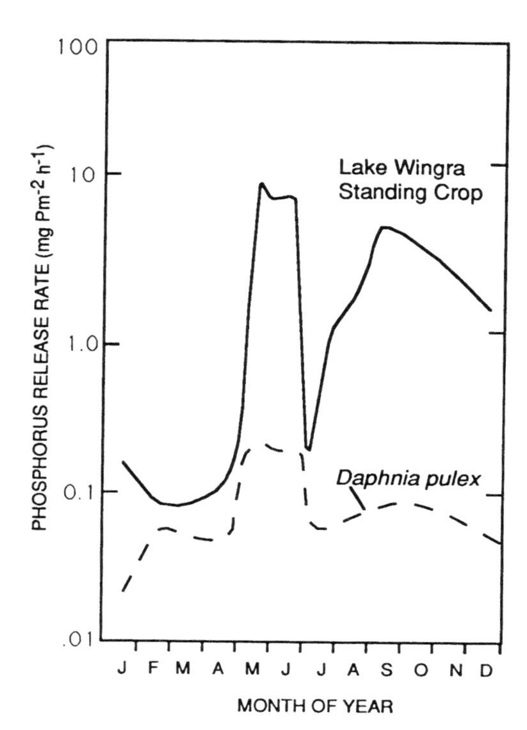

Figure 53 Calculated seasonal rates of phosphorus excretion, Lake Wingra, Wisconsin, U.S., zooplankton. Solid line indicates excretion of zooplankton, including *Daphnia* <1.0 mm in mean size. Broken line indicates hypothetical phosphorus release by equivalent biomass of larger zooplankton >1.75 mm. (From Kitchell, J. F. et al., *BioScience,* 29, 28, 1979. With permission.)

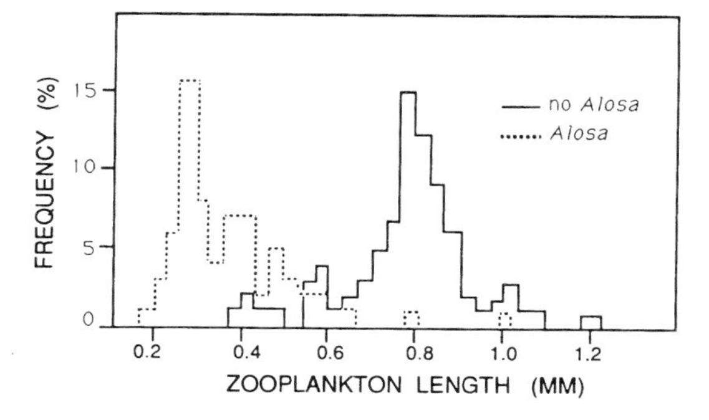

Figure 54 Mean size of zooplankton in Crystal Lake, Vermont, U.S., 10 years before the introduction of the planktivorous fish *Alosa aestivalis* (Clupeidae) compared to 10 years after introduction. (From Brooks, J. L. and Dodson, S. I., *Science,* 150, 28, 1965. With permission.)

Numerous papers on the control of excessive algae growth using silver carp were published in the former U.S.S.R., but these papers dealt largely with the potential of this fish or were based on circumstantial evidence (e.g., see Vovk,[225] Yefimova and Nikanorov,[489] and Verigin[631]). More substantial information came from mesocosm experiments in which the environmental influence of silver carp was compared to a fishless control group. Vybornov[632] conducted tests in concrete basins filled with pond water and stocked with 0, 250, 400, and 700 g of silver carp per cubic meter. In all the stocking densities silver carp decreased total algal biomass and increased primary production and the abundance of small phytoplankton (nannoplankton). A considerable reduction in zooplankton biomass was also observed.

In Poland[633] polyethylene enclosures, impermeable to water, were used in an experiment conducted in a eutrophic lake. An increase in water transparency and a decrease in algal biomass were found in this experiment when silver carp were present. Water quality improvement was due to the elimination of algae from the water column. Nevertheless, the authors were careful when extrapolating the results to the whole-lake situation. They alluded to the uncertain fate of the sedimented feces, which in a well-mixed aquatic system or in the presence of bottom-feeding fishes may provide nutrients for planktonic algae development, especially as the zooplankton were also eliminated by the fishes in this experiment.

Eastern German lakes and ponds were intensively stocked with silver carp to increase fish yield and improve water quality.[634,635] Despite high stocking rates (up to 10,000 fishes per hectare) the effect of silver carp on lake water quality was less than expected. The most distinctive effect was a shift in the phytoplankton structure from *Microcystis*-dominated to *Oscillatoria*-dominated associations. *Oscillatoria* became dominant because they were not eaten and digested by silver carp. Zooplankton were heavily preyed upon, and zooplankton biomass declined sharply in the pelagic zone, adversely affecting the survival of perch *(Perca fluviatilis)* and pike-perch *(Stizostedion lucioperca)* fry. The decline of zooplankton in the pelagic zone was offset somewhat by their abundance in the littoral zone, which created good feeding conditions for fish fry living there. Changes in the pond biota became visible beginning at a stocking density of about 1000 3- to 4-year-old silver carp per hectare. Zooplankton biomass decreased, while phytoplankton increased, as did primary production.

Extensive studies on the effect of silver carp on the environment of carp ponds were conducted in Poland.[406,636–638] Ponds stocked with the same density of common carp and different densities of silver carp were compared to ponds stocked with common carp only (Figure 48). The presence of silver carp resulted in a number of environmental and biological changes (Figure 55). Ponds stocked with silver carp had considerably more net primary production than ponds stocked with common carp and more different groups of bacteria in the water and bottom sediments. A smaller but significant increase was observed in biological oxygen demand (BOD_5), dissolved oxygen, phytoplankton biomass, chlorophyll-*a* concentration, gross primary production and destruction, and in numbers and biomass of benthos. When silver carp were present the biomass and numbers of zooplankton and, to a smaller extent, the concentrations of ammonium nitrogen and phosphates, were decreased. Changes in some parameters, however, were not always directly correlated with silver carp stocking densities, as increased stocking densities did not always cause gradual intensification of changes in all parameters (Figure 55).

In addition to increased algal biomass, the structure of the algal community was changed in silver carp ponds. As silver carp stocking densities were increased, the numbers and biomass of small green algae (Chlorophyceae) decreased. The larger algal species, notably belonging to the Bacillariophyceae, Euglenophyceae, and Cryptophyceae, became more numerous than in the ponds with common carp only. Green algae biomass declined in the consecutive groups of ponds with increasing silver carp stocking densities. The declines were 18, 35, and 37% as compared to the control ponds. This was due to a three- to fivefold drop in the biomass of the dominant nannoplankton alga *Chlorella minutissima.* This is one of the smallest algae, having a diameter of about 3 µm. Because of its small size it easily passes through the filtering apparatus of silver carp. The biomass of diatoms increased by an average of 32, 42, and 66% in the ponds with consecutively increasing densities of silver carp. A small diatom *(Nitzschia palea)* was found in small numbers in control ponds, but was the dominant species in silver carp ponds. It is

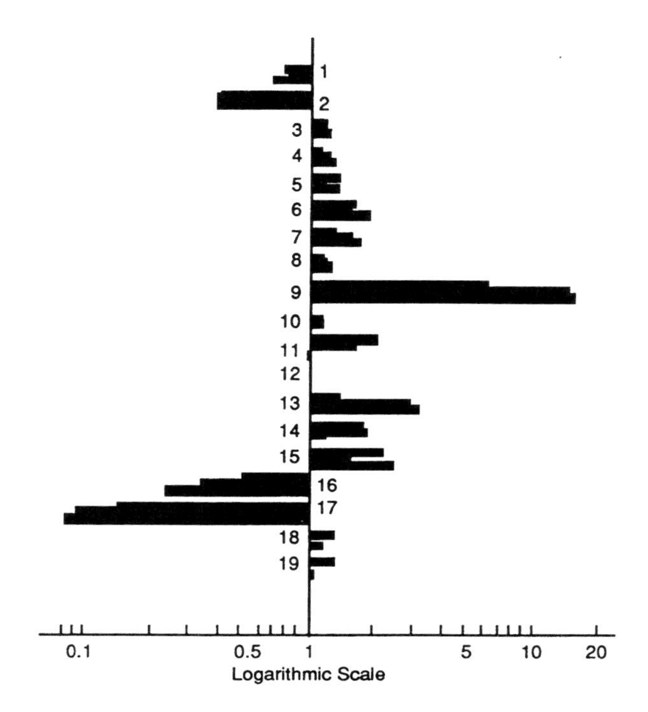

Figure 55 The influence of stocking densities with silver carp on average seasonal values of some environmental and biocenotic factors. Logarithmic scale: 1 corresponds to the value of a given factor in the control group of ponds with common carp only. The three bars correspond to the three consecutive increasing stocking densities of silver carp (the highest bar shows the particular value at lowest stocking density of silver carp). (1) ammonium nitrogen, (2) phosphates, (3) BOD_5, (4) soluble oxygen, (5) phytoplankton biomass, (6) chlorophyll, (7) gross primary production, (8) destruction, (9) net primary production, (10) total number of bacteria, (11) number of proteolytic bacteria, (12) number of ammonificating bacteria, (13) total number of bacteria in sediments, (14) number of proteolytic bacteria in sediments, (15) number of denitrifying bacteria in sediments, (16) zooplankton numbers, (17) zooplankton biomass, (18) benthos numbers, (19) benthos biomass. (From Opuszynski, K., *Ekol. Pol.,* 27, 117, 1979. With permission.)

interesting to note that bighead carp affected algal communities in a different way. When this fish was stocked, the percentage of blue-green algae (Cyanophyceae) increased.[637]

The shift in algal size structure toward dominance of larger species was unexpected. Filter feeding fishes usually cause the development of smaller algae through feeding pressure on the larger algae. It is presumed that the changes in the algae community resulted from the indirect effect of silver carp acting through alteration of the pond environment. The pH level, which is related to free carbon dioxide (CO_2) and bicarbonate (HCO_3^-) concentrations in the water, is of special importance for algae. Because of increased algae development in the silver carp ponds, pH was higher than in the control group, and averaged 8.5 and 8.0.

It is well known that as the pH increases the amount of carbon dioxide declines, and bicarbonate increases. At a pH of 8.4 the inorganic carbon in the water is represented by bicarbonates, and as pH increases above 8.4 also by carbonates. Nannoplankton algae are not capable of utilizing bicarbonates as a source of carbon at a high pH.[639] It appears that algae in oligotrophic systems utilize only free CO_2 as a source of carbon for photosynthesis and may be unable to assimilate it at low concentrations. For example, *Chlorella pyrenoidosas* cannot utilize bicarbonates or utilizes them only slightly. On the other hand, algae in eutrophic systems (*N. palea* is included here) can directly utilize bicarbonates or free CO_2 even at very low concentrations.[640] These data show that the differences in the pH values caused by the presence of silver carp were sufficient to change the structure of phytoplankton communities. It should also be noted that nutrient uptake is accomplished through the algal cell surface. Because smaller cells have a larger surface to volume ratio than larger cells, the small algae have an advantage at lower nutrient levels. Although the average nutrient concentration was lower in the silver carp ponds, there was a greater standing crop and higher nutrient uptake by the algae. Changes in zooplankton community structure were also not typical in comparison to those that usually occur under the influence of planktivorous fishes. The percentage of large cladoceran species (mainly *Daphnia* sp. and *Moina rectirostris*) increased, and the small species decreased (mainly due to the disappearance of *Bosmina longirostris*). The influence of the fishes on rotifers was relatively small. Phytoplankton control may have occurred due to zooplankton grazing. However, the zooplankton biomass was reduced enough by silver carp (over tenfold, see Figure 55) to make zooplankton an ineffective algal control agent.

Several factors could have accounted for the increase in phytoplankton production and biomass in these experiments. The most important were low algal consumption by silver carp due to detrital feeding, elimination of zooplankton, and increased nutrient cycling due to silver carp feeding activity. The fishes remove dead and living plant material from the water column, process it, and transport it rapidly as fish feces to the bottom. The fish feces are colonized by bacteria and undergo rapid mineralization.

Fishes affect not only water quality and the structure and function of ecosystems, but these changes also affect other organisms living there. Considerable changes have been observed in fish communities and in fish standing crop with increased eutrophication. These changes have been monitored by fishery biologists in many countries.[641,642] Generally, in temperate regions the ichthyofauna of clear and nutrient-poor (oligotrophic) lakes is predominated by salmonid species (for example, lake trout, *Salvelinus namaycush*). As eutrophication proceeds, these species are initially replaced by coregonids (for example, whitefish, *Coregonus clupeaformis*) and finally by cyprinids. This group is rich in species that occupy many different ecological niches. Therefore, they exist for long periods during succession, and many changes in species composition occur during this time. These changes consist of the replacement of the larger, less abundant species, which have important roles in the fishery, by small fishes with little economic importance. The cyprinids coexist with the piscivorous fishes such as pike or pike-perch; as eutrophication progresses, the piscivorous fishes disappear. Total fish biomass along with progressing eutrophication increases up to a maximum limit and then decreases if eutrophication continues to increase.

Two sets of data pertaining to the effect of fishes on the ecosystem and changes in the fish community due to ecosystem changes led to the hypothesis of ichthyo-eutrophication.[638,643] This hypothesis assumes that a relationship exists based on the feedback mechanism between environmental changes and fish community changes within a body of water (Figure 56). According to this hypothesis, management to counteract eutrophication

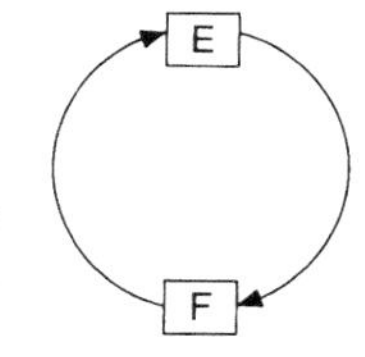

Figure 56 The hypothesis of ichthyo-eutrophication: eutrophication processes (E) result in changes in the fish community (F), which in turn speed up eutrophication processes.

should consist of controlling planktivorous and benthic fishes and protect piscivorous fishes. If the hypothesis of ichthyo-eutrophication is valid, the use of filter feeding phytophagous fishes in temperate climates to counteract eutrophication is ill-advised.

III. NORTH AMERICAN STUDIES

The environmental impact of a native filter feeding fish (gizzard shad) has been studied in the U.S. This fish is common in the southern states, where it may constitute >50% of total fish biomass in eutrophic lakes. Because of high biomass and high filtration rate, gizzard shad populations may filter a great volume of water in a short time. For example, in Barkley Lake, Kentucky, and Patten Lake, Texas, it was calculated that these fishes could filter volumes equivalent to those of the entire lake in 56 and 130 h of feeding, respectively.[644] Gizzard shad change their feeding behavior from visual particulate feeding to filter feeding at a size of about 2.5 cm. As the fishes grow, the distances between the gill rakers increase, resulting in a shift in food particle size selection. For example, gizzard shad of 5, 15, and 25 cm total length may selectively feed on particles larger than 19, 40, and 63 μm, respectively, which explains why phytoplankton become relatively less important in the diet as the fishes grow.[645]

Because of finely spaced gill rakers, gizzard shad may filter all rotifer and crustacean zooplankton. However, gill raker spacing and prey escape ability governs the feeding selectivity of this fish. Gizzard shad have the highest feeding rates for the most easily captured rotifers, cladoceran, and copepod nauplii. In pond experiments gizzard shad suppressed the rotifer *Keratella,* a small species which has limited swimming ability; however, the large copepod *Diaptomus,* which is one of the most evasive zooplankters, increased.[646] Because increases in the herbivorous *Diaptomus* may partially offset suppression of cladocerans, gizzard shad zooplanktivory may not result in the dramatic increases in phytoplankton found in some experiments with other planktivorous fishes.

Gizzard shad at stocking densities of 233 and 257 kg/ha caused increased water turbidity in ponds, but the algal biomass was not significantly altered. Changes in the algal community consisted of reduced numbers of *Ceratium* and increased numbers of nannoplankton.[647] *Ceratium,* a large dinoflagellate with a relatively slow growth rate, may be vulnerable to fish feeding pressure because it does not survive passage through the gizzard shad digestive tract.

In another experiment Drenner et al.[648] showed that the effect of gizzard shad on phytoplankton may be caused not only by direct grazing, but by other factors. Using outdoor tanks, they attempted to determine the direct and indirect effects of gizzard shad on phytoplankton by using a zooplanktivorous atherinid fish *(Menidia beryllina)* to suppress zooplankton. In all experiments gizzard shad caused increased amounts of phytoplankton as measured by algal chlorophyll-*a,* Secchi depth, turbidity, Coulter® counts of suspended particles, and counts of algal cells. *Menidia* suppressed zooplankton biomass to a greater extent than gizzard shad without causing an increase in phytoplankton; therefore, they rejected the hypothesis that the increase in phytoplankton due to gizzard shad was an indirect effect of zooplankton suppression. Instead, Drenner and coworkers formed an alternative hypothesis that gizzard shad caused increased phytoplankton by increased nutrient cycling. Unfortunately, testing this hypothesis was not included in the experimental design. An important inference from this study was that large gizzard shad were ineffective in controlling algal abundance when the community was dominated by filamentous blue-green algae or nannoplankton.

Even more complex effects of gizzard shad on the plankton community were revealed in a changeover experimental design in which the persistence of fish-derived changes was examined for periods of 38 to 45 d after the fishes were removed from the tanks.[649] Enhancement of phytoplankton by gizzard shad in the prechangeover period was a persistent effect in the December to February and July to September experiments. In the April to June experiment, however, phytoplankton responded primarily to the postchangeover gizzard shad treatment and showed few residual (due to the presence or absence of shad in tanks before the changeovers) effects. A distinct feature of zooplankton composition in the April to June experiment was the presence of two large-body *Daphnia* species *(D. pulicaria* and *D. galeata mendotae)* that were absent in the other experiments. In the other experiments, in which the enhancement of phytoplankton by gizzard shad was not altered by the changeover, zooplankton were composed of smaller *Daphnia, Diaphanosoma,* copepods, and rotifers. These experiments show the importance of zooplankton community structure in biomanipulation procedures and clearly indicate one of the reasons for different biomanipulation results in various bodies of water.

The analysis of trophic-level interactions in subtropical Florida lakes led Crisman[650] and Crisman and Beaver[651] to the conclusion that biomanipulation schemes for subtropical and tropical waters should be different from those in temperate waters. The most substantial difference between temperate and subtropical lakes is the absence of large *Daphnia* in the latter. These species are effective phytoplankton grazers. Subtropical cladocerans are not only smaller bodied and represented by fewer species, but unlike in the temperate zone, they are largely absent during periods of peak algal biomass. Hence, zooplankton in subtropical systems may not be as important for controlling algae as has been demonstrated for temperate lakes. This was confirmed by enclosure experiments conducted in Florida. These experiments indicated that even small-bodied cladocerans were subject to heavy predation from gizzard shad because when the fishes were absent, zooplankton populations increased. Algal biomass was not reduced by increased zooplankton populations, but actually increased by 9 to 31%, unlike the result in temperate lakes.

The changes in fish communities, along with progressing eutrophication, are similar in subtropical lakes to those already described for temperate lakes. In subtropical lakes, however, increasing eutrophication results in the enhancement of predominantly herbivorous filter feeding fishes (Figure 57). Gizzard shad or threadfin shad are absent or relatively less important in temperate climates. Gizzard shad prey upon small-bodied macrozooplankton in Florida lakes, and although relatively inefficient, they also feed on larger algae that would otherwise not be consumed because of the lack of large-bodied macrozooplankton. Crisman and Beaver[651] argue that if biomanipulation is to be successful in the subtropics, the emphasis should be shifted from zooplankton to the role played by planktivorous fishes. They are not specific, however, as to the kind of biomanipulations needed with planktivorous fishes.

The possibility of using sterile triploid bighead carp was studied to alleviate eutrophication in hypereutrophic Lake Apopka, Florida.[256] A large green alga, *Botryococcus*

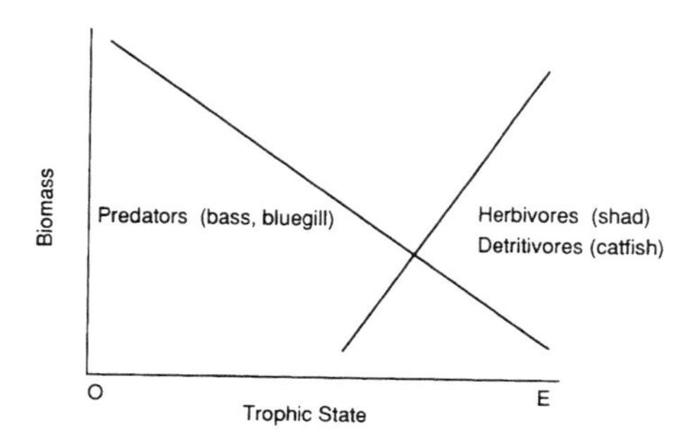

Figure 57 Changes in fish biomass and community structure in subtropical Florida lakes as a function of increasing eutrophication. O = oligotrophic; E = eutrophic lakes. (Data from Reference 650.)

braunii, is a nuisance species in this lake. Bighead carp were used because they possess gill rakers spaced widely enough to allow smaller algae to pass, but to retain *B. braunii.* The evaluation of the potential impact of bighead carp was conducted in flow-through ponds receiving water from Lake Apopka. Although bighead carp fed selectively on *B. braunii* they did not reduce its abundance. The possible reason was that fish biomass was too low, ranging from 60 to 97 kg/ha in individual ponds. The growth rate of the fish was low and mortality was high, indicating that *B. braunii* was not the proper food for bighead carp or that the triploid bighead carp was inferior to diploid bighead. The low numbers of zooplankton in the ponds were probably due to fish feeding. Despite low biomass, the fishes exerted a significant impact on the algal community, decreasing the ratio of blue-green to green algae in the ponds (Figure 58).

Boyd[451] summarized a number of experiments that were conducted at Auburn University (Auburn, AL) to evaluate the potential of plankton-feeding fishes to limit phytoplankton development in catfish ponds. Blue tilapia, silver carp, and bighead carp used in various stocking combinations actually caused increased phytoplankton production as well as increased chemical oxygen demand values. These results were confirmed by Burke et al.[260] Stocking catfish ponds with silver or bighead carp reduced the density of zooplankton, significantly increased phytoplankton biomass, and decreased ammonia and nitrite concentrations in the water.

Attempts have been made to overcome the detrimental effect of filter feeding fishes on zooplankton by using zooplankton refuges. Smith[652] conducted experiments with catfishes and silver carp stocked in tanks without a partition or with a partition of plastic mesh that excluded all fishes from one half of the tank. Tanks with no zooplankton refuge were dominated by small chydorid cladocerans and cyclopoid copepods, whereas large cladocerans, mostly *Ceriodaphnia* and *Daphnia* sp., dominated in refuge tanks. The refuge permitted the coexistence of high densities of large zooplankters with the filter feeding fishes. This combination resulted in decreased algal biomass by as much as 99%, increased phytoplankton diversity, and tended to improve silver carp growth. The idea of zooplankton refuge was tested on a larger scale in commercial prawn ponds in Hawaii.[653] Some ponds contained no fishes, some contained free swimming silver carp, and some contained silver carp confined to one half of the pond by a net placed across the pond. The results were not encouraging. Although silver carp (confined or not) always caused a dramatic decline in net phytoplankton (>10 μm), nannoplankton (<10 μm) increased. No significant differences were found in nannoplankton and total phytoplankton (expressed as chlorophyll-*a* concentrations) among the ponds with confined and free swimming

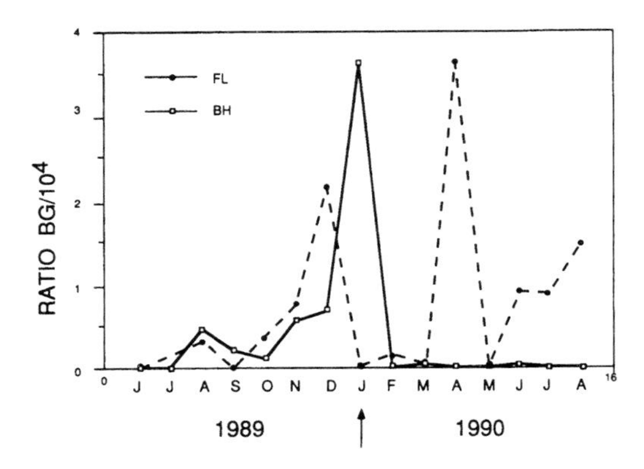

Figure 58 Ratio of blue-green to green algae (BG:G) in fishless (FL) and bighead carp (BH) ponds. The arrow indicates when BH ponds were stocked. (From Opuszynski, K. and Shireman, J. V., *J. Fish Biol.*, 42, 517, 1993. With permission.)

silver carp. Compared to the fishless control ponds, total phytoplankton biomass increased in the presence of 75 silver carp per hectare, but did not change significantly at 3750 silver carp per hectare.

In addition to Chinese carp the impact of another non-native planktivorous fish, blue tilapia, was studied.[42] The effect was similar to that of the other filter feeding species described earlier. They caused reductions in the populations of large phytoplankton and the zooplankton species with limited escape abilities. Small algae and the evasive zooplankter *Diaptomus* sp. increased in abundance.

IV. USE OF FILTER FEEDING FISHES IN ISRAEL

Israel is possibly the only country in which planktivores have been used on a mass scale in order to improve water quality. The results, however, are controversial and difficult to describe quantitatively. In the 1960s the Israel National Water Carrier (INWC) was built to deliver water from Lake Kinneret to the southern part of the country for drinking, agricultural, and industrial purposes. Lake Kinneret is the only freshwater lake in the country and provides about 35% of Israel's freshwater. The INWC consists of open reservoirs, channels, and 250 km of pipes for water distribution. After the water is pumped from the lake, it flows at a rate of approximately 1 million m^3/d to the Tsalmon Reservoir (20 ha), next to a settling reservoir (40 ha), and the Eshkol Reservoir (100 ha). These reservoirs were stocked from 1978 to 1981 with the species and quantities of fishes shown in Table 37.

The primary goal of the stocking was not for fish production, but for water quality improvement.[300] Blue tilapia and mullet, *Mugil capito,* were stocked to reduce the quantity of organic matter on the reservoir bottom, silver carp to reduce phytoplankton, and bighead carp to reduce zooplankton. To create a kind of "biological filter" in the Tsalmon Reservoir, the stocking density of silver and bighead carp was much higher there than in the Eshkol Reservoir.

Summarizing the results of fish introduction into the reservoirs, Leventer and Teltsch[654] maintain that the goal of the stocking program was achieved. During the 8-year study period, the concentration of suspended organic matter in the water flowing out of the Eshkol Reservoir was less than in the water pumped from Lake Kinneret. Most of the organic suspended matter in Lake Kinneret consisted of algae and zooplankton, whereas water leaving the Eshkol Reservoir contained mainly undefined organic matter. The average annual load of dry suspended organic matter from Lake Kinneret was 1634 tons, compared to 931 tons leaving the Eshkol Reservoir. The authors concluded that the

Table 37 **Fish species and numbers stocked into Tsalmon, Settling, and Eshkol Reservoirs (Israel), 1978–1981**

Species	Tsalmon	No./ha Settling	Eshkol	Total no.
Tilapia	1,500	—	—	30,000
Mullet	—	400	700	86,000
Silver carp	4,350	538	235	132,000
Bighead carp	800	125	235	44,500

From Leventer, H and Teltsh, B., *Hydrobiologia*, 191, 47, 1990. With permission.

annual reduction rate was 703 tons or approximately 5600 tons over the 8-year period. They maintained that this substantial reduction was due to organic matter removal "by the fish and other biological processes". Furthermore, they calculated that because 300 tons of silver carp were removed during the same period and because the fishes consumed about 8 kg of dry food to produce 1 kg of weight, the silver carp alone removed approximately 2500 tons of suspended organic matter. Although this figure is impressive, the authors conceded that up to 80% of the organic matter returned to the water settled to the bottom as fish feces. Other fishes, mostly tilapia, and benthic organisms consumed organic matter from the bottom and reduced the concentration.

The use of fishes to improve water quality after macrophyte removal by grass carp in the INWC constitutes an interesting case of the applicability of biological methods for water treatment. Unfortunately, the results are inconclusive. It is difficult to separate the effect of fishes from other processes that may have been responsible for the improvement of water quality in the reservoirs; e.g., mechanical sedimentation of the suspended matter was not taken into account. Moreover, besides biological methods, the water was intermittently treated with chlorine to control the large nuisance dinoflagellate *Peridinium* sp., and this treatment led to a reduction in the number of algae. Finally, these results were obtained in reservoirs with very short retention times and, therefore, may not be applicable to lake ecosystems.

The effect of silver carp on the water quality in Lake Kinneret was deemed negative in another study.[232] The fishes fed predominantly on zooplankton from September to January and suppressed crustaceans, which are needed to prevent algae blooms. Moreover, silver carp competed with the native fish, the Galilee cichlid, for the same zooplankton resource. This cichlid consumes *Peridinium* more efficiently, and has a higher commercial value.

The catch of the Galilee cichlid in Lake Kinneret possibly declined as the result of stocking blue tilapia. The impact that these two cichlids had on Lake Kinneret plankton was compared in laboratory experiments.[655] Crustaceans and rotifers were reduced when fishes were present. In the spring, chlorophyll concentration and the dominant phytoplankter *P. cinctum* were reduced. In the summer, however, when the smaller *P. elpatiewsky* was abundant, *Peridinium* was reduced in the presence of the Galilee cichlid. Chlorophyll concentrations were reduced when blue tilapia were present. Phytoplankton decreased and fewer small zooplankton increased in the presence of the Galilee cichlid. In another laboratory experiment conducted in outdoor tanks the effect of different densities of the Galilee cichlid was compared.[656] Crustaceans, rotifers, and *Peridinium* spp. declined as fish density increased, but the abundance of small nannoplankton was greatest at intermediate fish densities. Fewer *Peridinium* resulted in reduced total chlorophyll and gross primary production. Drenner et al. concluded that although the reduction of *Peridinium* spp. and algal biomass was a beneficial effect of fish activity, the increase in nannoplankton could have unforeseen implications for managing the water quality of Lake Kinneret.

The effect of fishes on the plankton community was also studied in ponds stocked with silver carp and bottom feeding fishes, including common carp and tilapia hybrids, Nile tilapia × blue tilapia.[657–659] Different fish densities and various species combinations were tested. The presence of silver carp resulted in an increase in total phytoplankton numbers and a decrease in phytoplankton size. This was due to the increase in the small green algae species *Scenedesmus* spp., *Chlorella* spp., *Selenastrum minutum,* and *Ankistrodesmus setigerus,* the small blue-green alga *Merismopedia minima,* and diatoms of the order Pennales. According to the authors, silver carp did not influence the abundance of small algae because they were passing freely through the filtering apparatus of the fishes. Other reasons for the rapid development of small algal species included the high turnover rate of small algae, the elimination of larger algae that competed with the smaller ones for nutrients, and decreased numbers and diversity of zooplankton predators. The presence of bottom feeding fishes partially offset the effect of silver carp on phytoplankton. The bottom feeding fishes caused the development of large phytoplankton possibly by stirring up bottom sediments, making settled nutrients available to large phytoplankton. As discussed previously, small algae have an advantage over large algae at low nutrient concentrations.

V. RESULTS IN OTHER REGIONS

In China a common opinion prevails among fishermen that silver and bighead carp, due to their grazing activity, decrease the development of phytoplankton. This opinion, however, is unsupported in recent studies. Chen et al.[660] showed that Chinese carp may intensify nitrogen and phosphorus cycling in a lake ecosystem. In another study conducted in cages located in a reservoir and in concrete tanks a positive correlation was found between stocking densities of silver and bighead carp and phytoplankton production and biomass.[661] A negative correlation occurred among fish stocking densities and zooplankton production and biomass. The food conversion rate of phytoplankton became less and the ratio of fish production to phytoplankton production decreased with the increase in fish stocking density. Shi et al. concluded that silver carp and bighead carp enhanced the eutrophication process in bodies of shallow water.

Promising results from using silver carp to improve water quality were obtained in a mesocosm experiment in Paranoá Reservoir, Brazil.[662] This large reservoir (4000 ha) was built in 1959 to provide water to the city of Brásilia. Water quality in the reservoir was deteriorating due to increased eutrophication and excessive algal blooms, caused mostly by a blue-green alga, *Cylindrospermopsis raciborskii.* The impact of fishes was compared in polyethylene cylinders that completely isolated a 2-m^3 column of water and were open to the atmosphere, but closed to the sediments. Four species of fishes were compared: Congo tilapia, bluegill, tambaqui *(Colossoma macropomum),* and silver carp. The first two species inhabited this reservoir, and the remaining two were recommended for introduction. *C. raciborskii,* which constituted 98% of the phytoplankton biomass during the experiments, was increased when bluegill, tilapia, and tambaqui were present, but significantly reduced by the presence of silver carp. In another mesocosm experiment conducted in Paranoá Reservoir,[663] bottomless tanks were dug about 30 cm into the soft bottom and stocked with silver carp. A decrease in net phytoplankton (>20 μm) abundance, total phytoplankton biomass, and net primary production occurred, but *C. raciborskii* numbers were not reduced.

Success in blue-green algae control, using silver carp in a whole-lake experiment, was reported in New Zealand.[664]

VI. STRATEGY AND EFFICACY OF USING FISHES: DIRECT OR INDIRECT ALGAL CONTROL?

This chapter has reviewed recent papers describing the use of phytoplanktivorous fishes to control the excessive development of phytoplankton, which is the most visible and noxious effect of increased eutrophication. Because of the practical importance of this problem, a great number of studies have been completed using various species of fishes in different climates and using different experimental designs. It is clear that the impact of phytoplanktivorous fishes on the algal community is more complex than may have been expected from a simple predator-prey model. Here, if predation pressure is strong enough, it results in a decline of the prey population. The possible effects planktivorous fishes have on algal communities include increases or decreases in algal numbers and biomass and a shift in community structure toward dominance by small-bodied (more often) or large-bodied algae (Table 38). It should be considered that most of the results were obtained in microcosm and mesocosm experiments which were extrapolated to whole-lake situations. These extrapolations may not be valid.

A strategy contrary to algae control using fish grazing was also discussed in this chapter. This strategy emphasized a reduction of planktivorous and benthic fishes, which proved to be successful in many cases (temperate regions) (Table 38). This strategy resulted in water quality improvement, even in large bodies of water, as shown by Scavia et al.[665] in Lake Michigan (U.S.). Salmonids heavily stocked into this lake reduced populations of the planktivorous alewife *(Alosa pseudoharengus),* which allowed the development of large *Daphnia.* Because *Daphnia* were more effective grazers than the dominant calanoid copepods, seston concentrations were reduced, causing dramatic increases in Secchi disk transparency.

It appears that in temperate climates the methods of algal control using nonpredatory fish reduction are superior to methods of phytoplanktivorous fish enhancement.

Notwithstanding the strategy to be selected, two important questions remain unanswered: how long lasting are the effects, and to what degree can water quality improvement be expected? Benndorf et al.[666] compared the response of a hypereutrophic German reservoir during the first 5 years of biomanipulation (enhancement of piscivorous and control of planktivorous fishes) to the 4-year pretreatment period. They found significant fluctuations in the measured parameters and concluded that the response of the ecosystem to the treatment was remarkably unstable. They postulated that effective eutrophication control in bodies of water with high external phosphorus loadings seems to be possible only via the combined use of biological and technical methods. In his concluding remarks at the International Workshop on Perspectives of Biomanipulation in Inland Waters, Golterman[667] admitted the validity of biomanipulation, but stated that its usefulness was restricted to situations in which essential technical measures were simultaneously taken. Recovery is quicker when biomanipulation is included, although this does not change the pattern of the recovery process (Figure 59).

State-of-the-art biomanipulation techniques do not allow formation of a judgment as to whether the above opinions underestimate the efficacy of biomanipulation. At present

Table 38 **Effect of phytoplanktivorous fish stocking or planktivorous and benthivorous fish removal on phytoplankton and water quality**

Experiment description, ref.	Results
Stocking of Phytoplanktivorous Fish	
Triploid bighead carp in ponds, Florida, U.S.[256]	Ratio of blue-green:green algae decreased
Gizzard shad in enclosures in Florida, U.S.[651]	No significant changes in chlorophyll lakes, concentration, and algal abundance, cladoceran abundance decreased
Gizzard shad in ponds, Kansas, U.S.[42]	Large alga *Ceratium* suppressed, nannoplankton enhanced
Gizzard shad, laboratory experiments, Texas, U.S.[648]	Algal chlorophyll, turbidity, Coulter® counts and microscopic algal counts increased, zooplankton suppressed
Silver carp, enclosures in a reservoir, Brazil[662]	Dominant blue-green alga and zooplankton suppressed
Silver and bighead carp in tanks and ponds, China[661]	Biomass and production of phytoplankton increased, zooplankton decreased
Silver carp in a lake, Germany[635]	No significant changes in water quality, phytoplankton dominance changed from *Microcystis* to *Oscillatoria*
Silver carp in lakes and ponds, Germany[634]	No significant changes in water quality, shift from *Microcystis* to *Oscilatoria* dominated phytoplankton, sharp decline of zooplankton
Silver carp in reservoirs, Israel[654]	Alleged significant reduction of algae, zooplankton, and suspended organic matter
Silver carp in ponds, Israel[657,658]	Increase in total phytoplankton numbers, decrease in their dominant size, decrease in zooplankton abundance
Silver carp in a lake, New Zealand[664]	Effective control of blue-green algal blooms
Silver carp in enclosures in a lake, Poland[633]	Phytoplankton biomass decreased, nannophytoplankton enhanced, zooplankton suppressed
Silver carp in ponds, Poland[638]	Phytoplankton biomass and primary production increased, large-sized algae enhanced
Silver and bighead carp in ponds, Poland[637]	Diatoms (silver carp) or blue-green algae (bighead carp) enhanced, shift toward large algal species
Silver or bighead carp in catfish ponds, Alabama, U.S.[260]	Phytoplankton biomass increased, zooplankton density decreased, lower ammonia and nitrite concentrations
Silver carp in tanks with zooplankton refuge, California, U.S.[652]	Large cladocerans present, reduced algal biomass, and increased species diversity

Table 38 **(continued) Effect of phytoplanktivorous fish stocking or planktivorous and benthivorous fish removal on phytoplankton and water quality**

Experiment description, ref.	Results
Silver carp in ponds, Hawaii, U.S.[653]	Net plankton decreased, nannoplankton increased, total phytoplankton biomass increased (low-density ponds) or not changed (high-density ponds)
Silver carp in tanks, former Uzbek SSR[632]	Biomass of zoo- and phytoplankton decreased, primary production increased
Tilapia *(T. galilaea),* laboratory experiments, Israel[656]	Zooplankton and *Peridinium* declined, nannoplankton increased, total chlorophyll and gross primary production reduced
Tilapia *(T. galilaea* and *T. aurea),* outdoor tanks, Israel[655]	*T. galilaea* suppressed more and enhanced fewer phytoplankton species than did *T. aurea*
Tilapia (*T. aurea*) in ponds, Kansas, U.S.[647]	Large-sized algae suppressed, small-sized algae enhanced
Removal of Planktivorous and Benthivorous Fishes	
Enhancement of piscivorous fishes in a hypertrophic reservoir, Germany[666]	*Dapnia* biomass and water transparency increased, total algal biomas unchanged, *Microcystis blooms* enhanced
Piscivorous fishes stocked, planktivorous and benthivorous fishes removed from a lake, The Netherlands[762]	Phytoplankton abundance and chlorophyll concentration declined, transparency increased, zooplankton more abundant
Planktivorous and benthivorous fishes removed from a lake, Sweden[765]	Total phytoplankton biomass and total P and N decreased
Largemouth bass stocked, planktivorous minnows removed from a lake, Indiana, U.S.[763]	*Daphnia* enhanced, algal community changed, and biomass declined
Planktivorous and benthivorous fishes removed from a lake, Minnesota, U.S.[764]	Water transparency increased, chlorophyll concentration and algal densities decreased, large crustaceans enhanced

it is impossible to decisively determine the preferable strategy for using herbivorous filter feeding fishes to counteract eutrophication. Possibly it is impracticable to use a single management strategy for the control of algae by fishes. Instead, various strategies should be considered in conjunction with temperate or tropical conditions, regional differences in structure and function of ecosystems (species and abundance of fishes, zooplankton and phytoplankton community structures, the major "nuisance" algal species), and finally, with the kind of body of water to be treated. A different strategy may be effective in small artificial bodies of water with short water residence time (reservoirs, canals, or

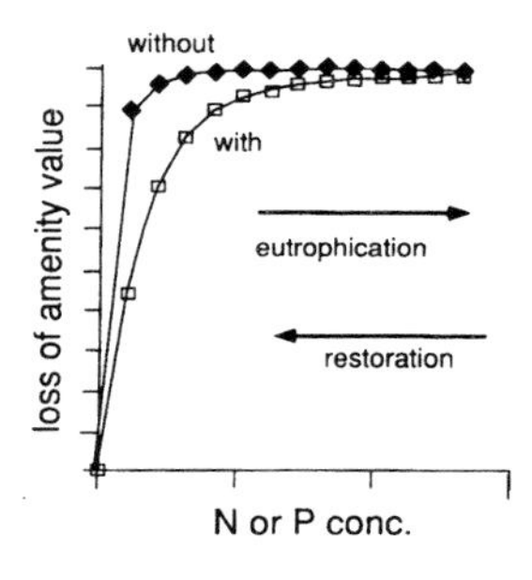

Figure 59 Lake restoration process with or without biomanipulation with fishes. Losses of amenity value = 1/ water quality. (From Golterman, H. L., *Hydrobiologica,* 191, 319, 1990. With permission.)

ponds) than in natural larger lakes. To distinguish between these two strategies, Gophen[620] proposed two different terms. "Biological control" is the term used when fishes are used to improve water quality in small bodies of water, and "biomanipulation" is the term used to describe long-term trophic-level manipulations in large natural systems. As biological control is concerned, the use of filter feeding fishes for wastewater treatment looks especially interesting.

REFERENCES

1. **Berg, L. S.,** Freshwater fishes of the USSR and adjacent countries (Ryby presnykh vod SSSR i sopredel'nykh stran), *IPST Catalog No. 742,* Jerusalem, Israel Program for Scientific Translations (translated 1964), 2, 134, 1949.
2. **Nelson, J. S.,** *Fishes of the World,* 1st ed., John Wiley & Sons, New York, 1976, 416.
3. **Howes, G.,** Anatomy and phylogeny of the Chinese major carps *Ctenopharyngodon* Steind., 1866 and *Hypophthalmichthys* Blkr., 1860, *Bull. Br. Mus. (Nat. Hist.) Zool.,* 41, 1, 1981.
4. **Kryzanovskij, S. G.,** Sistema semeistva karpovych ryb (Cyprinidae), *Zool. Zh.,* 26, 53, 1947.
5. **Oshima, M.,** Contributions to the study of fresh water fishes of the island of Formosa, *Ann. Carnegie Mus.,* 12, 169, 1919.
6. **Gosline, W. A.,** Unbranched dorsal-fin rays and subfamily classification of the fish family Cyprinidae, *Occ. Pap. Mus. Zool. Univ. Mich.,* 684, 1, 1978.
7. **Robins, C. R., Bailey, R. M., Bond, C. E., Brooker, J. R., Lachner, E. A., Lea, R. N., and Scott, W. B.,** Fishes important to North Americans, exclusive of species from continental waters of the U. S. and Canada, *Common Names of Fishes,* American Fisheries Society, Bethesda, MD, 1989.
8. **Cuvier, G. and Valenciennes, A.,** *Histoire Naturelle des Poissons,* Vol. 17, P. Bertrand, Paris, 1844, 362.
9. **Antalfi, A. and Tölg, I.,** *The Herbivorous Fishes,* Panstwowe Wydawnictwo Rolnicze i Lesne, Warsaw, 1975, 270 (in Polish).
10. **Berry, P. V. and Low, M. P.,** Comparative studies on some aspects of the morphology and histology of *Ctenopharyngodon idella, Aristichthys nobilis,* and their hybrid (Cyprinidae), *Copeia,* 4, 708, 1970.
11. **Shireman, J. V. and Smith, C. R.,** Synopsis of biological data on the grass carp, *Ctenopharyngodon idella* (Cuvier and Valenciennes, 1844), *FAO Fish. Synop.,* 135, 86, 1983.
12. **Ojima, Y., Hayashi, M., and Ueno, K.,** Cytogenetic studies in lower vertebrates. X. Karyotype and DNA studies in 15 species of Japanese cyprinidae, *Jpn. J. Genet.,* 47, 431, 1972.
13. **Manna, G. K. and Khuda-Bukhsh, A. R.,** Karyomorphology of cyprinid fishes and cytological evaluation of the family, *Nucleus,* 20, 119, 1977.
14. **Márián, T. and Krasznai, Z.,** Comparative karological studies on Chinese carp, *Aquaculture,* 18, 325, 1979.
15. **Beck, M. L., Biggers, C. J., and Dupree, H. K.,** Karyological analysis of *Ctenopharyngodon idella, Aristichthys noblis* and their F_1 hybrid, *Trans. Am. Fish. Soc.,* 109, 433, 1980.
16. **Utter, F. and Folmar, L.,** Protein systems of grass carp: allelic variants and their applications to management of introduced populations, *Trans. Am. Fish. Soc.,* 107, 129, 1978.
17. **Magee, S. M. and Philipp, D. P.,** Biochemical genetic analyses of the grass carp ♀ × bighead carp ♂ F_1 hybrid and the parental species, *Trans. Am. Fish. Soc.,* 111, 593, 1982.
18. **Okoniewska, Z. and Okoniewski, Z.,** Nutritional and organoleptic properties of herbivorous fish flesh — white amur and white tolpyga, *Prezemysl Spozywezy* (translation from Polish by Office of Foreign Fisheries, Translations, U.S. Dept. of the Interior, Washington, D.C.), 22, 304, 1968.

19. **Slechtova, V., Slechta, V., Do Doan Hiep, and Valenta, M.,** Biochemical genetic comparison of bighead *(Hypophthalmichthys molitrix),* silver carp *(Aristichthys nobilis)* and their hybrids reared in Czechoslovakia, *J. Fish. Biol.,* 39(Suppl. A), 349, 1991.
20. **Pasteur, R. and Herzberg, A.,** Proximate composition and freshness determination of the silver carp *Hypophthalmichthys molitrix* (Val.), *Fish Fishbreed. Isr.,* 14, 14, 1979.
21. **Richardson, J.,** Ichthyology, in *The Zoology of the Voyage of Hms "Sulphur", Under the Command of Captain Sir Edward Belcher, During the Years 1836–42,* Vol. 2, Hinds, R. B, Ed., 1845, 71.
22. **Jennings, D. P.,** Bighead carp *(Hypophthalmichthys nobilis):* biological synopsis, *U.S. Fish Wildl. Serv., Biol. Rep.,* 29, 35, 1988.
23. **Blaxter, J. H. S. and Hunter, J. R.,** The biology of clupeoid fishes, *Adv. Mar. Biol.,* 20, 1, 1982.
24. **Horn, M. H.,** Biology of marine herbivorous fishes, *Oceanogr. Mar. Biol. Annu. Rev.,* 27, 167, 1989.
25. **Nelson, J. S.,** *Fishes of the World,* 2nd ed., John Wiley & Sons, New York, 1984, 523.
26. **Choat, J. H.,** Fish feeding and the structure of benthic communities in temperate waters, *Annu. Rev. Ecol. Syst.,* 13, 423, 1982.
27. **Gaines, S. D. and Lubchenco, J.,** A unified approach to marine plant-herbivore interactions. II. Biogeography, *Annu. Rev. Ecol. Syst.,* 13, 111, 1982.
28. **Estes, J. A. and Steinberg, P. D.,** Predation, herbivory, and kelp evolution, *Paleobiology,* 14, 19, 1988.
29. Fishery Statistics, Catches and Landing, FAO Yearbook, Food and Agriculture Organization of the United Nations, Rome, 1991.
30. **Philippart, J.-C. L. and Ruwet, J.-C. L.,** Ecology and distribution of tilapias, in *The Biology and Culture of Tilapias,* Pullin, R. S. V. and Lowe-McConnell, R. H., Eds., International Center for Living Aquatic Resources Mgmt, Manila, the Philippines, 1983, 15.
31. **Okeyo, D. O.,** Herbivory in freshwater fishes: a review, *Bamidgeh,* 41, 79, 1989.
32. **Bowen, S. H.,** Detritivory and herbivory, in *Biology and Ecology of African Freshwater Fishes,* Leveque, C., Bruton, M. N., and Ssentongo, G., Eds., Office de la Recherche Scientifique et Technique Outre-Mer, Paris, 1988, chap. 11.
33. **Horn, M. H.,** Herbivorous fishes: feeding and digestive mechanisms, in *Plant-Animal Interactions in the Marine Benthos,* Systematics Association Spec. Vol. No. 46, John, D. M., Hawkins, S.J., and Price, J. H., Eds., Clarendon Press, Oxford, 1992, 339.
34. **Lazzaro, X.,** A review of planktivorous fishes: their evolution, feeding behaviours, selectivities, and impacts, *Hydrobiologia,* 146, 97, 1987.
35. **Boruckij, E. W.,** Materials on feeding of *Ctenopharyngodon idella* and *Plagiognathops microlepis* in the basin of Amur River, *Proc. Amur Ichthyol. Expedition 1945–1949,* 3, 505, 1952 (in Russian).
36. **Hickling, C. F.,** *Tropical Inland Fisheries,* Longmans, London, 1961, 65.
37. **Gerking, S. D.,** Assimilation and maintenance ration of an herbivorous fish, *Sarpa salpa,* feeding on green alga, *Trans. Am. Fish. Soc.,* 113, 378, 1984.
38. **Bowen, S. H.,** Feeding, digestion and growth — qualitative consideration, in *The Biology and Culture of Tilapias,* Pullin, R. S. V. and Lowe-McConnell, R. H., Eds., ICLARM, Manila, the Philippines, 1983, 141.
39. **Inaba, D. and Nomura, M.,** On the digestive system and feeding habits of young Chinese carps collected in the River Tone, *J. Tokyo Univ. Fish.,* 42, 17, 1956.

40. **Hickling, C. F.,** On the feeding process in the white amur, *Ctenopharyngodon idella, J. Zool. (London),* 148, 408, 1966.
41. **Prejs, A.,** Herbivory by temperate freshwater fishes and its consequences, *Environ. Biol. Fish.,* 10, 281, 1984.
42. **Drenner, R. W., Mummert, J. R., deNoyelles, F., Jr., and Kettle, D.,** Selective particle ingestion by a filter-feeding fish and its impact on phytoplankton community structure, *Limnol. Oceanogr.,* 29, 941, 1984.
43. **Voropaev, N. V.,** Morphological characteristics, diet and certain fish farming attributes of silver carp and bighead carp and their hybrids, in *New Investigations in the Ecology and Breeding of Herbivorous Fish,* Nikolskii, G. V., Ed., Nauka Press, Moscow, 1968, 206.
44. **Spataru, P., Wohlfarth, G. W., and Hulata, G.,** Studies on the natural food of different fish species in intensively manured polyculture ponds, *Aquaculture,* 35, 283, 1983.
45. **Drenner, R. W., Vinyard, G. L., Hambright, K. D., and Gophen, M.,** Particle ingestion by *Tilapia galilaea* is not affected by removal of gill rakers and microbranchiospines, *Trans. Am. Fish. Soc.,* 116, 272, 1987.
46. **Cremer, M. C. and Smitherman, R. O.,** Food habits and growth of silver and bighead carp in cages and ponds, *Aquaculture,* 20, 57, 1980.
47. **Borutskiy, Y. V.,** The food of the bighead carp and the silver carp in the natural waters and ponds of the USSR, in *The Trophology of Aquatic Animals,* Nikolskij, G. V. and Prioznikov, P. L., Eds., Nauka Press, Moscow, 1973, 299 (in Russian).
48. **Durbin, A. G. and Durbin, E. G.,** Grazing rates of the Atlantic menhaden as a function of particle size and concentration, *Mar. Biol.,* 33, 265, 1975.
49. **Al-Hussaini, A. H.,** The feeding habits and the morphology of the alimentary tract of some teleosts living in the neighborhood of the Marine Biological Station, Ghardaqa, Red Sea, *Publ. Mar. Biol. Stn.,* 5, 1, 1947.
50. **Fryer, G. and Iles, T. D.,** *The Cichlid Fishes of the Great Lakes of Africa: Their Biology and Evolution,* T. F. H. Publisher, Neptune, NJ, 1972.
51. **Odum, W. E.,** Utilization of the direct grazing and plant detritus food chains by the striped mullet *Mugil cephalus,* in *Marine Food Chains,* Steele, J. H., Ed., University of California Press, Berkeley, 1970, 222.
52. **Nikolski, G. V.,** *Fishes of the Amur Basin,* Academy of Science U. S. S. R., Moscow, 1956 (in Russian).
53. **Verigin, B. V.,** The development of young silver carp with relation to its biology, *Proc. Amur Ichthyol. Expedition, 1945–1949,* 1, 303, 1950.
54. **Girgis, S.,** On the anatomy and histology of the alimentary tract of an herbivorous bottom-feeding cyprinoid fish, *Labeo horie* (Cuvier), *J. Morphol.,* 90, 317, 1952.
55. **Montgomery, W. L. and Pollak, P. E.,** Gut anatomy and pH in a Red Sea surgeonfish, *Acanthurus nigrofuscus, Mar. Ecol. Prog. Ser.,* 44, 7, 1988.
56. **Matthes, H.,** A comparative study of the feeding mechanisms of some African cyprinids, *Bijdr Dierk,* 33, 3, 1963.
57. **Lobel, P. S.,** Trophic biology of herbivorous reef fishes: alimentary pH and digestive capabilities, *J. Fish Biol.,* 19, 365, 1981.
58. **Osman, A. H. K. and Caceci, T.,** Histology of the stomach of *Tilapia nilotica* (Linnaeus, 1758) from the River Nile, *J. Fish Biol.,* 38, 211, 1991.
59. **Collins, M. R.,** The feeding periodicity of striped mullet, *Mugil cephalus* L., in two Florida habitats, *J. Fish Biol.,* 19, 307, 1981.

60. **Pillay, T. V. R.,** Food, feeding habits and alimentary tract of grey mullet, *Proc. Nat. Inst. Sci. India,* 19, 777, 1953.
61. **Rimmer, D. W. and Wiebe, W. J.,** Fermentative microbial digestion in herbivorous fishes, *J. Fish Biol.,* 31, 229, 1987.
62. **Migita, M. and Hashimoto, Y.,** On the digestion of higher carbohydrates by Zsuanhi (*Ctenopharyngodon idella* Cuv. and Val.), *Bull. Jpn. Soc. Sci. Fish.,* 15, 259, 1949.
63. **Fish, G. R.,** Digestion in *Tilapia esculenta, Nature,* 167, 900, 1951.
64. **Barrington, E. J. W.,** The alimentary canal and digestion, in *The Physiology of Fishes,* Brown, M. E., Ed., Academic Press, New York, 1957, 109.
65. **Stickney, R. R. and Shumway, S. E.,** Occurrence of cellulase activity in the stomachs of fish, *J. Fish Biol.,* 6, 779, 1974.
66. **Prejs, A. and Blaszczyk, M.,** Relationship between food and cellulase activity in freshwater fish, *J. Fish Biol.,* 11, 447, 1977.
67. **Niederholzer, R. and Hofer, R.,** The adaptation of digestive enzymes to temperature, season and diet in roach *Rutilus rutilus* L. and rudd *Scardinius erythrophthalmus* L., *J. Fish Biol.,* 15, 411, 1979.
68. **Lindsay, G. T. H. and Harris, T. E.,** Carboxymethyl cellulase activity in the digestive tract of fish, *J. Fish Biol.,* 16, 219, 1980.
69. **Shcherbina, M. A. and Kazlauskene, O. P.,** The reaction of the medium and the rate of absorption of nutrients in the intestine of carp, *J. Ichthyol.,* 11, 81, 1971.
70. **Lesel, R., Fromageot, C., and Lesel, M.,** Cellulose digestibility in grass carp, *Ctenopharyngodon idella,* and in goldfish, *Carassius auratus, Aquaculture,* 54, 11, 1986.
71. **Das, K. M. and Tripathi, S. D.,** Studies on the digestive enzymes of grass carp, *Ctenopharyngodon idella* (Val.), *Aquaculture,* 92, 21, 1991.
72. **Moriarty, D. J. W.,** The physiology of digestion of blue-green algae in the cichlid fish, *Tilapia nilotica, J. Zool. (London),* 171, 25, 1973.
73. **Payne, A. I.,** Gut pH and digestive strategies in estuarine grey mullet (Mugilidae) and tilapia (Cichlidae), *J. Fish Biol.,* 13, 627, 1978.
74. **Bitterlich, G.,** The nutrition of stomachless phytoplanktivorous fish in comparison with tilapia, *Hydrobiologia,* 121, 173, 1985.
75. **Jonás, E., Ragyanszki, M., Oláh, J., and Boross, L.,** Proteolytic digestive enzymes of carnivorous (*Silurus glanis* L.), herbivorous (*Hypophthalmichthys molitrix* Val.) and omnivorous (*Cyprinius carpio* L.) fishes, *Aquaculture,* 30, 145, 1983.
76. **Anderson, T. A.,** Mechanisms of digestion in the marine herbivore, the luderick, *Girella tricuspidata* (Quoy and Gaimard), *J. Fish Biol.,* 39, 535, 1991.
77. **Nikolski, G. V.,** *The Ecology of Fishes,* Academic Press, New York, 1963, 352.
78. **Kirilenko, N. S. and Chigrinzkaya, Yu. N.,** Activity of digestive enzymes in the silver carp, *Hypophthalmichthys molitrix* (Cyprinidae), feeding on blue-green algae, *J. Ichthyol.,* 23, 79, 1983.
79. **Jancarik, A.,** Die Verdauung der Hauptnährstoffe beim Karpfen, *Z. Fisch. Deren Hilfswiss.,* 12, 601, 1964.
80. **Dhage, K. P.,** Studies of the digestive enzymes in the three species of the major carps of India, *J. Biol. Sci.,* 11, 63, 1968.
81. **Fänge, R. and Grove, D.,** Digestion, in *Fish Physiology,* Vol. 8, Hoar, W. S., Randall, D., and Brett, J. R., Eds., Academic Press, New York, 1979, chap. 4.
82. **Tanaka, S. and Abe, T.,** *Zusetsu Yuyo Gyorui Senshu,* Morikita Shuppan, Showa, Tokyo, 1955.

83. **Buddington, R. K., Chen, J. W., and Diamond, J.,** Genetic and phenotypic adaptation of intestinal nutrient transport to diet in fish, *J. Physiol. (London),* 393, 261, 1987.
84. **Reshkin, S. J., Vilella, S., Ahearn, G. A., and Storelli, C.,** Basolateral inositol transport by intestines of carnivorous and herbivorous teleosts, *Am. J. Physiol.,* 256(3), G509, 1989.
85. **Vilella, S., Reshkin, S. J., Storelli, C., and Ahearn, G. A.,** Brush-border inositol transport by intestines of carnivorous and herbivorous teleosts, *Am. J. Physiol.,* 256, G501, 1989.
86. **Wissing, T. E.,** Energy transformation by young-of-the-year white bass *Morone Chrysops* (Rafinesque) in Lake Mendota, Wisconsin, *Trans. Am. Fish. Soc.,* 103, 32, 1974.
87. **Webb, P. W.,** Partitioning of energy into metabolism and growth, in *Ecology of Freshwater Fish Production,* Gerking, S. D., Ed., John Wiley & Sons, New York, 1978, chap. 8.
88. **White, T. C. R.,** When is a herbivore not a herbivore?, *Oecologia (Berlin),* 67, 596, 1985.
89. **Neighbors, M. A. and Horn, M. H.,** Nutritional quality of macrophytes eaten and not eaten by two temperate-zone herbivorous fishes: a multivariate comparison, *Mar. Biol.,* 108, 471, 1991.
90. **Pandian, T. J. and Vivekanandan, E.,** Energetics of feeding and digestion, in *Fish Energetics — New Perspectives,* Tyler, P. and Calow, P., Eds., The Johns Hopkins University Press, Baltimore, 1985, 99.
91. **Horn, M. H. and Neighbors, M. A.,** Protein and nitrogen assimilation as a factor in predicting the seasonal macroalgal diet of the monkeyface prickleback, *Trans. Am. Fish. Soc.,* 113, 388, 1984.
92. **Horn, M. H., Murray, S. N., Fris, M. B., and Irelan, C. D.,** Diurnal feeding periodicity of an herbivorous blenniid fish, *Parablennius sanguinolentus,* in the western Mediterranean, in *Trophic Relationships in the Marine Environment,* Barnes, M. and Gibson, R. N., Eds., Aberdeen University Press, Aberdeen, Scotland, 1990, 170.
93. **Fischer, Z. and Lyakhnovich, V. P.,** Biology and bioenergetics of grass carp (*Ctenopharyngodon idella* Val.), *Pol. Arch. Hydrobiol.,* 4, 521, 1973.
94. **Panov, D. A., Sorokin, Yu. I., and Motenkova, A. G.,** Assimilation of plant and animal food by young grass carp and silver carp, in *Sbornik Poprudovmu Rybovodstvu,* VNIRO, Moscow, 1969, 153.
95. **Menzel, D. W.,** Utilization of algae for growth by the Angelfish, *Holacanthus bermudensis, J. Conserv., Cons. Int. Explor. Mer.,* 24, 308, 1959.
96. **Vaughan, F. A.,** *Bull. Mar. Sci.,* 28, 527, 1978.
97. **Krupauer, V.,** Food selection of two-year-old grass carp, *Bul. Vyzk. Ust. Ryb. Vodnany,* 3, 7, 1967 (in Czech with English summary).
98. **Opuszynski, K.,** Feeding of *Ctenopharyngodon idella* Val. on aquatic plants under aquarium conditions, *Rocz. Nauk Roln.,* 90-H-3, 453, 1967.
99. **Shireman, J. V., Colle, D. E., and Rottmann, R. W.,** Growth of grass carp fed natural and prepared diets under intensive culture, *J. Fish Biol.,* 12, 457, 1978.
100. **Brett, J. R. and Groves, T. D. D.,** Physiological energetics, in *Fish Physiology,* Vol. 8, Hoar, W. S., Randall, D. J., and Brett, J. R., Eds., Academic Press, New York, 1979, chap. 6.
101. **Hseih, C.,** *Atlas of China,* McGraw-Hill, New York, 1973, 282.

102. **Kuronuma, K.,** Do Chinese carps spawn in Japanese waters?, *Proc. IPFC,* 5, 126, 1954.
103. **Nikolsky, G. V.,** Aquaculture development: U.S.S.R. culture of herbivorous Chinese carps in the USSR, *FAO Aquacult. Bull.,* 3, 12, 1971.
104. **Guillory, V. and Gasaway, R. D.,** Zooeography of the grass carp in the United States, *Trans. Am. Fish. Soc.,* 107, 105, 1978.
105. **Crossman, E. J., Nepszy, S. J., and Krause, P.,** The first record of grass carp, *Ctenopharyngodon idella,* in Canadian waters, *Can. Field-Nat.,* 101, 584, 1987.
106. **Dukravets, G. M.,** White amur in the Ili River basin, *Izv. Akad. Nauk Kas. SSSR (Biol.),* 1, 52, 1972 (in Russian).
107. **Nezdoliy, V. K. and Mitrofanov, V. P.,** On the natural reproduction of grass carp *Ctenopharyngodon idella* Val. in the Ili River, *Vopr. Ikhtiol.,* 15, 1039, 1975 (translated from Russian in *J. Ichthyol.,* 15, 927, 1975).
108. **Aliev, D. S.,** Breeding of white amur *(Ctenopharyngodon idella),* silver carp *(Hypophthalmichthys molitrix)* and bighead carp *(Aristichthys nobilis)* in the Amur Darya basin, *Vopr. Ikhtiol.,* 5, 593, 1965.
109. **Aliev, D. S.,** What's new in the use of biological method for preventing the overgrowth and siltation of collecting and drainage network canals, in *The Hydrobiology of Canals of the USSR and Biological Intervention in Their Operation,* Naukova Dumka, Moscow, 1976, 297, (translated from Russian).
110. **Martino, K. V.,** The natural reproduction of the grass carp in the waters of the lower Volga, *Gidrobiol. Zh.,* 10, 91, 1974 (translated from Russian in *Hydrobiol. J.,* 10, 76, 1974).
111. **Vinogradov, V. K. and Zolotova, Z. K.,** The effects of the grass carp on the ecosystems of waters, *Gidrobiol. Zh.,* 10, 90, 1974 (translated from Russian in *Hydrobiol. J.,* 10, 72, 1974).
112. **Motenkov, Y.,** Reproduction of silver carp in the Kuban, *Rybovod. Rybolov.,* 1, 18, 1966 (translated from Russian by Stanley, J. G., 1973).
113. **Djisalov, N.,** Some information on the occurrence of young herbivorous fish in the Tisa River in the territory of SFR Yugoslavia, *Investii (Belgrade),* in press.
114. **Stanley, J. G., Miley, W. W., II, and Sutton, D. L.,** Reproductive requirements and likelihood for naturalization of escaped grass carp in the United States, *Trans. Am. Fish. Soc.,* 197, 119, 1978.
115. **Inaba, D., Nomura, M., and Nakamura, M.,** Preliminary report on the spawning of grass carp and silver carp in the Tone River, Japan, and the development of their eggs, *J. Tokyo Univ. Fish.,* 43, 81, 1957.
116. **Tsuchiya, M.,** Natural reproduction of grass carp in the Tone River and their pond spawning, in *Proc. Grass Carp Conf.,* Shireman, J. V., Ed., University of Florida Press, Gainesville, 1979, 185.
117. **Tang, Y. A.,** Report of the Investigation on Spawning of Chinese Carps in Ah Kung Tian Reservoir, Tainan, Taiwan, Taiwan Fisheries Research Institute, Taiwan Fish Culture Station, Tainan, 1960, 11.
118. **Tang, Y. A.,** Reproduction of the Chinese carps, *Ctenopharyngodon idellus* and *Hypophthalmichthys molitrix* in a reservoir in Taiwan, *Jpn. J. Ichthyol.,* 8, 1, 1960.
119. **Anon.,** North American reproduction of grass carp, *Bull. Sport Fish. Inst.,* No. 269, 5, 1975.
120. **Rosas, M.,** Now the grass carp is born in Mexico (in Spanish), *Tec. Pecu. Mex.,* 9, 45, 1976.

121. **Conner, J. V., Gallagher, R. P., and Chatry, M. F.,** Larval Evidence for Natural Reproduction of the Grass Carp *(Ctenopharyngodon idella)* in the Lower Mississippi River (Louisiana and Arkansas), Office of Biological Services, United States Fish and Wildlife Service, Washington, D.C., 1980, 19.
122. **Zimpfer, S. P. and Bryan, C. F.,** Grass Carp in the Lower Mississippi, Res. Info. Bull., 85–31, U.S. Fish and Wildlife Service, Washington, D.C., 1985, 1.
123. **Courtenay, W. R., Jr., Jennings, D. P., and Williams, J. D.,** Exotic fishes, in *Common and Scientific Names of Fishes from the United States and Canada,* Spec. Publ. 20, 5th ed., Robins, C. R., Bailey, R. M., Bond, C. E., Brooker, J. R., Lachner, E. A., Lea, R. N., and Scott, W. B., Eds., American Fisheries Society, Bethesda, MD, 1991, 97.
124. **Brown, D. J. and Coon, T. G.,** Grass carp larvae in the Lower Missouri River and its tributaries, *N. Am. J. Fish. Manag.,* 11, 62, 1991.
125. **Pflieger, W. L.,** Natural reproduction of bighead carp *(Hypophthalmichthys nobilis)* in Missouri, *Am. Fish. Soci., Introduced Fish Sect. Newsl.,* 9(4), 9, 1989.
126. **Jhingran, V. G. and Pullin, R. S. V.,** *A Hatchery Manual for the Common, Chinese, and Indian Major Carps,* Asian Development Bank and International Center for Living Aquatic Resources Management, Manila, the Philippines, 1988, 191.
127. **Li, S. and Wang, R.,** Maturity speed and genetic analysis of silver carp and bighead carp from Changjiang and Zhujiang River systems, *J. Fish. China,* 14, 189, 1990.
128. **Chen, F. Y., Chow, M., and Sim, B. K.,** Induced spawning of the three major Chinese carps in Malacca, Malaysia, *Malay. Agric. J.,* 47, 211, 1969.
129. **Kamilov, B. G.,** Gonad condition of female silver carp, *Hypophthalmichthys molitrix* Val., in relation to growth rate in Uzbekistan, *J. Ichthyol.,* 27, 135, 1987.
130. **Kuronuma, K.,** New systems and new fishes for culture in the Far East, *FAO Fish. Rep.,* 5, 123, 1968.
131. **Chang, Y. F.,** Culture of freshwater fish in China, in *Chinese Fish Culture,* Tech. Rep. A-79, Gangstad, E. O., Ed., U.S. Army Waterways Experiment Station, (draft translated by T. S. Y. Koo, 1980), 1966.
132. **Makeeva, A. P.,** The maturation of grass carp and silver carp females and the reproduction of these species in the Amur basin, in *Problems of the Fisheries Exploitation of Plant-Eating Fishes in the Water Bodies of the USSR,* Ashkhabad, Akademia Nauk Turkmenistan SSSR, 1963, 148 (in Russian).
133. **Alikunhi, K. H., Sukumaran, K. K., and Parameswaran, S.,** Induced spawning of the Chinese grass carp, *Ctenopharyngodon idellus* (C. and V.), and the silver carp, *Hypophthalmichthys molitrix* (C. and V.), in ponds at Cuttack, India, *Curr. Sci.,* 32, 103, 1963.
134. **Sukhanova, A. I.,** Development of the bighead *Aristichthys nobilis, Vopr. Ikhtiol.,* 6, 39, 1966.
135. **Vinogradov, V. K.,** Techniques of raising phytophagous fishes, *FAO Fish. Rep.,* 5, 227, 1968.
136. **Nicolau, A. and Steopoe, I.,** The oogenesis of phytophagous fish species *(Ctenopharyngodon idella, Hypophthalmichthys molitrix* and *Aristichthys nobilis),* reared in controlled units, in Romanian waters, *Bul. Cercet. Piscic.,* 29, 5, 1970 (in Romanian with English summary).
137. **Shelton, W. L. and Jensen, G. L.,** Production of reproductively limited grass carp for biological control of aquatic weeds, *Bull. Water Resources, Res. Inst. Auburn Univ.,* 1979, 173.

138. **Smith, C. R. and Shireman, J. V.,** *Grass Carp Biblography,* University of Florida Center for Aquatic Weeds, Gainesville, 1981, 177.
139. **Woynarovich, E. and Horváth, L.,** The artificial propagation of warm-water finfishes — a manual for extension, *FAO Fish. Tech. Pap.,* 201, 183, 1980.
140. **Emel'yanova, N. G.,** Seasonal changes in the cytoplasm of oocytes during previtellogenesis of silver carp, *Hypophthalmichthys molitrix, J. Ichthyol.,* 25, 73, 1985.
141. **Duvarova, A. S.,** Oocyte maturation in the silver carp, *Hypophthalmichthys molitrix* (Cyprinidae), after hormonal stimulation, *J. Ichthyol.,* 22, 76, 1982.
142. **Gorbach, E. I.,** Fecundity of the grass carp (*Ctenopharyngodon idella* Val.), in the Amur basin, *J. Ichthyol.,* 12, 616, 1972.
143. **Qu, W.-L. and Pan, W.-Z.,** Study on the annual variation of the ovary of silver carp in the lake in Heilongjiang Province, *J. Fish. China,* 9, 143, 1985.
144. **Hickling, C. F.,** On the biology of herbivorous fish, the white amur or grass carp, *Ctenopharyngodon idella* Val., *Proc. R. Soc. Edinburgh (B),* 70, 62, 1967.
145. **Bobrova, Y. P.,** Gonadal development and the fertilization process in grass carp (O ravitii gonad i protsesse oplodotvoreniya u belogo amura), in *Genetics, Selection and Hybridization of Fish (Genetika, selektsiya i gibridizatsiya ryb),* Cherfas, B. I., Ed., Israel Program for Scientific Translations, Jerusalem, 1972, 116.
146. **Courtenay, W. R., Jr. and Miley, W. W.,** II, Sex Determination in the Grass Carp, *Ctenopharyngodon idella,* Annual Report, Bureau of Aquatic Plant Research and Control, Florida Department of Natural Resources, Tallahassee, 1973, 15.
147. **Gorbach, E. I.,** Age composition, growth and age of onset of sexual maturity of the white *Ctenopharyngodon idella* (Val.) and the black *Mylopharyngodon piceus* (Rich.) amurs in the Amur River basin, *Vopr. Ikhtiol.,* 1, 119, 1961 (translated from Russian by R. M. Howland, 1971).
148. **Aliev, D. S. and Sukhanova, A. I.,** Fecundity of the grass carp, *Ctenopharyngodon idella* (Val.) and the silver carp, *Hypophthalmichthys molitrix* (Val.) in the Kara Kum canal and its reservoirs, *Izv. Akad. Nauk Turkm. SSR. Serv. Biol. Nauk,* 4, 77, 1974 (in Russian with English summary).
149. **Alikunhi, K. H., Sukumaran, K. K., and Parameswaran, S.,** Observations on growth, maturity and breeding of induced-bred, pond-reared silver carp, *Hypophthalmichthys molitrix,* and grass carp, *Ctenopharyngodon idellus,* in India during July 1962 to August 1963, *Bull. Cent. Inst. Fish. Educ.,* 19, 1965.
150. **Vinogradov, V. K., Erokhina, L. V., Savin, G. I., and Konradt, A. G.,** Methods of artificial breeding of herbivorous fishes, *Biol. Abstr.,* 48, 1966.
151. **Santiago, C. B., Camacho, A. S., and Laron, M. A.,** Growth and reproductive performance of bighead carp *(Aristichthys nobilis)* reared with or without feeding in floating cages, *Aquaculture,* 96, 109, 1991.
152. **Belova, N. V.,** Ecological-physiological properties of semen of some cyprinid fishes. IV. Physiological-biochemical properties of testes, *J. Ichthyol.,* 23, 75, 1983.
153. **Belova, N. V.,** Ecological-physiological peculiarities of semen of pond carps. II. Change in the physiological parameters of spermatozoids of some carps under the influence of environmental factors, *J. Ichthyol.,* 21, 70, 1981.
154. **Lin, S. Y.,** Life-history of waan ue, *Ctenopharyngodon idellus* (Cuv. and Val.), *Lingnan Sci. J.,* 14, 129. 1935.
155. **Lin, S. Y.,** Life-history of waan ue, *Ctenopharyngodon idellus* (Cuv. and Val.), *Lingnan Sci. J.,* 14, 271, 1935.

156. **Krykhtin, M. L. and Gorbach, E. I.**, Reproductive ecology of the grass carp, *Ctenopharyngodon idella,* and silver carp, *Hypophthalmichthys molitrix,* in the Amur Basin, *J. Ichthyol.,* 21, 109, 1981.
157. **Aliev, D. S.**, The role of phytophagous fish in the reconstruction of commercial ichthyofauna and biological melioration of water reservoirs, *J. Ichthyol.,* 16, 216, 1976 (translated from Russian).
158. **Leslie, A. J., Van Dyke, J. M., and Nall, L. E.**, Current velocity for transport of grass carp eggs, *Trans. Am. Fish. Soc.,* 3, 99, 1982.
159. **Anon.**, Manual on the Biotechnology of the Propagation and Rearing of Phytophagous Fishes, Fisheries Ministry of the USSR, All-Union Scientific Research Institute of Pond Fishery, 1970, 49, Moscow, (translated from Russian by R. M. Howland, 1971).
160. **Soin, S. G. and Sukhanova, A. I.**, Comparative morphological analysis of the development of the grass carp, the black carp, the silver carp and the bighead *(Cyprinidae), J. Ichthyol.,* 12, 61, 1972.
161. **Stott, B. and Cross, D. G.**, A note on the effect of lowered temperatures on the survival of eggs and fry of the grass carp, *Ctenopharyngodon idella* (Val.), *J. Fish Biol.,* 5, 649, 1973.
162. **Opuszynski, K.**, Comparison of temperature and oxygen tolerance in grass carp (*Cteonpharyngodon idella* Val.), silver carp (*Hypophthalmichthys molitrix* Val.), and mirror carp (*Cyprinus carpio* L.)., *Ekol. Pol. (A),* 15, 385, 1967.
163. **Negonovskaya, I. T. and Rudenko, G. P.**, Oxygen threshold and characteristics of respiratory metabolism in young herbivorous fish grass carp *Ctenopharyngodon idella* and bighead *Aristichthys nobilis, Vopr. Ikhthiol.,* 14, 1111, 1974 (translated from Russian in *J. Ichthyol.,* 14, 965, 1974).
164. **Singh, S. B., Banerjee, S. C., and Chakrabarti, P. C.**, Preliminary observations on response of young Chinese carps to various physicochemical factors of water, *Proc. Natl. Acad. Sci. India (B Biol. Sci.),* 37, 320, 1967.
165. **Doroshev, S. I.**, The survival of white amur and tolstolobik fry in Sea of Azov and Aral Sea water of varing salinity, in *Problems of the Fisheries Exploitation of Plant-Eating Fishes in the Water Bodies of the USSR,* Akademii Nauk Turkmenistan SSSR, Ashkhabad, 1963, 144 (in Russian).
166. **Chervinski, J.**, Note on the adaptability of silver carp *(Hypopthalmichthys molitrix)* and grass carp *(Ctenopharyngodon idella)* to various saline concentrations, *Aquaculture,* 11, 179, 1977.
167. **Maceina, M. J. and Shireman, J. V.**, Grass carp: effects of salinity on survival, weight loss, and muscle tissue water content, *Prog. Fish-Cult.,* 41, 69, 1979.
168. **Maceina, M. J. and Shireman, J. V.**, Effects of salinity on vegetation consumption and growth in grass carp, *Prog. Fish-Cult.,* 42, 50, 1980.
169. **Appelbaum, S. and Uland, B.**, Intensive rearing of grass carp larvae *Ctenopharyngodon idella* (Valenciennes, 1844) under controlled conditions, *Aquaculture,* 17, 175, 1979.
170. **Bessmertnaya, R. E.**, Problems in grass carp larvae feeding, in *New Studies on Ecology and Propagation of Herbivorous Fishes,* Nauka, Moscow, 1968, 154.
171. **Kornenko, G. S.**, The role of infusoria in feeding of larvae of herbivorous fishes, *Vopr. Ikhtiol.,* 11, 241, 1971.
172. **Kornenko, G. S.**, The importance of ciliata as a food for larval Asiatic herbivorous fish, *Vopr. Ikhtiol.,* 2, 303, 1971.
173. **Rozmanova, M. D.**, The feeding of white amur larvae kept in fish-cages, *J. Biol. Acad. USSR,* 166, 729, 1966.

174. **Sobolev, Yu. A.,** Food interrelationships of young grass carp, silver carp and carp reared jointly in ponds in Belorussia, *J. Ichthyol.,* 10, 528, 1970.
175. **Linchevskaya, M. D.,** The role of phytoplankton in the diet of grass carp during early stages of its development, in *Biological Bases of the Fishing Industry in Central Asian and Kazakhstan Waters,* Nauka, Alma-Ata, 1966, 255.
176. **Gaevskaya, N. S.,** *The Role of Higher Aquatic Plants in the Nutrition of the Animals of Fresh-Water Basins,* Vol. 1, National Lending Library for Science and Technology, Yorkshire, England, 1969, chap. 1 and 2.
177. **De Silva, S. S. and Weerakoon, D. E. M.,** Growth, food intake and evacuation rates of grass carp, *Ctenopharyngodon idella,* fry, *Aquaculture,* 25, 67, 1981.
178. **Opuszynski, K.,** *Fundamentals of Fish Biology,* PWRiL, Warsaw, 1979, 364 (in Polish).
179. **Watkins, C. E., Shireman, J. V., Rottmann, R. W., and Colle. D. E.,** Food habits of fingerling grass carp, *Prog. Fish-Cult.,* 43, 95, 1981.
180. **Scheer, D., Jähnichen, H., and Grahl, K.,** Beobachtungen bei der Haltung von ein und zweisömmigen graskarpten *(Ctenopharyngodon idella)* in Kleinen Teichen in Gebiet von Karl-Marx-Stadt, *Dtsch. Fisch. Z.,* 14, 141, 1967.
181. **Fedorenko, A. Y. and Frazer, F. J.,** Review of Grass Carp Biology, Fisheries and Marine Service Tech. Rep. No. 786, Fish and Wildlife Branch, Ministry of Recreation and Conservation, Victoria, British Columbia, Canada, 1978, 15.
182. **Edwards, D. J.,** Weed preference and growth of young grass carp in New Zealand, *N. Z. J. Mar. Freshwat. Res.,* 8, 341, 1974.
183. **Stroganov, N. S.,** The food selectivity of the grass carp, in *Problems of the Fisheries Exploitation of Plant-Eating Fishes in the Water Bodies of the USSR,* Akademia Nauk Turkmenistan SSSR, Ashkhabad, 1963, 181.
184. **Aliev, D. S.,** Trials of using grass carp for aquatic weed control, in *Problems of the Fisheries Exploitation of Plant-Eating Fishes in the Water Bodies of the USSR,* Akademia Nauk Turkmenistan SSSR, Ashkhabad, 1963, 203.
185. **Verigin, B. V., Viet, N., and Dong, N.,** Data on the food selectivity and the daily ration of white amur, in *Problems of the Fisheries Exploitation of Plant-Eating Fishes in the Water Bodies of the USSR,* Akademia Nauk Turkmenistan SSSR, Ashkhabad, 1963, 192.
186. **Bonar, S. A., Sehgal, H. S., Pauley, G. B., and Thomas, G. L.,** Relationship between the chemical composition of aquatic macrophytes and their consumption by grass carp, *Ctenopharyngodon idella, J. Fish Biol.,* 36, 49, 1990.
187. **Pine, R. T. and Anderson, L. W. J.,** Effects of static versus flowing water on aquatic plant preferences of triploid grass carp, *Trans. Am. Fish. Soc.,* 118, 336, 1989.
188. **Avault, J. W.,** Preliminary studies with grass carp for aquatic weed control, *Prog. Fish-Cult.,* 27, 207, 1965.
189. **Penzes, B. and Tölg, I.,** Aquaristische Untersuchung des Pflanzenverbrauches des graskarpten (*Ctenopharyngodon idella* Cuv. et Val.), *Z. Fish. Deren Hilfswiss.,* 1/2, 131, 1966.
190. **Ilin, W. M. and Solovieva, L. M.,** Production and wintering of yearlings of phytophageous fish, *Vopr. Prud. Ryb.,* 13, 11, 1965.
191. **Opuszynski, K.,** Weed control and fish production, in *Proc. Grass Carp Conf.,* Shireman, J. V., Ed., Aquatic Weeds Research Center, University of Florida, Gainesville, 1979, 103
192. **Prowse, G. A.,** Experimental criteria for studying grass carp feeding in relation to weed control, *Prog. Fish-Cult.,* 33, 128, 1971.

193. **Bailey, W. M. and Boyd, R. L.,** Some observations on the white amur in Arkansas, *Ctenopharygodon idella,* aquatic weeds control, *Hyacinth Control J.,* 10, 20, 1972.
194. **Colle, D. E., Shireman, J. V., and Rottmann, R. W.,** Food selection by grass carp fingerlings in a vegetated pond, *Trans. Am. Fish. Soc.,* 107, 149, 1978.
195. **Kilgen, R. H.,** Food habits of white amur, largemouth bass, bluegill, and redear sunfish receiving supplemental feed, *Proc. 27th Annu. Conf. Southeast Assoc. Game and Fish Comm.,* Alabama, 1973, 620.
196. **Kilgen, R. H. and Smitherman, R. O.,** Food habits of the white amur stocked in ponds alone and in combination with other species, *Prog. Fish-Cult.,* 33, 123, 1971.
197. **Forester, J. S. and Avault, J. W., Jr.,** Effects of grass carp on freshwater red swamp crayfish in ponds, *Trans. Am. Fish. Soc.,* 107, 157, 1978.
198. **Terrell, J. W. and Fox, A. C.,** Food Habits, Growth and Catchability of Grass Carp in the Absence of Aquatic Vegetation, paper presented at Annu. Meet., Southern Div., Am. Fish. Soc., White Sulphur Springs, WV, November 1974, 15.
199. **Edwards, D. J.,** Aquarium studies on the consumption of small animals by O-group grass carp, *Ctenopharyngodon idella* (Val.), *J. Fish Biol.,* 5, 599, 1973.
200. **Singh, S. B., Dey, R. K., and Reddy, P. V. G. K.,** Some additional notes on the piscivorous habits of the grass carps *(Ctenopharyngodon idella), Aquaculture,* 9, 195, 1976.
201. **Fischer, Z.,** The elements of energy balance in grass carp (*Ctenopharyngodon idella* Val.). IV. Consumption rate of grass carp fed on different types of food, *Pol. Arch. Hydrobiol.,* 20, 309, 1973.
202. **Wiley, M. J. and Wike, L. D.,** Energy balances of diploid, triploid, and hybrid grass carp, *Trans. Am. Fish. Soc.,* 115, 853, 1986.
203. **Chapman, V. J. and Coffey, B. J.,** Experiments with grass carp in controlling exotic macrophytes in New Zealand, *Hydrobiologia (Bucharest),* 12, 313, 1971.
204. **Shireman, J. V. and Maceina, M. J.,** Recording Fathometer for Hydrilla Distribution and Bimass Studies. Annu. Rep. Corps of Engineers, Aquatic Plant Control Research Program, Waterways Experiment Station, Vicksburg, MS, University of Florida, School of Forest Resources and Conservation, Gainesville, 1980, 69.
205. **Opuszynski, K.,** Use of phytophagous fish to control aquatic plants, *Aquaculture,* 1, 61, 1972.
206. **Sutton, D. L.,** Utilization of hydrilla by white amur, *Hyacinth Control J.,* 12, 66, 1974.
207. **Kilambi, R. V. and Robison, W. R.,** Effects of temperature and stocking density on food consumption and growth of grass carp *Ctenopharyngodon idella* Val., *J. Fish Biol.,* 15, 337, 1979.
208. **Cassani, J. R. and Caton, W. E.,** Feeding behaviour of yearling and older hybrid grass carp, *J. Fish Biol.,* 22, 35, 1983.
209. **Shireman, J. V., Colle, D. E., and Rottmann, R. W.,** Intensive culture of grass carp, *Ctenopharyngodon idella,* in circular tanks, *J. Fish Biol.,* 11, 267, 1977.
210. **Stanley, J. G.,** Annu. Rep. U.S. Army Corps of Engineers, Stuttgart, AZ, Fish Farming Experimental Station, Bureau of Sports Fisheries and Wildlife, 1972, 11.
211. **Tooby, T. E., Lucey, J., and Stott, B.,** The tolerance of grass carp, *Ctenopharyngodon idella* Val., to aquatic herbicides, *J. Fish Biol.,* 16, 591, 1980.
212. **Stanley, J. G.,** Energy balance of white amur fed egeria, *Hyacinth Control J.,* 12, 62, 1974.
213. **Fischer, Z.,** The elements of energy balance in grass carp (*Ctenopharyngodon idella* Val.). I. *Pol. Arch. Hydrobiol.,* 17, 421, 1970.

214. **Hajra, A.,** Biochemical investigations on the protein-calorie availability in grass carp (*Ctenopharyngodon idella* Val.) from an aquatic weed (*Ceratophyllum demersum* Linn.) in the tropics, *Aquaculture,* 61, 113, 1987.
215. **Cassani, J. R., Caton, W. E., and Hansen, T. H.,** Culture and diet of hybrid grass carp fingerlings, *J. Aquat. Plant Manage.,* 20, 30, 1982.
216. **Cai, Z. and Curtis, L. R.,** Effects of diet and temperature on food consumption, growth rate and tissue fatty-acid composition of triploid grass carp, *Aquaculture,* 88, 313, 1990.
217. **Sutton, D. L.,** Utilization of duckweed by the white amur, in *Proc. 4th Int. Symp. Biological Control of Weeds,* Freeman, T. E., Ed., University of Florida Press, Gainesville, 1977, 257.
218. **Stott, B. and Orr, L. D.,** Estimating the amount of aquatic weed consumed by grass carp, *Prog. Fish-Cult.,* 1, 51, 1970.
219. **Fowler, M. C.,** Experiments on food conversion ratios and growth rates of grass carp (*Ctenopharyngodon idella* Val.) in England, in *Proc. 2nd Int. Symp. Herbivorous Fish, Novi Sad, Yugoslavia,* European Weed Research Society, Wageningen, The Netherlands, 1982, 107.
220. **Lupaceva, L. I.,** Food relationships of silver carp and grass carp reared together in ponds, *Ryb. Choz. Kiev,* 11, 34, 1970.
221. **Sobolev, Yu. A. and Abramovitch, L. V.,** Culture of larvae of herbivorous fishes in heated water effluence from a power station, *Vopr. Ryb. Choz. Belorussi,* 10, 129 1974.
222. **Boruckij, E. W.,** Materials on feeding of silver carp (*Hypophthalmichthys molitrix* Val.), *Proc. Amur Ichthyol. Expedition, 1945–1949,* 1, 319, 1950.
223. **Kopylova, T. S.,** Elements of nitrogen balance and food rations of silver carp fingerlings, *Tr. VNIPRX,* 18, 76, 1971.
224. **Nabereznyj, A. I., Zelemnin, A. M., and Jalovickaja, N. I.,** On trophical relations of herbivorous fishes in waters of Moldavia, in *Biology and Culture of Herbivorous Fishes,* Kozokaru, E. V., Ed., Stünca, Kisinev, 1972, 27.
225. **Vovk, P. S.,** The possibility of using the silver carp *(Hypophthalmichthys molitrix)* to increase the fish production of the Dnieper reservoirs and to decrease eutrophication, *J. Ichthyol.,* 14, 351, 1974.
226. **Lupaceva, L. I.,** Feeding of silver carp in ponds of Cjurupinskos Fish Breeding Farm, *Ryb. Choz. Kiev,* 9, 35, 1969.
227. **Salar, V. M.,** The phytoplankton in ponds of Moldavia and its significance in the feeding of phytophagous fishes, in *Proc. 7th Meet. Pol. Hydrobiol. Assoc.,* Warsaw, 1967, 119.
228. **Opuszynski, K.,** Silver carp, *Hypopthalmichthys molitrix* (Val.), in carp ponds. II. Rearing of fry, *Ekol. Pol.,* 27, 93, 1979.
229. **Opuszynski, K.,** Comparison of the usefulness of the silver carp and the bighead carp as additional fish in carp ponds, *Aquaculture,* 25, 223, 1981.
230. **Adamek, Z. and Spittler, P.,** Particle size selection in the food of silver carp, *Hypophthalmichthys molitrix, Folia Zool.,* 33, 363, 1984.
231. **Kajak, Z., Spodniewska, I., and Wisniewski, R. J.,** Studies on food selectivity of silver carp, *Hypophthalmichthys molitrix* (Val.), *Ekol. Pol.,* 25, 227, 1977.
232. **Spataru, P. and Gophen, M.,** Feeding behaviour of silver carp *Hypophthalmichthys molitrix* Val. and its impact on the food web in Lake Kinneret, Israel, *Hydrobiologia,* 120, 53, 1985.
233. **Shapiro, J.,** Food and intestinal contents of the silver carp, *Hypophthalmichthys molitrix* (Val.), in Lake Kinneret between 1982–1984, *Bamidgeh,* 37, 3, 1985.

234. **Guenther, V.,** Silver carp in Lake Duemmer, *Fischwirt,* 29, 11, 1979.
235. **Zhou, J. and Lin, F.,** The feeding habit of silver carp and bighead and their digestion of algae, *Acta Hydrobiol. Sin.,* 14, 170, 1990.
236. **Iwata, K., Chen, S.-L., and Liu, X.-F.,** Nitrogen balance in the silver and bighead carps. I. Estimation of several parameters in relation to the nitrogen balance during the growing season (summer) of fish, *Acta Hydrobiol. Sin.,* 10, 297, 1986.
237. **Tarasova, M. O.,** Feeding of silver carp in ponds of experimental fish farm "Nivka", *Rybn. Khoz. Kiev,* 13, 51, 1971.
238. **Barthelmes, D. and Janichen, H.,** Food choice and ration size in third and fourth summer silver carp, *Z. Binnenfisch. DDR,* 25, 331, 1978.
239. **Ivlev, V. S.,** *The Experimental Ecology of the Feeding of Fishes,* Yale University Press, New Haven, CT, 1961, 302.
240. **Spittler, P.,** Zur Bestimmung der Filtrierrafen vorgestreckter Silberkarpten *(Hypophthalmichthys molitrix), Wiss. Z. Wilhelm-Pieck-Univ. Rostock,* 30, 109, 1981.
241. **Herodek, S., Tatrai, I., Olah, J., and Vörös, L.,** Feeding experiments with silver carp (*Hypophthalmichthys molitrix* Val.) fry, *Aquaculture,* 83, 331, 1989.
242. **Wang, J.-Q., Flickinger, S. A., Be, K., Liu, Y., and Xu, H.,** Daily food consumption and feeding rhythm of silver carp *(Hypophthalmichthys molitrix)* during fry to fingerling period, *Aquaculture,* 83, 73, 1989.
243. **Muchamedova, A. F. and Sarsembaiev, Z. G.,** On daily feeding activity and rations of silver carp fingerlings, *Hypophthalmichthys molitrix* Val., *Tr. Volg. Otg. GNIORRX,* 3, 45, 1967.
244. **Herrmann, J.,** Experimental studies on the filtration efficiency and the energy balance of young silver carp (*Hypophthalmichthys molitrix* Val.), *Arch. Hydrobiol.,* 65, 268, 1983.
245. **Chen, S., Hu, C., Tiam, L., and Shun, X.,** Digestion of silver carp *(Aristichthys nobilis)* for *Daphnia pulex, Collect. Treatises Ichthyol.,* 4, 163, 1985.
246. **Omarov, M. O.,** Daily ration of silver carp, *Hypophthalmichthys molitrix* (Val.), *Vopr. Ikhthiol.,* 10, 580, 1970.
247. **Bialokoz, W. and Krzywosz, T.,** Feeding intensity of silver carp (*Hypophthalmichthys molitrix* Val.) from the Paprotechic Lake in the annual cycle, *Elid. Pol.,* 29, 53, 1981.
248. **Chen, S.-L., Hu, C.-L., and Zhang, S.-Y.,** Feeding intensity of silver carp and bighead under natural conditions. I. Feeding intensity of fingerlings of silver carp and bighead in summer, *Acta Hydrobiol. Sin.,* 10, 277, 1986.
249. **Savina, R. A.,** Feeding of silver carp in ponds of RSSR, in *New Research in Ecology and Breeding of Herbivorous Fishes,* Nikoslkij, G. V., Ed., Nauka, Moscow, 1968, 116.
250. **Lin, W. L., Liu, X. Z., and Liu, J. K.,** Detritus formation of two plankters and query on the role of their detritus in the nutrition of the silver carp and bighead, *Verh. Int. Verein. Limnol.,* 21, 1287, 1981.
251. **Zhu, H. and Deng, W.-J.,** Studies on the digestion of algae by fish. II. *Microcystis aernginosa* and *Euglera* sp. digested and absorbed by silver carp and bighead, *Trans. Chin. Ichthyol. Soc.,* 3, 77, 1983.
252. **Ekpo, I. and Bender, J.,** Digestibility of a commercial fish feed, wet algae, and dried algae by *Tilapia nilotica* and silver carp, *Prog. Fish-Cult.,* 51, 83, 1989.
253. **Hamada, A., Maeda, W., Iwasaki, J., and Kumanaru, A.,** Conversion efficiency of phytoplankton in the growth of silver carp, *Hypophthalmichthys molitrix, Jpn. J. Limnol.,* 44, 321, 1983.

254. **Dabrowski, K. and Bardega, R.,** Mouth size and predicted food size preference of larvae of three cyprinid fish species, *Aquaculture*, 40, 41, 1984.
255. **Lazareva, L. P., Omarov, M. O., and Lezina, A. N.,** Feeding and growth of the bighead, *Aristichthys nobilis,* in the waters of Dagestan, *J. Ichthyol.,* 17, 65, 1977.
256. **Opuszynski, K. and Shireman, J. V.,** Food habits, feeding behavior and impact of triploid bighead carp, *Hypophthalmichthys nobilis* (Richardson), in experimental ponds, *J. Fish Biol.,* 42, 517, 1993.
257. **Opuszynski, K., Shireman, J. V., and Cichra, C. E.,** Food assimilation and filtering rate of bighead carp kept in cages, *Hydrobiologia,* 220, 49, 1991.
258. **Nickolsky, G. V. and Aliev, D. D.,** Role of far eastern herbivorous fish in ecosystems of natural water bodies used for acclimatization, *Vopr. Ikhtiol.,* 14, 974, 1974 (translated from Russian in *J. Ichthyol.,* 14, 842, 1974).
259. **Danchenko, E. V.,** The role of zooplankton in the food of second-year grass carp and bighead reared jointly with carp in ponds of the Sinyukhinskiy fish farm in Krasnodarsk Province, in *Materials from a Scientific Conference on Intensive Fisheries Exploitation of the Inland Waters of the Northern Caucasus, Kradnodar,* 1970, 53 (translated from Russian).
260. **Burke, J. S., Bayne, D. R., and Rea, H.,** Impact of silver and bighead carps on plankton communities of channel catfish ponds, *Aquaculture*, 55, 59, 1986.
261. **Wang, Y.,** The food organism of pond-cultured bighead *Aristichthys nobilis* (Rich.), *J. Fish. China,* 12, 43, 1988.
262. **Sifa, L., Hequen, Y., and Weimin, L.,** Preliminary reasearch on diurnal feeding rhythm and the daily ration for silver carp, bighead carp and grass carp, *J. Fish. China,* 4, 275, 1980.
263. **Opuszynski, K. and Shireman, J. V.,** Food passage time and daily ration of bighead carp, *Aristichthys nobilis*, kept in cages, *Environ. Biol. Fish.*, 30, 387, 1991.
264. **Panov, D., Vinogradov, V., and Chromov, L.,** Larval fish rearing, *Rybovod. Rybolov.,* 1, 8, 1969 (in Russian).
265. **Sharma, K. P. and Kulshrestha, S. D.,** Preliminary studies on food and growth of white amur fry and fingerlings at Kota, Rajasthan, India, *Hyacinth Control J.*, 12, 55, 1974.
266. **Dabrowski, K. and Poczyczynski, P.,** Comparative experiments on starter diets for grass carp and common carp, *Aquaculture*, 69, 317, 1988.
267. **Dabrowski, K.,** Protein requirements of grass carp fry (*Ctenopharyngodon idella* Val.), *Aquaculture*, 12, 63, 1979.
268. **Dabrowski, K. and Kozak, B.,** The use of fishmeal and soybean meal as a protein source in the diet of grass carp fry, *Aquaculture*, 18, 107, 1979.
269. **Meske, C. and Pfeffer, E., Ahrensburg and Göttingen,** Growth experiments with carp and grass carp, *Arch. Hydrobiol. Beih. Ergebn. Limnol.*, 11, 98, 1978.
270. **Opuszynski, K., Myszkowski, L., Okoniewska, G., Opuszynska, W., Szlaminska, M., Wolnicki, J., and Wozniewski, M.,** Rearing of common carp, grass carp, silver carp, and bighead carp larvae using zooplankton and/or different dry feeds, *Pol. Arch. Hydrobiol.*, 36, 217, 1989.
271. **Rottmann, R. W., Shireman, J. V., and Lincoln, E. P.,** Comparison of three live foods and two dry diets for intensive culture of grass carp and bighead carp larvae, *Aquaculture*, 96, 269, 1991.
272. **Tan, Y. T.,** Composition and nutritrive value of some grasses, plants and aquatic weeds tested as diets, *J. Fish Biol.*, 2, 253, 1970.

273. **Sutton, D. L. and Blackburn, R. D.,** Feasibility of the Amur *(Ctenopharyngodon idella)* as a Biocontrol of Aquatic Weeds, in *Herbivorous Fish for Aquatic Plant Control,* Gangstad, E. O., Ed., Aquatic Plant Control Program, Tech. Rep. 4, U.S. Army Corps of Engineers Water Exp. Stn., Vicksburg, MS, 1973, D1.
274. **Shireman, J. V., Colle, D. E., and Maceina, M. J.,** Grass carp growth rates in Lake Wales, Florida, *Aquaculture,* 19, 379, 1980.
275. **Venkatesh, B. and Shetty, H. P. C.,** Studies on the growth rate of the grass carp *Ctenopharyngodon idella* (Valenciennes) fed on two aquatic weeds and a terrestrial grass, *Aquaculture,* 13, 45, 1978.
276. **Tal, S. and Ziv, I.,** Culture of exotic fishes in Israel, in *Culture of Exotic Fishes Symp. Proc.,* Smitherman, R. O., Shelton, W. L., and Grover, J. H., Eds., Atlanta, 1978, 1.
277. **Tal, S. and Ziv, I.,** Culture of exotic species in Israel, *Bamidgeh,* 30, 3, 1978.
278. **Gasaway, R. D.,** Growth, survival and harvest of grass carp in Florida lakes, in *Culture of Exotic Fishes Symp. Proc.,* Smitherman, R. O., Shelton, W. L., and Grover, J. H., Eds., Atlanta, 1978, 167.
279. **Chow, T.,** Growth characteristics of four species of pondfish in Hong Kong, *Hong Kong Univ. Fish. J.,* 2, 29, 1958.
280. **Hoa, D. T.,** Variability of juvenile grass carp (*Ctenopharyngodon idella* Val.) and pond carp (*Cyprinus carpio* L.) reared at a hatchery in the southern Ukraine, *J. Ichthyol.,* 13, 305, 1973.
281. **Cassani, J. R. and Caton, W. E.,** Growth comparisons of diploid and triploid grass carp under varing conditions, *Prog. Fish-Cult.,* 48, 184, 1986.
282. **Shireman, J. V.,** Predation, Spawning and Culture of White Amur *(Ctenopharyngodon idella),* Annual report to the Florida Department of Natural Resources, Gainesville, 1975, 40.
283. **Mitzner, L.,** Vital statistics of white amur in Red Haw Lake, in Evaluation of Biological Control of Nuisance Aquatic Vegetation by White Amur, *Stud. Iowa Conserv. Comm. Fish. Sect.,* 504–1, 51, 1975.
284. **Shireman, J. V. and Hoyer, M. V.,** Assessment of grass carp for weed management in an 80-hectare Florida lake, *Am. Fish. Soc Symp.,* 1986, 469.
285. **Peirong, S. and Binchen, X.,** Pen culture in Wuli Lake, Jiangsu, China, *Aquaculture,* 71, 301, 1988.
286. **Stanley, J. G.,** Production of Monosex White Amur for Aquatic Plant Control. Final report to Aquatic Plant Control Program, Waterways Experiment Station, Vicksburg, MS, U.S. Army Corps of Engineers, Washington, DC, 1975, 49.
287. **Blackburn, R. D. and Sutton, D. L.,** Growth of the white amur (*Ctenopharyngodon idella* Val.) on selected species of aquatic plants, *Proc. Eur. Weeds Res. Counc. Int. Symp. Aquatic Weeds,* 3, 87, 1971.
288. **Shelton, W. L., Smitherman, R. O., and Jensen, G. L.,** Density related growth of grass carp (*Ctenopharyngodon idella* Val.) in managed small impoundments in Alabama, *Fish. Soc. Br. Isles,* 22, 45, 1981.
289. **Antalfi, A. and Tölg, I.,** *Graskarpfen,* Donau Verlag, Günzburg, 1971 (translated into German from Hungarian by I. Bogsch).
290. **Opuszynski, K.,** Method of phytophagous fish culture used in China and countries of Eastern Europe, in *Proc. 2nd Int. Symp. Herbivorous Fish,* Novi Sad, Yugoslavia, European Weed Research Soc., Wageningen, The Netherlands, 1982, 121.
291. **Van Zon, J. C. J.,** Grass carp (*Ctenopharyngodon idella* Val.) in Europe, *Aquat. Bot.,* 3, 143, 1977.

292. **Kilambi, R. V.,** Food consumption, growth and survival of grass carp, *Ctenopharyngodon idella* (Val.), at four salinities, *J. Fish Biol.*, 17, 613, 1980.
293. **Savina, R. A.,** Some features of growth and feeding of silver carp larvae, *Vopr. Prud. Ryb.*, 15, 86, 1967.
294. **Prinsloo, J. F. and Schoonbee, H. J.,** Comparison of the early larval growth rates of the Chinese grass carp *Ctenopharyngodon idella* and the Chinese silver carp *Hypophthalmichthys molitrix* using live and artificial feed, *Water S. A.*, 12, 229, 1986.
295. **Dabrowski, K.,** Influence of initial weight during the change from live to compound feed on the survival and growth of four cyprinids, *Aquaculture*, 40, 29, 1984.
296. **Mumtazuddin, M. and Khaleque, M. A.,** Observations on the relative growth potential of carp hatchlings in relation to fertilization and supplemental feeding, *Bangladesh J. Zool.*, 15, 71, 1987.
297. **Konradt, A. G.,** Methods of breeding the grass carp, *Ctenopharyngodon idella,* and silver carp, *Hypophthalmichthys molitrix,* in *Proc. World Symp. Warm-Water Pond Fish Culture, Rome, 18–25 May 1966,* FAO Fish. Rep. No. 44, Vol. 4, Pillay, T. V. R., Ed., 1968, 195.
298. **Karamchandani, S. J. and Mishra, D. N.,** Preliminary observations on the status of silver carp in relation to catla in the culture fishery of Kulgarhi Reservoir, *J. Bombay Nat. Hist. Soc.*, 77, 261, 1980.
299. **Kozianowski, A. and Schmidt, K.,** The development of fisheries management with the silver and bighead carp in GDR., *Fortschr. Fisch., Adv. Fish. Sci.*, 3, 149, 1984.
300. **Leventer, H.,** *Biological Control of Reservoirs by Fish,* MEKOROTH Water Co., Jordan District Central Laboratory of Water Quality, Nazareth Elit, Israel, 1979, 71.
301. **Shefler, D. and Reich, K.,** Growth of silver carp *(Hypophthalmichthys molitrix)* in Lake Kinneret in 1967–1975, *Bamidgeh*, 29, 3, 1977.
302. **Li, S., Lu, W., Peng, C., and Zhao, P.,** A genetic study of the growth performance of the silver carp from the Changjiang and Zhujiang Rivers, *Aquaculture*, 65, 93, 1987.
303. **Gaigher, I. G. and Krause, J. B.,** Growth rates of Mozambique tilapia *(Oreochromis mossambicus)* and silver carp *(Hypophthalmichthys molitrix)* without artificial feeding in floating cages in plankton-rich waste water, *Aquaculture*, 31, 361, 1983.
304. **Moav, R., Wohlfarth, G., Schroeder, G. L., Hulata, G., and Barash, H.,** Intensive polyculture of fish in freshwater ponds. I. Substitution of expensive feeds by liquid cow manure, *Aquaculture*, 10, 25, 1977.
305. **Buck, D. H., Baur, R. T., and Rose, C. R.,** Utilization of swine manure in a polyculture of Asian and North American fishes, *Trans. Am. Fish. Soc.*, 107, 216, 1978.
306. **Dimitrov, M.,** Intensive polyculture of common carp (*Cyprinus carpio* L.) and herbivorous fish [silver carp, *Hypopthalmichthys molitrix* (Val.), and grass carp, *Ctenopharyngodon idella* (Val.)], *Aquaculture*, 38, 241, 1984.
307. **Santiago, C. B. and Reyes, O.S.,** Optimum dietary protein level for growth of bighead carp *(Aristichthys nobilis)* fry in a static water system, *Aquaculture*, 93, 155, 1991.
308. **Dah-Shu, L.,** The Method of Cultivation of Grass Carp, Black Carp, Silver Carp and Bighead Carp, China, Aquatic Biology Reasearch Institute, Academica Sinica, 1957, 90 (translated from Chinese by Language Services Branch, U.S. Dept. of Commerce, Washington, D.C.).
309. **Baltadgi, R. A.,** The artifical reproduction, feeding and growth of the herbivorous fishes in the reservoirs with ordinary and higher thermal regimes, in *Symp. Biol. Manage. Herbivorous Freshwater Fishes in the Pacific Area,* Pacific Science Assoc. 14th Pacific Science Congress, Khabarovsk, U.S.S.R., 1979, 49.

310. **Negonovskaya, I. T.,** On the results and prospects of the introduction of phytophagous fishes into waters of the USSR, *J. Ichthyol.*, 20, 101, 1980.
311. **Krzywosz, T., Bialokoz, W., and Brylinski, E.,** Growth of the bullhead carp *Aristichthys nobilis* in Lake Dgal Wielki, *Rocz. Nauk Roln. Ser. H Rybactwo,* 98, 103, 1977.
312. **Woynarovich, E.,** New systems and new fishes for culture in Europe, *FAO Fish. Rep.*, 5, 162, 1968
313. **Newton, S. H.,** Catfish farming with Chinese carps, *Ark. Farm Res.*, 29, 8, 1980.
314. **Li, S., Lu, W., Peng, C., and Zhao, P.,** Growth performance of different populations of silver carp and bighead, in *Proc. World Symp. Selection, Hybridization and Genetic Engineering in Aquaculture,* Bordeaux, France, May 27–30, 1986, 243.
315. **Green, B. W. and Smitherman, R. O.,** Relative growth, survival and harvestability of bighead carp, silver carp, and their reciprocal hybrids, *Aquaculture,* 37, 87, 1984.
316. **Maddox, J. J., Behrends, L. L., Madewell, C. E., and Pile, R. S.,** Algae-swine manure system for production of silver carp, bighead carp and tilapia, in *Culture of Exotic Fishes Symp. Proc.,* Vol. 24, Smitherman, R. O., Shelton, W. L., and Grover, J. H., Eds., American Fisheries Society, Auburn, AL, 1978, 109.
317. **Henderson, S.,** An Evaluation of Filter-Feeding Fishes for Removing Excessive Nutrients and Algae from Waste Water, USEPA Rep. 600/S2-83-019, U.S. Environmental Protection Agency, Washington, D.C., 1983, 5.
318. *Manual on the Biotechnology of the Propagation and Rearing of Phytophagous Fishes,* Fishery Ministry of the USSR, Moscow, 1970 (free translation from Russian by R. M. Howland, distributed by Division of Fishery Research Bureau of Sport Fisheries and Wildlife, Washington, D.C., 1971, 49).
319. **Rottmann, R. W. and Shireman, J. V.,** *Hatchery Manual for Grass Carp and Other Riverine Cyprinids*, Bull. No. 244, Florida Cooperative Extension Service and Institute of Food and Agricultural Sciences, Gainesville, 1990, 27.
320. **Schoonbee, H. J. and Prinsloo, J. F.,** Use of pituitary glands of the sharptooth catfish *Clarias gariepinus* in the induced spawning of the European common carp *Cyprinus carpio* and the Chinese grass carp *Ctenopharyngodon idella, Water S. A.,* 12, 235, 1986.
321. **Jaco, Z., Epler, P., and Bieniarz, K.,** Effect of the time of storage on gonadotropic activity of carp hypophyses fixed in acetone, *Pol. Arch. Hydrobiol.,* 36, 373, 1989.
322. **Thalathiah, S., Ahmad, A. O., and Zaini, M. S.,** Induced spawning techniques practised at Batu Berendam, Melaka, Malaysia, *Aquaculture,* 74, 23, 1988.
323. **Kumarasiri, W. S. A. A. L. and Seneviratne, P.,** Induced multiple spawnings of Chinese carps in Sri Lanka, *Aquaculture*, 74, 57, 1988.
324. **Hussain, M. G.,** Development of induced spawning procedures for grass carp, *Ctenopharyngodon idella* in Syria, *Asian Fish. Sci.*, 2, 115, 1988.
325. **Burlakov, A. B., Belova, N. V., Godvich, P. L., and Tsibezov, V. V.,** Role of endocrine system in the formation of egg quality of silver carp *Hypophthalmichthys molitrix* under artificial reproduction, *J. Ichthyol.*, 32, 83, 1992.
326. **Chondar, S. L.,** Mass scale breeding of silver carp in "Bangla bundh" through human chorionic gonadotropin and its combination with pituitary, in *Carp Seed Production Technology,* Spec. Publ. 2, Keshavanath, P. and Radhakrishnan, K. V., Eds., Asian Fisheries Society, Mangalore, India, 1990, 17.
327. **Dwivedi, S. N., Chaturvedi, C. S., and Varshney, P. K.,** Breeding of silver carp in combination of human chronionic gonadotrophin and constant dosage of pituitary hormone, *Agric. Biol. Res.*, 2, 8, 1986.

328. **Banerjee, S., Saha, D., and Podder, S.,** Induced spawning of silver carp, *Hypophthalmichthys molitrix*, by human chorionic gonadotrophic hormone (HCG) administration, *Environ. Ecol.*, 2, 153, 1984.
329. **Ghosh, A. and Roy, P. K.,** Mass breeding of silver carp in a cement cistern with crude H.C.G., *J. Indian Fish. Assoc.*, 16/17, 5, 1987.
330. **Fermin, A. C.,** LHRH-a and domperidone-induced oocyte maturation and ovulation in bighead carp, *Aristichthys nobilis* (Richardson), *Aquaculture*, 93, 87, 1991.
331. **Rottmann, R. W. and Shireman, J. V.,** The use of synthetic LH-RH analogue to spawn Chinese carp, *Aquat. Fish. Manage.*, 1, 19, 1985.
332. **Kouril, J., Barth, T., Hamackova, J., Slaninova, J., Servitova, L., Machacek, T., and Flegel, M.,** Application LH-RH and its analog for reaching ovulation in female tench, grass carp, carp and sheat fish, *Bul. Vyzk. Ustav. Ryb. Hydrobiol. Vodn.*, 19, 3, 1983.
333. **Ngamvongchon, S., Pawaputanon, O., Leelapatra, W., and Johnson, W. E.,** Effectiveness of an LHRH analogue for the induced spawning of carp and catfish in Northeast Thailand, *Aquaculture*, 74, 35, 1988.
334. **Lin, H.-R., Kraak, G. V. D., Liang, J.-Y., Peng, C., Li, G.-Y., Lu, L.-Z., and Zhou, X.-J.,** The effect of LHRH analogue and drugs which block the effects of dopamine on gonadotropin secretion and ovulation in fish cultured in China, in *Aquaculture of Cyprinids*, Billard, R. and Marcel, J., Eds., Institut National de la Recherche Agronomique, Paris, 1985, 139.
335. **Popek, W., Bieniarz, K., Epler, P., Mikolajczyk, T., Motyka, K., Malczewski, B., and Krakowie, A.,** Hormonal stimulation of induced spawn of grass carp without use of hypophysis, *Gosp. Rybna*, 1, 7, 1987, (in Polish).
336. **Huisman, E. A.,** The culture of grass carp (*Ctenopharyngodon idella* Val.) under artificial conditions, *Schr. Bundesforschungs. Fisch. Hamburg*, 14/15, 491, 1979.
337. **Shireman, J. V., Colle, D. E., and Rottmann, R. W.,** Manipulation of temperature and photoperiod for inducing maturation in grass carp, in *Symp. Culture of Exotic Fishes, Aquaculture/Atlanta/'78*, Smitherman, R. O., Shelton, W. L., and Grover, J. H., Eds., Atlanta, January 5, 1978, 156.
338. **Makeeva, A. P. and Emel'yanova, N. G.,** Cytological investigation of preovulatory oocytes of silver carp, *Hypophthalmichthys molitrix*, using biopsy, *J. Ichthyol.*, 30, 69, 1990.
339. **Rottmann, R. W. and Shireman, J. V.,** Tank spawning of grass carp, *Aquaculture*, 17, 257, 1979.
340. **Zonneveld, N.,** The spawning season and the relation between temperature and stripping time of grass carp (*Ctenopharyngodon idella* Val.) in Egypt, *Bamidgeh*, 36, 21, 1984.
341. **Zalepukhin, V. V.,** Egg quality of cultured grass carp, in *Physiology of Cultured Fishes*, Vol. 42, Shcherbina, M. A., Ed., VNIIPRKH, Moscow, 96, 1984.
342. **Makeeva, A. P., Emel'yanova, N. G., and Verigin, V. B.,** A quality of eggs produced by *Hypophthalmichthys molitrix, Aristichthys nobilis*, and *Ctenopharyngodon idella* under artificial culture, *J. Ichthyol.*, 28, 48, 1988.
343. **Zhukinskiy, V. N. and Alekseenko, V. R.,** Semen quality in common carp, *Cypinus carpio,* and white amur, *Ctenopharyngodon idella* (Cyprinidae), in different periods of the spawning season and as influenced by extraction methods, *J. Ichthyol.*, 23, 124, 1983.
344. **Durbin, H., Durbin, F. J., and Stott, B.,** A note on the cryopreservation of grass carp milt, *Fish. Manage.*, 13, 115, 1982.

345. **Gonzal, A. C., Aralar, E. V., and Pavico, J. Ma. F.,** The effect of water hardness on the hatching and viability of silver carp *(Hypophthalmichthys molitrix)* eggs, *Aquaculture*, 64, 111, 1987.
346. **Makeeva, A. P. and Sukhanova, A. I.,** Development of hybrids of herbivorous fishes, *Vopr. Ikhthiol.*, 6, 477, 1966 (in Russian).
347. **Márián, T. and Krasznai, A.,** Karyological investigations on *Ctenopharyngodon idella* and *Hypophthalmichthys nobilis* and their cross-breeding, *Aquacult. Hung. (Szarvas)*, 1, 44, 1978.
348. **Sutton, D. L., Stanley, J. G., and Miley, W. W., II,** Grass carp hybridization and observations of a grass carp and bighead hybrid, *J. Aquat. Plant Manage.*, 19, 37, 1981.
349. **Allen, S. K., Jr. and Stanley, J. G.,** Ploidy of hybrid grass carp x bighead carp determined by flow cytometry, *Trans. Am. Fish. Soc.*, 112, 431, 1983.
350. **Cassani, J. R., Caton, W. E., and Clark, B.,** Morphological comparisons of diploid and triploid hybrid grass carp, *Ctenopharyngodon idella* ♀ and *Hypophthalmichthys nobilis* ♂, *J. Fish Biol.*, 25, 269, 1984.
351. **Osborne, J. A.,** The potential of the hybrid grass carp as a weed control agent, *J. Freshw. Ecol.*, 1, 353, 1982.
352. **Kilambi, R. V. and Zdinak, A.,** Food intake and growth of hybrid carp [female grass carp, *Ctenopharyngodon idella* × male bighead, *Aristichthys (Hypophthalmichthys) nobilis*] fed on zooplankton and *Chara*, *J. Fish Biol.*, 21, 63, 1982.
353. **Shireman, J. V. and Hoyer, M. V.,** Long-Term Impact Assessment of Grass Carp and Field Evaluation of the Hybrid Grass Carp, final report to National Fisheries Research Laboratory, Gainesville, FL, 1984, 32.
354. **Opuszynski, K., Shireman, J. V., Aldridge, F. J., and Rottmann, R.,** Intensive culture of grass carp and hybrid grass carp larvae, *J. Fish Biol.*, 26, 563, 1985.
355. **Shelton, W. L.,** Broodstock development for monosex production of grass carp, *Aquaculture*, 57, 311, 1986.
356. **Mirza, J. A. and Shelton, W. L.,** Induction of gynogenesis and sex reversal in silver carp, *Aquaculture*, 68, 1, 1988.
357. **Van Eenennaam, J. P., Stocker, R. K., Thiery, R. G., Hagstrom, N. T., and Doroshov, S. I.,** Egg fertility, early development and survival from crosses of diploid female × triploid male grass carp *(Ctenopharyngodon idella)*, *Aquaculture*, 86, 111, 1990.
358. **Allen, S. K., Jr., Thiery, R. G., and Hagstrom, N. T.,** Cytological evaluation of the likelihood that triploid grass carp will reproduce, *Trans. Am. Fish. Soc.*, 115, 841, 1986.
359. **Allen, S. K., Jr. and Wattendorf, R. J.,** Triploid grass carp: status and management implications, *Fisheries*, 12, 20, 1987.
360. **Thompson, B. Z., Wattendorf, R. J., Hestand, R. S., and Underwood, J. L.,** Triploid grass carp production, *Prog. Fish-Cult.*, 49, 213, 1987.
361. **Cassani, J. R. and Caton, W. E.,** Efficient production of triploid grass carp *(Ctenopharyngodon idella)* utilizing hydrostatic pressure, *Aquaculture*, 55, 43, 1986.
362. **Wattendorf, R. J.,** Rapid identification of triploid grass carp with a Coulter counter and channelyzer, *Prog. Fish-Cult.*, 48, 125, 1986.
363. **Pine, R. T. and Anderson, L. W. J.,** Blood preparation for flow cytometry to identify triploidy in grass carp, *Prog. Fish-Cult.*, 52, 266, 1990.
364. **Cassani, J. R.,** A new method for early ploidy evaluation of grass carp larvae, *Prog. Fish-Cult.*, 52, 207, 1990.

365. **Aldridge, F. J., Marston, R. Q., and Shireman, J. V.,** Induced triploids and tetraploids in bighead carp, *Hypophthalmichthys nobilis*, verified by multi-embryo cytofluorometric analysis, *Aquaculture*, 87, 121, 1990.
366. **Chourrout, D., Chevassus, B., Krieg, F., Happe, A., Burger, G., and Renard, P.,** Production of second generation triploid and tetraploid rainbow trout by mating tetraploid males and diploid females — potential of tetraploid fish, *Theor. Appl. Genet.*, 72, 193, 1986.
367. **Cassani, J. R., Maloney, D. R., Allaire, H. P., and Kerby, J. H.,** Problems associated with tetraploid induction and survival in grass carp, *Ctenopharyngodon idella, Aquaculture,* 88, 273, 1990.
368. **Lirski, A. and Opuszynski, K.,** Lower lethal temperatures for carp (*Cyprinus carpio* L.) and the phytophagous fishes (*Ctenopharyngodon idella* Val., *Hypopthalmichthys molitrix* Val., *Aristichthys nobilis* Rich.) in the first period of life, *Rocz. Nauk Roln.*, H-101(4), 11, 1988 (in Polish with English summary).
369. **Lirski, A. and Opuszynski, K.,** Upper lethal temperatures for carp (*Cyprinus carpio* L.) and the phytophagous fishes (*Ctenopharyngodon idella* Val., *Hypopthalmichthys molitrix* Val., *Aristichthys nobilis* Rich.) in the first period of life, *Rocz. Nauk Roln.*, H-101(4), 31, 1988.
370. **Jobling, M.,** Temperature tolerance and final preferendum — rapid methods for the assessment of optimum growth temperatures, *J. Fish Biol.*, 19, 439, 1981.
371. **Opuszynski, K., Lirski, A., Myszkowski, L., and Wolnicki, J.,** Upper lethal and rearing temperatures for juvenile common carp, *Cyprinus carpio* L., and silver carp, *Hypopthalmichthys molitrix* Val., *Aquacult. Fish. Manage.*, 20, 287, 1989.
372. **Schlumpberger, W.,** Zur kombination der rinnen und Gasekafiganfzucht von Silberkarpfenbrut *(Hypopthalmichthys molitrix), Z. Binnenfisch. DDR,* 27, 172, 1980.
373. **Vovk, P. S.,** *Biology of Asiatic Herbivorous Fish and Their Use in Water Reservoirs of Ukraina*, Naukova Dumka, Kiev, 1976 (in Russian).
374. **Radenko, V. N. and Alimov, I. A.,** Significance of temperature and light for growth and survival of larvae of silver carp, *Hypopthalmichthys molitrix, J. Ichthyol.*, 32, 16, 1992.
375. **Wozniewski, M. and Opuszynski, K.,** Threshold oxygen content for juvenile stages of the cyprinids (*Ctenopharyngodon idella* Val., *Hypopthalmichthys molitrix* Val., *Aristichthys nobilis* Rich., *Cyprinus carpio* L.), *Rocz. Nauk Roln.*, H-101(4), 51, 1988.
376. **Wozniewski, M. and Myszkowski, L.,** Oxygen conditions during common carp larval rearing at high water temperature, *Gosp. Rybna*, 4, 14, 1987 (in Polish).
377. **Wozniewski, M., Littak, A., and Opuszynski, K.,** Production of the Silver Carp Stocking Material. II. Rearing of Larvae Under Controlled Conditions, No. 128, Inland Fisheries Institute, Olsztyn, Poland, 1980, 1 (in Polish).
378. **Carlos, M. H.,** Growth and survival of bighead carp *(Aristichthys nobilis)* fry fed different intake levels and feeding frequencies, *Aquaculture*, 68, 267, 1988.
379. **Krebs, J. R.,** Optimal foraging: decision rules for predators, in *Behavioral Ecology: An Evolutionary Approach*, Krebs, J. R. and Davies, N.B., Eds., Blackwell Scientific, Oxford, 1978, 23.
380. **Ito, J. and Suzuki, R.,** Feeding habits of a *Cyprinidae* loach fry in the early stages, *Bull. Freshw. Fish. Res. Lab.*, 27, 85, 1977.
381. **Van der Wind, J. J.,** Techniques of rearing phytophagous fishes, *FAO Fish. Rep.*, 44, 227, 1979.
382. **Okoniewska, G. and Wolnicki, J.,** Food preference of common carp (*Cyprinus carpio* L.) and grass carp (*Ctenopharyngodon idella* Val.) after starvation, *Rocz. Nauk Roln.*, H-101(4), 7, 1988 (in Polish with English summary).

383. **Okoniewska, G. and Opuszynski, K.,** Rearing of grass carp larvae (*Ctenopharyngodon idella* Val.) in ponds receiving heated effluents. IV. Food, *Rocz. Nauk Roln.*, H-101(4) 161, 1988 (in Polish with English summary).
384. **Ciborowska, J.,** Food of Asiatic herbivorous fish (*Ctenopharyngodon idella* Val., *Hypopthalmichthys molitrix* Val., *Aristichthys nobilis* Rich.) raised together with common carp in fry ponds, *Rocz. Nauk Roln.*, H-94, (2), 41, 1972 (in Polish with English summary).
385. **Khadka, R. B. and Rao, T. R.,** Prey size selection by common carp (*Cyprinus carpio* var. *communis*) larvae in relation to age and prey density, *Aquaculture*, 54, 89, 1986.
386. **Dabrowski, K. and Culver, D.,** The physiology of larval fish: digestive tract and formulation of starter diets, *Aquacult. Mag.*, 17, 49, 1991.
387. **Charlon, N. and Bergot, P.,** Rearing system for feeding fish larvae on dry diets. Trial with carp (*Cyprinus carpio* L.) larvae, *Aquaculture*, 41, 1, 1984.
388. **Charlon, N., Durante, H., Escaffre, A. M., and Bergot, P.,** Alimentation artificielle des larves de carpe (*Cyprinus carpio* L.), *Aquaculture*, 54, 83, 1986.
389. **Alami-Durante, H., Charlon, N., Escaffre, A., and Bergot, P.,** Supplementation of artificial diets for common carp (*Cyprinus carpio* L.) larvae, *Aquaculture*, 93, 167, 1991.
390. **Bryant, P. L. and Matty, A. J.,** Adaptation of carp (*Cyprinus carpio*) larvae to artificial diets. I. Optimum feeding rate and adaptation age for a commercial diet, *Aquaculture*, 23, 275, 1981.
391. **Pillay, T. V. R.,** *Aquaculture: Principles and Practices*, Fishing News Books, Oxford, 1990.
392. **Fritzche, S. and Taege, M.,** Damages at the rearing of fry, (*Hypopthalmichthys molitrix*, *Cyprinus carpio* and *Coregonus albula*) in industrial production plants because of strong occurrence of copepods (Copepoda), *Z. Binnenfisch. DDR*, 26, 304, 1979.
393. **Wolny, P.,** Results of the six-year study on the effectiveness of nursery pond fertilization, *Rocz. Nauk Roln.*, 91-H(4), 565, 1970 (in Polish with English summary).
394. **Olah, J. and Farkas, J.,** Effect of temperature, pH, antibiotics, formalin and malachite green on the growth and survival of *Saprolegnia* and *Achlya* parasitics on fish, *Aquaculture*, 3, 273, 1978.
395. **Opuszynski, K.,** Fresh-water pond ecosystems managed under moderate European climate, in *Managed Aquatic Ecosystems*, Michael, R. G., Ed., Elsevier, Amsterdam, 1987, 63.
396. **Lannan, J. E., Smitherman, R. O., and Tchobanoglous, G.,** *Principles and Practices of Pond Aquaculture*, Oregon State University Press, Corvallis, 1986.
397. **Grygierek, E. and Wasilewska, B.,** Regulation of Fish Pond Biocoenosis, in *Cultivation of Fish Fry and Its Live Food*, Spec. Publ. 4, Styczynska-Jurewicz, E., Backiel, T., Jaspers, E., and Persoone, G., Eds., European Mariculture Society, Bredene, Belgium, 1979.
398. **Opuszynski, K. and Okoniewska, Z.,** Survival and changes in fat and protein content of *Ctenopharyngodon idella* Val., *Hypopthalmichthys molitrix* Val., and *Cyprinus carpio* L. during wintering in ponds, *Rocz. Nauk Roln.*, H-4, 657, 1969 (in Polish with English summary).
399. **Opuszynski, K.,** Production of herbivorous fishes (*Ctenopharyngodon idella* and *Hypophthalmichthys molitrix*) in carp ponds, *Rocz. Nauk. Roln.*, H-91, 219, 1969 (in Polish with English summary).
400. **Opuszynski, K.,** Silver carp, *Hypopthalmichthys molitrix* (Val.), in carp ponds. I. Fishery production and food relations, *Ekol. Pol.*, 27, 71, 1979.

401. **Wolny, P.,** The effect of intensification measures on growth, survival and production of Asiatic herbivorous fish, *Rocz. Nauk Roln.*, H-92, 97, 1970 (in Polish with English summary).
402. **Fijan, N.,** Problems in carp pond fertilization, *FAO Fish. Rep.*, 44, 114, 1967.
403. **Januszko, M., Bednarz, T., Broad, M., and Grzelewska, E.,** Fertilization of enclosed parts of ponds with nitrogen and phosphorus. IV. Phytoplankton, *Rocz. Nauk Roln.*, H-98 (1), 75, 1977 (in Polish with English summary).
404. **Vinberg, G. G. and Lakhnovitch, V. P.,** *Pond Fertilization*, Pischtchevaia Promyslennost, Moscow, 1965 (in Russian).
405. **Aleksandrijskaia, A.,** Contemporary practices in pond fertilization, *Rybov. Rybol.*, 6, 15, 1978 (in Russian).
406. **Piotrowska-Pouszynska, W.,** The influence of nitrogen fertilizers on physico-chemical conditions in nursery ponds, *Rocz. Nauk Roln.*, H-100 (4), 111, 1984 (in Polish with English summary).
407. **Culver, D. A., Maden, S. P., and Qin, J.,** Percid pond production techniques: timing, enrichment, and stocking density manipulation, *J. Appl. Aquat.*, 2, 1992.
408. **Anderson, R. O.,** Problems and solutions for production of fingerling striped bass, *J. Appl. Aquat.*, 2, 1992.
409. **Opuszynski, K. and Shireman, J. V.,** Pond environmental manipulation to stimulate rotifers for larval fish rearing, *Rocz. Nauk Roln.*, H-101(4), 183, 1988.
410. **Kane, A. S. and Johnson, D. L.,** Use of TFM (3-trifluoromethyl-4-nitrophenol) to selectively control frog larvae in fish production ponds, *Prog. Fish-Cult.*, 51, 207, 1989.
411. **Gabbadon, P. W.,** Use of the lampricide TFM (3-trifluoromethyl-4-nitrophenol) to control frog larvae in warm-water ornamental fish ponds, M.S. thesis, University of Florida, Gainesville, 1992, 110.
412. **Chen, L.-C.,** *Aquaculture in Taiwan,* Fishing News Books, Oxford, 1990.
413. **Lirski, A., Ononszkiewicx, B., Opuszynski, K., and Wozniewski, M.,** Rearing of cyprinid larvae in new type flow-through cages placed in carp ponds, *Pol. Arch. Hydrobiol.*, 26, 545, 1979.
414. **Kossmann, H.,** A warm water recycling system for grass carp production, in *Aquaculture in Heated Effluents and Recirculation Systems,* Tiews, K., Ed., Heenemann Verlagsgesellschaft, Berlin, 1981, 431.
415. **Lincoln, E. P., Hall, T. W., and Koopman, B.,** Zooplankton control in mass algae cultures, *Aquaculture,* 32, 331, 1983.
416. **Schluter, M. and Groeneweg, J.,** Mass production of freshwater rotifers on liquid wastes. I. The influence of some environmental factors on population growth of *Brachionus rubens* Ehrenberg 1938, *Aquaculture,* 25, 17, 1981.
417. **Tamas, G.,** Rearing of common carp fry and mass cultivation of its food organisms in ponds, in *Cultivation of Fish Fry and Its Live Food,* Spec. Publ. 4, Styczynska-Jurewicz, E., Backiel, T., Jaspers, E., and Persoone, G., Eds., European Mariculture Society, Bredene, Belgium, 1979.
418. **Komis, A., Candreva, P., Franiceric, V., Moreau, V., Van Bauaer, E., Leger, Ph., and Sorgeloos, P.,** Successful Application of a New Combined Culture and Enrichment Diet for the Mass Cultivation of the Rotifer *Brachionus plicatilis* at Commercial Hatchery Scale in Monaco, Yugoslavia, France, and Thailand, Spec. Publ. 15, European Aquaculture Society, Ghent, Belgium, 1991, 102.
419. **Dhout, T., Lavens, P., and Sorgeloos, P.,** Development of a Lipid-Enrichment Technique for *Artemia* Juveniles Produced in an Intensive System for Use in Marine Larviculture, Spec. Publ. 15, European Aquaculture Society, Ghent, Belgium, 1991, 51.

420. **Leger, P., Bengtson, D. A., Simpson, K. L., and Sorgeloos, P.,** The use and nutritional value of *Artemia* as food source, *Ocean. Mar. Biol., Annu. Rev.*, 24, 521, 1986.
421. **Campton, D. E.,** A simple procedure for decapsulating and hatching brine shrimp, *Prog. Fish-Cult.*, 51, 176, 1989.
422. **Huisman, E. A.,** Integration of hatchery, cage, and pond culture of common carp (*Cyprinus carpio* L.) and grass carp (*Ctenopharyngodon idella* Val.) in the Netherlands, in *Proc. Bio-Engineering Symp. Fish Culture,* Publ. 1, Allen, L. J. and Kinney, E. C., Eds., American Fisheries Society, Bethesda, MD, 1981, 266.
423. **Okoniewska, G., Piotrowska-Opuszynska, W., and Opuszynski, K.,** Rearing of grass carp larvae (*Ctenopharyngodon idella* Val.) in ponds receiving heated effluents. I. Scheme of the experiment, temperature, water flow in the ponds and fish production, *Rocz. Nauk Roln.*, H-101(4), 111, 1988 (in Polish with English summary).
424. **Saeuberlich, E.,** Increased production of I+ herbivorous fishes by utilization of warm water in the aquaculture, *Z. Binnenfisch. DDR,* 28, 365, 1981.
425. **Tapiador, D. D., Henderson, H. F., Delmendo, M. N., and Tsutsui, H.,** Freshwater fisheries and aquaculture in China, *FAO Fish. Tech. Pap.*, 168, 84, 1977.
426. Freshwater aquaculture development in China, *FAO Fish. Tech. Pap.*, 215, 125, 1983.
427. **Chang, W. Y. B.,** Fish culture in China, *Fisheries,* 12, 11, 1987.
428. **Marcel, J.,** Aquaculture in China, in *Aquaculture,* Vol. 2, Barnabé, G., Ed., Ellis Horwood, Chichester, England, 1990, 962.
429. **Li, S.,** Energy structure and efficiency of a typical Chinese integrated fish farm, *Aquaculture,* 65, 105, 1987.
430. **Zhu, Y., Yang, Y., Wan, J., Hua, D., and Mathias, J. A.,** The effect of manure application rate and frequency upon fish yield in integrated fish farm ponds, *Aquaculture,* 91, 233, 1990.
431. **Lin, S. Y. and Chen, T. P.,** Increase of production in fresh-water fish ponds by the use of inorganic fertilizers, in *Proc. World Symp. on Warm Water Pond Fish Culture, Rome, 18–25 May 1966,* Vol. 44-3, Pillay, T. V. R, Ed., Food and Agriculture Organization, Rome, 1967, 210.
432. **Wang, Y., Feng, B., Tan, D., Zhang, X., Wu, X., and Chen, H.,** Evaluation of the leaf of *Zizania latifolia* as a substitute for cereal stuff in practical diets for herbivorous fishes, *Acta Hydrobiol. Sin.*, 14, 247, 1990.
433. **Hu, B.-T., Qian, W.-M., and Xi, S.-R.,** Studies on stocking density in cage culture of silver carp fingerlings in the Bailianhe Reservoir, *J. Fish. China,* 7, 45, 1983.
434. **Liu, H.-Q., Xie, H.-G., Huang, S.-W., and Deng, B.-L.,** On the scales annuli formation of silver and bighead carps in Lake Doug Hu, with special reference to the problem of rational size of "seedlings" at the time of stocking, *J. Fish. China,* 6, 129, 1982.
435. **Sharma, B. K. and Olah, J.,** Integrated fish-pig farming in India and Hungary, *Aquaculture,* 54, 135, 1986.
436. **Sukhoverkhor, F. M.,** Results of the investigations and the perspectives of grass carp, silver carp, and bighead carp in ponds of the European part of the RSFSR, in *Problems of the Fisheries Exploitation of Plant-Eating Fish in the Water Bodies of the USSR,* Academy of Science of Turkmenistan SSSR, Ashabad, 1963, 48 (in Russian).
437. **Verigin, B. V.,** The present state and the perspectives of the utilization of the silver carp and the grass carp in the water bodies of the USSR, in *Problems of the Fisheries Exploitation of Plant-Eating Fishes in the Water Bodies of the USSR,* Academy of Sciences at Turkmerristan SSSR, Ashabad, 1963, 20 (in Russian).

438. **Verigin, B. V.,** Ecological aspects of culture herbivorous fishes in ponds, in *2nd Int. Symp. Herbivorous Fish, Svietlost Fojnica, Novi Sad,* 1982, 176 (in Russian with English summary).
439. **Babayan, K. E.,** A new stage in the culture of plant-eating fishes, *Ryb. Khoz.,* 42, 4, 1966 (in Russian).
440. **Janecek, V., Prikryl, I., and Kepr, T.,** Experimental rearing of three-year-old common carp in polyculture with silver carp and grass carp, *Bul. Vyzk. Ustav. Ryb. Hydrobiol. Vodn.,* 21, 3, 1985.
441. **Sarodnik, W., Braeuer, V., Nagel, L., Greim, K.-H.,** Results of intensive pond rearing of three-year-old grass carp, *Z. Binnenfisch.,* 37, 279, 1990.
442. **Yashouv, A.,** Interaction between the common carp *(Cyprinus carpio)* and the silver carp *(Hypopthalmichthys molitrix)* in fish ponds, *Bamidgeh,* 23, 85, 1971.
443. **Reich, K.,** Multispecies fish culture (polyculture) in Israel, *Bamidgeh,* 27, 85, 1975.
444. **Milstein, A.,** Ecological aspects of fish species interactions in polyculture ponds, *Hydrobiologia,* 231, 177, 1992.
445. **Hepher, B., Milstein, A., Leventer, H., and Teltsch, B.,** The effect of fish density and species combination on growth and utilization of natural food in ponds, *Aquacult. Fish. Manage.,* 20, 59, 1989.
446. **Schroeder, G. L., Wohlfarth, G., Alkon, A., Halevy, A., and Krueger, H.,** The dominance of algal-based food webs in fish ponds receiving chemical fertilizers plus organic manures, *Aquaculture,* 86, 219, 1990.
447. **Mgbenka, B. O. and Lovell, R. T.,** Intensive feeding of grass carp in ponds, *Prog. Fish-Cult.,* 48, 238, 1986.
448. **Pfeiffer, T. J. and Lovell, R. T.,** Responses of grass carp, stocked intensively in earthen ponds, to various supplemental feeding regimes, *Prog. Fish-Cult.,* 52, 213, 1990.
449. **Henderson, S.,** Production potential of catfish grow-out ponds supplementally stocked with silver and bighead carp, *Proc. Annu. Conf. S.E. Assoc. Fish Wildl. Agencies,* 33, 584, 1979.
450. **Dunseth, D. R. and Smitherman, R. O.,** Polyculture of catfish, tilapia and silver carp, *Proc. Annu. Commercial Fish. Workshop,* 6, 25, 1977.
451. **Boyd, C. E.,** *Water Quality in Warmwater Fish Ponds,* Auburn University Agricultural Experiment Station, Auburn, AL, 1979, 168.
452. **Behrends, L. L., Kingsley, J. B., and Price, A. H.,** III, Polyculture of freshwater prawns, tilapia, channel catfish and Chinese carp, *J. World Maricult. Soc.,* 16, 437, 1986.
453. **Jhingran, V. G.,** Aquaculture in India, in *Aquaculture,* Vol. 2, Barnabé, G., Ed., Ellis Horwood, Chichester, England, 1990, chap. 5.
454. **Jhingran, V. G.,** *Fish and Fisheries of India,* Hindustan Publishing, Delhi, 1975, 466.
455. **Ling, S.-W.,** *Aquaculture in Southeast Asia,* University of Washington Press, Seattle, 1977, chap. 3.
456. **Tripathi, S. D. and Ranadhir, M.,** An economic analysis of composite fish culture in India, in *Aquaculture Economics Research in Asia,* Ottawa, Ontario, Canada, 1982, 90.
457. **De Silva, S. S. and Gunasekera, R. M.,** An evaluation of the growth of Indian and Chinese major carps in relation to the dietary protein content, *Aquaculture,* 92, 237, 1991.
458. **Tripathi, S. D. and Mishra, D. N.,** Synergistic approach in carp polyculture with grass carp as a major component, *Aquaculture,* 54, 157, 1986.

459. **Mohire, K. V., Devaraj, K. V., and Das, T. K.,** Utilization of waste mulberry leaves as food for the grass carp, *Ctenopharyngodon idella* (Val.), in *Exotic Aquatic Species in India,* Joseph, M. M., Ed., Asian Fisheries Society, Mangalore, India, 1989, 113.
460. **Prasad, G. S.,** Performance of grass carp, *Ctenopharyngodon idella* (Val.), in different aquaculture practices, in *Exotic Aquatic Species in India,* Joseph, M. M., Ed., Asian Fisheries Society, Mangalore, India, 1989, 105.
461. **Sen, P. R., Roa, N. G. S., Chakrabarty, R. D., Kar, S. L., and Jena, S.,** Effect of additions of fertilizer and vegetation on growth of major Indian carps in ponds containing grass carp, *Prog. Fish-Cult.,* 40, 69, 1978.
462. **Rabanal, H. R.,** Stock manipulation and other biological methods of increasing production of fish through pond fish culture in Asia and the Far East, in *Proc. FAO World Symp. Warm-Water Pond Fish Culture,* Food and Agriculture Organization, Rome, FAO Fish. Rep., 44, Pillay, T. V. R., Ed., 1968, 274.
463. **Ji, L. S.,** Status and potential for development of freshwater fish culture in Malaysia, *Indo-Pacific Fish Counc. Proc.,* 17, 96, 1976.
464. **Law, A. T.,** Digestibility of low cost ingredients in pelleted feed by grass carp (*Ctenopharyngodon idella* C. et V.), *Aquaculture,* 51, 97, 1986.
465. **Le Mare, D. W.,** Pig rearing, fish farming and vegetable growing, *Malay. Agric. J.,* 35, 156, 1952.
466. **Swar, D. B. and Gurung, T. B.,** Introduction and cage culture of exotic carps and their impact on fish harvested in Lake Begnas, Nepal, *Hydrobiologia,* 166, 277, 1988.
467. **Arrignon, J.,** Aquaculture in Africa, in *Aquaculture,* Vol. 2, Barnabé, G., Ed., Ellis Horwood, Chichester, England, 1990, chap. 7.
468. **Schoonbee, H. J., Nakani, V. S., and Prinsloo, J.,** The use of cattle manure and supplementary feeding in growth studies of the Chinese silver carp in Transkei, *S. Afr. J. Sci.,* 75, 489, 1979.
469. **Doumenge, F.,** Aquaculture in other Far Eastern countries, in *Aquaculture,* Vol. 2, Barnabé, G., Ed., Ellis Horwood, Chichester, England, 1990, chap.4.
470. **Bauer, O. N., Musselius, V. A., and Strelkow, Y. A.,** Die parasiten und drankheiten von *Ctenopharyngodon idella, Hypophthalmichthys molitrix* und *Aristichthys nobilis* bei der autzucht in Teichwirtschaffen der USSR, *Z. Fischer.* 17, 205, 1969.
471. **Bohl, M.,** Disease control and reproduction of grass carp in Germany, in *Proc. Grass Carp Conf.,* Shireman, J. V., Ed., Institute of Food and Agricultural Sciences, University of Florida, Gainesville, 1979, 243.
472. **Molnar, K.,** Parasite Range Extension by Introducing of Fish to Hungary, EIFAC Tech. Pap./Doc. Tech. CECPI, 42 Suppl. Vol. 2, 1984, 283.
473. **Riley, D. M.,** Parasites of grass carp and native fishes in Florida, *Trans. Am. Fish. Soc.,* 107, 207, 1978.
474. **McDaniel, D. W.,** *Fish Health Blue Book: Procedures for the Detection and Identification of Certain Fish Pathogens,* American Fisheries Society, Bethesda, MD, 1985, 114.
475. **Ferguson, H. W.,** *Systemic Pathology of Fish,* Iowa State University Press, Ames, 1989, 263.
476. **Roberts, R. J., Ed.,** *Fish Pathology,* Bailliere Tindall, London, 1989, 467.
477. **Sarig, S.,** *Disease of Fishes. Book 3: The prevention and Treatment of Diseases of Warmwater Fishes Under Subtropical Conditions, with Special Emphasis on Intensive Fish Farming,* T. F. H. Publications, Neptune, NJ, 1971, 127.
478. **Hoffman, G. L. and Meyer, F. P.,** *Parasites of Freshwater Fishes, A Review of Their Control and Treatment,* T. F. H. Publishers, Neptune City, NJ, 1974.

479. **Pieterse, A. H.,** Introduction, in *Aquatic Weeds: The Ecology and Management of Nuisance Aquatic Vegetation,* Pieterse, A. H. and Murphy, K. J., Eds., Oxford University Press, New York, 1990, chap. 1.
480. **Spencer, W. and Bowes, G.,** Ecophysiology of the world's most troublesome aquatic weeds, in *Aquatic Weeds: The Ecology and Management of Nuisance Aquatic Vegetation,* Pieterse, A. H. and Murphy, K. J., Eds., Oxford University Press, New York, 1990, chap. 4.
481. **Ramaprabhu, T., Ramachandran, V., and Reddy, P. V. G. K.,** Some aspects of the economics of aquatic weed control in fish culture, *J. Aquat. Manage.,* 20, 41, 1982.
482. **Wade, P. M.,** Physical control of aquatic weeds, in *Aquatic Weeds: The Ecology and Management of Nuisance Aquatic Vegetation,* Pieterse, A. H. and Murphy, K. J., Eds., Oxford University Press, New York, 1990, chap. 7.
483. **Livermore, D. F. and Koegel, R. G.,** Mechanical harvesting of aquatic plants: an assessment of the state of the art, in *Aquatic Plants, Lake Management and Ecosystem Consequences of Lake Harvesting,* Breck, J., Prentki, R., and Loucks, O., Eds., Institute for Environmental Studies, University of Wisconsin, Madison, 1979.
484. **Ramey, V.,** Mechanical control of aquatic plants, *Aquaphyte,* 2, 3, 1982.
485. **Gallagher, J. E. and Haller, W. T.,** History and development of aquatic weed control in the United States, *Rev. Weed Sci.,* 5, 115, 1990.
486. **Murphy, K. J. and Barrett, P. R. F.,** Chemical control of aquatic weeds, in *Aquatic Weeds: The Ecology and Management of Nuisance Aquatic Vegetation,* Pieterse, A. H. and Murphy, K. J., Eds., Oxford University Press, New York, 1990, chap. 8.
487. **Cooke, G. D.,** Lake-level drawdown as a macrophyte control technique, *Water Resources Bull.,* 16, 317, 1980.
488. **Krupauer, V.,** The use of herbivorous fishes for ameliorative purposes in central and eastern Europe, in *Proc. Eur. Weed Res. Counc. 3rd Int. Symp. Aquatic Weeds,* 1971, 95.
489. **Yefimova, A. T. and Nikanorov, Yu. I.,** Prospects for the introduction of phytophagous fishes into Ivankovskoye reservoir, *J. Ichthyol.,* 17, 634, 1977.
490. **Zolotova, Z. K.,** Biological weed control in irrigation channels with the aid of grass carp, Sb. Nauchno-Issled. Rabot Vses. Nauchno-Issled.-Inst. Prudov., *Rybn. Khoz.,* 18, 112, 1970 (in Russian).
491. **Lupaceva, L. I.,** Higher aquatic vegetation in ponds of the Tsyrupinsk spawning-breeding farm, *Rybn. Khoz. Respub. Mezhved. Temat. Nauch. Sb.,* 6, 98, 1968 (in Russian).
492. **Zolotova, Z. K. and Khromov, L. V.,** The weeding role of grass carp, *Rybovod. Rybolov.,* 18, 8, 1970 (in Russian).
493. **Opuszynski, K.,** Carp polyculture with plant-feeding fish: grass carp (*Ctenopharyngodon idella* Val.) and silver carp (*Hypopthalmichthys molitrix* Val.), *Bull. Ac. Pol. Scin.,* 16, 677, 1968.
494. **Opuszynski, K.,** Rearing grass carp larvae in ponds fed with warm water discharged by power stations, in *Proc. 2nd Int. Symp. Herbivorous Fish. Novi Sad,* 1982, 140.
495. **Wolnicki, J. and Opuszynski, K.,** Rearing of grass carp larvae (*Ctenopharyngodon idella* Val.) in ponds receiving heated effluents. III. Food conditions, *Rocz. Nauk Roln.* 101, 149, 1988 (in Polish with English summary).
496. **Pinter, K.,** Exotic fishes in Hungarian waters, their importance in fishery utilization of natural water bodies and fish farming, *Fish. Manage.,* 11, 163, 1980.

497. **Krasznai, Z., Márián, T., Buris, L., and Ditrói, F.,** Production of sterile hybrid grass carp (*Ctenopharyngodon idella* Val. × *Aristichthys nobilis* Rich.) for weed control, in *Proc. 2nd Int. Symp. Herbivorous Fish.* Novi Sad, Yugoslavia, European Weed Research Soc., Waginingen, The Netherlands, 1982, 55.
498. **Krasznai, Z., Márián, T., Buris, L., and Ditrói, F.,** Production of sterile hybrid grass carp (*Ctenopharyngodon idella* Val. × *Aristichthys nobilis* Rich.) for weed control, *Aquacult. Hung.*, 4, 33, 1984.
499. **Márián, T. and Krasznai, Z.,** Comparative karyological and serological studies on Chinese major carps, *Aquacult. Hung.*, 2, 5, 1980.
500. **Krupauer, V.,** Experience gained in the rearing of herbivorous fish in Czechoslovakia, *Bul. Vyzk. Ust. Ryb. Vodn.*, 4, 3, 1968.
501. **Van Zon, J. C. J.,** The grass carp in Holland, in *Proc. EWRS 4th Symp. Aquatic Weeds,* 1974, 128.
502. **Zonneveld, N. and van Zon, H.,** The biology and culture of grass carp (*Ctenopharyngodon idella* Val.) with special reference to their utilisation for weed control, *Rec. Adv. Aquacult.*, 2, 120, 1985.
503. **Van der Zweerde, W.,** Biological control of aquatic weeds by means of phytophagous fish, in *Aquatic Weeds: The Ecology and Management of Nuisance Aquatic Vegetation,* Pieterse, A. H. and Murphy, K. J., Eds., Oxford University Press, New York, 1990, chap. 9(d).
504. **Van der Eijk, M.,** Notes on the experimental introduction of grass carp *(C, idella)* into the Netherlands, in *Proc. EWRS 5th Symp. Aquatic Weeds,* 1978, 245.
505. **Stott, B. and Robson, T. O.,** Efficacy of grass carp (*Ctenopharyngodon idella* Val.) in controlling submerged water weeds, *Nature,* 226, 870, 1970.
506. **Fowler, M. C. and Robson, T. O.,** The effects of the food preferences and stocking rates of grass carp (*Ctenopharyngodon idella* Val.) on mixed plant communities, *Aquat. Bot.*, 5, 261, 1978.
507. **Von Menzel, A.,** Gewaesserreinhaltung durch pflanzenfressende Fische (Water purification using herbivorous fish), *Oesterr. Fisch.*, 30, 12, 1977, in German).
508. **Anon.,** Grass carp doubts, *Sports Fish. Inst. Bull.*, 205, 2, 1969.
509. **Anon.,** From the research institutes: weed control in ponds, *FAO Fish Cult. Bull.*, 1, 6, 1969.
510. **Riechert, C. and Trede, R.,** Preliminary experiments on utilization of water hyacinth by grass carp, *Weed Res.*, 17, 357, 1977.
511. **Bailey, W. M.,** Operational experiences with the white amur in weed control programs, in *Proc. Symp. Water Quality and Management Through Biological Control,* Gainesville, 1975, 75.
512. **Bailey, W. M.,** A comparison of fish populations before and after extensive grass carp stocking, *Trans. Am. Fish. Soc.*, 107, 181, 1978.
513. **Avault, J. W., Jr.,** Biological weed control with herbivorous fish, *Proc. 7th Weed Control Conf.*, 18, 590, 1965.
514. **Swingle, H. S., Prather, E. E., Swingle, H. A., Hill, T. K., Jeffrey, N. B., and Addison, J. H.,** *Biological Control of Filamentous Algae and Other Pond Weeds,* Job No. 2, Alabama Department of Conservation, Auburn, 1967, 6.
515. **Lembi, C. A., Ritenour, B. C., Iverson, E. M., and Forss, E. C.,** The effects of vegetation removal by grass carp on water chemistry and phytoplankton in Indiana ponds, *Trans. Am. Fish. Soc.*, 107, 161, 1978.

516. **Buck, D. H., Baur, R. J., and Rose, C. R.,** Comparison of the effects of grass carp and the herbicide diuron in densely vegetated pools containing golden shiners and bluegills, *Prog. Fish-Cult.,* 37, 185, 1975.
517. **Stevens, V.,** Operational use of white amur in Kansas, *Proc. South. Weed Sci. Soc.,* 33, 205, 1980.
518. **Henderson, S.,** Review of the use of the grass carp in fisheries management in Arkansas, in *Symp. Culture of Exotic Fishes,* Smitherman, R. O., Shelton, W. L., and Grover, J. H., Eds., 1978, 165.
519. **Mitzner, L. R.,** The use of the white amur in Iowa (abstract), *Proc. South. Weed Sci. Soc.,* 33, 204, 1980.
520. **Kobylinski, G. J., Miley, W. W., II, Van Dyke, J. M., and Leslie, A. J., Jr.,** The Effects of Grass Carp (*Ctenopharyngodon idella* Val.) on Vegetation, Water Quality, Zooplankton, and Macroinvertebrates of Deer Point Lake, Bay County, Florida, Florida Department of Natural Resources, Tallahassee, 1980, 114.
521. **Theriot, R. F. and Decell, J. L.,** Large-scale operations management test of the use of the white amur to control aquatic plants, in *Symp. Culture of Exotic Fishes,* Smitherman, R. O., Shelton, W. L., and Grover, J. H., Eds., Atlanta, 1978, 220.
522. **Shireman, J. V. and Maceina, M. J.,** The utilization of grass carp *(Ctenopharyngodon idella)* for hydrilla control in Lake Baldwin, *J. Fish Biol.,* 19, 629, 1981.
523. **Shireman, J. V., Rottmann, R. W., and Aldridge, F. J.,** Consumption and growth of hybrid grass carp fed four vegetation diets and trout chow in circular tanks, *J. Fish Biol.,* 22, 685, 1983.
524. **Stanley, J. G., Biggers, C. J., and Schultz, D. E.,** Isozymes in androgenetic and gynogenetic white amur, gynogenetic carp, and carp amur hybrids, *J. Hered.,* 67, 129, 1976.
525. **Stanley, J. G. and Jones, J. B.,** Morphology of androgenetic and gynogenetic grass carp, *Ctenopharyngodon idella* (Valenciennes), *J. Fish Biol.,* 9, 523, 1976.
526. **Stanley, J. G.,** Female homogamety in grass carp (*Ctenopharyngodon idella* Val.) determined by gynogenesis, *J. Fish. Res. Bd. Can.,* 33, 1372, 1976.
527. **Stanley, J. G.,** Production of hybrid, androgenetic, and gynogenetic grass carp and carp, *Trans. Am. Fish. Soc.,* 105, 10, 1976.
528. **Stanley, J. G. and Thomas, A. E.,** Absence of sex reversal in unisex grass carp fed methyltestosterone, in *Symp. Culture of Exotic Fishes,* Smitherman, R. O., Shelton, W. L., and Grover, J. H., Eds., Atlanta, 1978, 194.
529. **Cassani, J. R. and Caton, W. E.,** Induced triploidy in grass carp, *Ctenopharyngodon idella* Val., *Aquaculture,* 46, 37, 1985.
530. **Leslie, A. J., Jr., Van Dyke, J. M., Hestand, R. S., and Thompson, B. Z.,** Management of aquatic plants in multi-use lakes with grass carp *(Ctenopharyngodon idella), Lake Reservoir Manage.,* 3, 266, 1987.
531. **Clugston, J. P. and Shireman, J. V.,** Triploid Grass Carp for Aquatic Plant Control, Fish and Wildlife Leafl. No. 8, U.S. Department of the Interior, Fish and Wildlife Service, Washington, D.C., 1987, 3.
532. **Gaudet, J. J.,** *Salvinia* infestation on Lake Naivasha in East Africa (Kenya), in *Aquatic Weeds in South East Asia,* Varshney, C. K. and Rzoska, J., Eds., W. Junk, The Hague, 1976, 193.
533. **Anon.,** Fish and shellfish introductions: introduction of carp into Sudan, *FAO Aquacult. Bull.,* 7, 25, 1974–75.
534. **Anon.,** Fish and shellfish introductions: *Ctenopharyngodon idella* in Ethiopia, *FAO Aquacult. Bull.,* 7, 24, 1975.

535. **Bailey, W. M.,** Fish and shellfish introductions: introduction of *Ctenopharyngodon idella* into Egypt, *FAO Aquacult. Bull.,* 8, 27, 1977.
536. Aquatic Weed Control Project, final report, International Land Development Consultants, 1978, 123.
537. **Spataru, P.,** Gut contents of silver carp — *Hypophthalmichthys molitrix* and some trophic relations to other fish species in a polyculture system, *Aquaculture,* 11, 137, 1977.
538. **Singh, S. B., Sukumaran, K. K., Pillai, K. K., and Chakrabarti, P. C.,** Observations on efficacy of grass carp, *Ctenopharyngodon idella* (Val.) in controlling and utilizing aquatic weeds in ponds in India, *Proc. Indo-Pacific Fish. Counc.,* 12, 220, 1967.
539. **Singh, C. S.,** Aquatic weeds in the fisheries ponds and their control measures, in *Aquatic Weeds in South East Asia,* Varshney, C. K. and Rzoska, J., Eds., W. Junk, The Hague, 1976, 331.
540. **Mehta, I., Sharma, R. K., and Tuank, A. P.,** The aquatic weed problem in the Chambal irrigated area and its control by grass carp, in *Aquatic Weeds in Southeast Asia,* Varshney, C. K. and Rzoska, J., Eds., W. Junk, The Hague, 1976, 307.
541. **Ahmed, N.,** Review of research work done by the directorate of fisheries, West Pakistan, *J. Agric.,* 19, 557, 1968.
542. **Naik, I. U.,** Introducing grass carp (*Ctenopharyngodon idellus* Cuvier and Valenciennes) in Pakistan, *Pak. J. Sci.,* 24, 45, 1972.
543. **Gopal, B.,** Aquatic weed problems and management in Asia, in *Aquatic Weeds: The Ecology and Management of Nuisance Aquatic Vegetation,* Pieterse, A. H. and Murphy, K. S., Eds., Oxford University Press, New York, 1990, chap. 16.
544. **De Zylva, R.,** New Zealand reviews use of grass carp to clear weeds, *Fish Farm. Int.,* 2, 40, 1993.
546. **Cope, O. B., Wood, E. M., and Wallen, G. H.,** Some chronic effects of 2,4-D on the bluegill *(Lepomis macrochirus), Trans. Am. Fish. Soc.,* 99, 1, 1970.
547. **Eller, L. L.,** Some pathology in redear sunfish exposed to Hydrothol 191, *Abstr. 1968 Weed Sci. Soc. Meeti.,* New Orleans, 1968.
548. **McBride, J. R., Dye, H. M., and Donaldson, E. M.,** Stress response of juvenile sockeye salmom *(Onchorhynhus nerka)* to the butoxyethanol ester of 2,4-dichlorophenoxyacetic acid, *Bull. Environ. Contam. Toxicol.,* 27, 877, 1981.
549. **McCraren, J. P., Cope, O. B., and Eller, L.,** Some chronic effects of diuron on bluegills, *Weed Sci.,* 17, 497, 1969.
550. **Folmar, L. C.,** Overt avoidance reaction of rainbow trout fry in nine herbicides, *Bull. Environ. Contam. Toxicol.,* 15, 509, 1976.
551. **Beitinger, T. L. and Freeman, L.,** Behavioural avoidance and selection responses of fishes to chemicals, *Residue Rev.,* 90, 35, 1983.
552. **McCorkle, F. M., Chambers, J. E., and Yarbrough, J. D.,** Acute toxicities of selected herbicides to fingerling channel catfish *Ictalurus punctatus, Bull. Environ. Contam. Toxicol.,* 18, 267, 1977.
553. **Brooker, M. P. and Edwards, R. W.,** Review paper: aquatic herbicides and the control of water weeds, *Water Res.,* 9, 1, 1975.
554. **McNabb, C. D., Jr. and Teirney, D. P.,** Growth and Mineral Accumulation of Submerged Vascular Hydrophytes in Pleiocutrophic Environs, Tech. Rep., 26th Inst. Water Res., Michigan State University, East Lansing, 1972.
555. **Bristow, J. M. and Whitcombe, W.,** The role of roots in the nutrition of aquatic vascular plants, *Am. J. Bot.,* 58, 8, 1971.

556. **DeMarte, J. A. and Hartman, R. T.,** Studies on absorption of ^{32}P, ^{59}Fe, and ^{45}Ca by water-milfoil (*Myriophyllum exalbescen* Fernald), *Ecology,* 55, 188, 1974.
557. **Bole, J. B. and Allen, J. R.,** Uptake of phosphorus from sediment by aquatic plants, *Myriophyllum spicatum* and *Hydrilla verticillata, Water Res.,* 12, 353, 1978.
558. **Carpenter, S. R. and Adams, M. S.,** Environmental Impacts of Mechanical Harvesting on Submersed Vascular Plants, IES Rep. 77, Institute for Environmental Studies, University of Wisconsin, Madison, 1977.
559. **Nichols, S. A.,** The effects of harvesting aquatic macrophytes on algae, *Trans. Wisc. Acad. Sci.,* 61, 165, 1973.
560. **Pearson, R. G. and Jones, N. V.,** The effects of dredging operations on benthic community of a chalk stream, *Biol. Conserv.,* 8, 273, 1975.
561. **Collett, L. C., Collina, A. J., Gibbs, P. J., and West, R. J.,** Shallow dredging as a strategy for the control of sublittoral macrophytes: a case study in Tuggerah Lakes, New South Wales, *Aust. J. Mar. Freshw. Res.,* 32, 563, 1981.
562. **Toner, E. D., O'Riordan, A., and Twomey, E.,** The effects of arterial drainage works on the salmon stock of a tributary of the River Moy, *Ir. Fish. Invest., Ser. A (Freshwater),* 1, 36, 1965.
563. **Haller, W. T., Shireman, J. V., and Durant, D. F.,** Fish harvest resulting from mechanical control of hydrilla, *Trans. Am. Fish. Soc.,* 109, 517, 1980.
564. **Mikol, G. F.,** Effects of harvesting on aquatic vegrtation and juvenile fish populations at Saratoga Lake, New York, *J. Aquat. Plant Manage.,* 23, 59, 1985.
565. **Lewis, W. M.,** Observations on the grass carp in ponds containing fingerling channel catfish and hybrid sunfish, *Trans. Am. Fish. Soc.,* 107, 153, 1978.
566. **Kilgen, R. H. and Smitherman, R. O.,** Food habits of the white amur *(Ctenopharyngodon idella)* stocked in ponds alone and in combination with other species, in *Herbivorous Fish For Aquatic Plant Control,* U.S. Army Corps of Engineers Tech. Rep. 4, Gangstad, E. O., Ed., Aquatic Plant Control Program, Waterways Exp. Stn., Vicksburg, MS, 1973, F1.
567. **Terrell, J. W. and Terrell, T. T.,** Macrophyte control and food habits of the grass carp in Georgia ponds, *Verh. Int. Ver. Theor. Angew. Limnol.,* 19, 2515, 1975.
568. **Klussmann, W. K., Noble, R. L., Martyn, R. D., Clark, W. J., Betsill, R. K., Bettoli, P. W., Cichra, M. F., and Campbell, J. M.,** Control of Aquatic Macrophytes by Grass Carp in Lake Conroe, Texas, and the Effects on the Reservoir Ecosystem, Misc. Publ. 1664, Texas Agricultural Exp. Stn., College Station, 1988.
569. **Shireman, J. V.,** Grass carp for weed control in Florida, *Proc. 4th Br. Freshw. Fish. Conf.,* 1985, 60.
570. **Blancher, E. C.,** Lake Conway, Florida: Nutrient Dynamics, Trophic State, Zooplankton Relationships, Ph.D. thesis, University of Florida, Gainesville, 1979.
571. **Noonan, T. A.,** Crustacean Zooplankton and Chlorophyll-*a* Relationships in Some Iowa Lakes and Reservoirs, M.S. thesis, Iowa State University, Ames, 1979.
572. **Gerking, S. D.,** A method of sampling the littoral macrofauna and its application, *Ecology,* 38, 219, 1957.
573. **Martin, R. G. and Shireman, J. V.,** A quantitative sampling method for hydrilla inhabiting macroinvertebrates, *J. Aquat. Plant Manage.,* 14, 16, 1976.
574. **Schramm, H. L., Jr., Jirka, L. J., and Hiyer, M. V.,** Epiphytic macroinvertebrates on dominant macrophytes in two central Florida lakes, *J. Freshw. Ecol.,* 4, 151, 1987.
575. **Watkins, C. E., II, Shireman, J. V., and Haller, W. T.,** The influence of aquatic vegetation upon zooplankton and benthic macroinvertebrates in Orange Lake, Florida, *J. Aquat. Plant Manage.,* 21, 78, 1983.

576. **Hargeby, A.,** Macrophyte associated invertebrates and the effect of habitat permanence, *Oikos,* 57, 338, 1990.
577. **Leslie, A. J. and Kobylinski, G. J.,** Benthic macroinvertebrates response to aquatic vegetation removal by grass carp in North-Florida reservoir, *Fla. Sci.,* 48, 220, 1985.
578. **Prowse, G. A.,** The role of cultured pond fish in the control of eutrophication in lakes and dams, *Verh. Int. Ver. Angew. Limnol.,* 17, 714, 1969.
579. **Canfield, D. E., Jr., Shireman, J. V., Colle, D. E., Haller, W. T., Watkins, C. E., II, and Maceina, M. J.,** Prediction of chlorophyll *a* concentrations in Florida lakes: importance of aquatic macrophytes, *Can. J. Fish. Aquat. Sci.,* 41, 497, 1984.
580. **Leslie, A. J., Jr., Nall, L. E., and van Dyke, J. M.,** Effects of vegetation control by grass carp on selected water-quality variables in four Florida lakes, *Trans. Am. Fish. Soc.,* 112, 777, 1983.
581. **Mitchell, C. P., Fish, G. R., and Burnet, A. M. R.,** Limnological changes in a small lake stocked with grass carp, *N.Z. J. Mar. Freshw. Res.,* 18, 103, 1984.
582. **Mitzner, L.,** Evaluation of biological control of nuisance aquatic vegetation by grass carp, *Trans. Am. Fish. Soc.,* 107, 135, 1978.
583. **Canfield, D. E., Jr., Maceina, M. J., and Shireman, J. V.,** Effects of hydrilla and grass carp on water quality in a Florida lake, *Water Res. Bull.,* 19, 773, 1983.
584. **Shireman, J. V., Hoyer, M. V., Maceina, M. J., and Canfield, D. E., Jr.,** The water quality and fishery of Lake Baldwin, Florida, 4 years after macrophyte removal by grass carp, in *Lake and Reservoir Management: Practical Applications,* North American Lake Management Society, 1985, 201.
585. **Killgore, K. J., Dibble, E. D., and Hoover, J. J.,** *Relationships between Fish and Aquatic Plants: A Plan of Study,* U.S. Army Corps of Engineers, Misc. Paper A-93–1, Waterways Exp. Stn., Vicksburg, MS, 1993.
586. **Swingle, H. S.,** Relationships and dynamics of balanced and unbalanced fish populations, *Ala. Agric. Exp. Stn. Bull.,* 274, 1, 1950.
587. **Shireman, J. V., Haller, W. T., Colle, D. E., Watkins, C. E., II, DuRant, D. F., and Canfield, D. E.,** Ecological Impact of Integrated Chemical and Biological Aquatic Weed Control, USEPA 600/3, U.S. Environmental Protection Agency, Washington, D.C., 1983, 333.
588. **Barnett, B. S. and Schneider, R. W.,** Fish populations in dense subersed plant communities, *Hyacinth Control J.,* 12, 12, 1974.
589. **Colle, D. E. amd Shireman, J. V.,** Coefficients of condition for largemouth bass, bluegill, and redear sunfish in hydrilla-infested lakes, *Trans. Am. Fish. Soc.,* 109, 521, 1980.
590. **Hoyer, M. V., Canfield, D. E., Jr., Shireman, J. V., and Colle, D. E.,** Relationship between abundance of largemouth bass and submerged vegetation in Texas reservoirs: a critique, *N. Am. J. Fish. Manage.,* 5, 613, 1985.
591. **Jones, J. R. and Hoyer, M. V.,** Sportfish harvest predicted by summer chlorophyll *a* concentration in midwestern lakes and reservoirs, *Trans. Am. Fish. Soc.,* 111, 176, 1982.
592. **Bays, J. S. and Crisman, T. L.,** Zooplankton and trophic state relationships in Florida lakes, *Can. J. Fish. Aquat. Sci.,* 40, 1813, 1983.
593. **Dunst, R. C., Born, S. M., Uttormark, P. D., Smith, S. A., Knauer, S. A., Serns, S. L., Winter, D. R., and Wirth, T. L.,** Survey of Lake Rehabilitation Techniques and Experiences, Tech. Bull. 75, Wisconsin Department of Natural Resources, Madison, 1974.

594. **Ware, F. J. and Gasaway, R. D.,** Effects of grass carp on native fish populations in two Florida lakes, *Proc. Southeast. Assoc. Game Fish Comm.,* 30, 324, 1976.
595. **Savino, J. F. and Stein, R. A.,** Predator-prey interaction between largemouth bass and bluegills as influenced by simulated, submersed vegetation, *Trans. Am. Fish. Soc.,* 11, 255, 1982.
596. **Crowder, L. B. and Cooper, W. E.,** Effects of macrophyte removal on the feeding efficiency and growth of sunfish: evidence from pond studies, in *Proc. Aquatic Plants, Lake Management, and Ecosystem Consequences of Lake Harvesting,* Creck, J. E., Prentki, R. T., and Louks, O. L., Eds., Institute for Environmental Studies, University of Wisconsin, Madison, 1979, 251.
597. **Canfield, D. E., Jr. and Hoyer, M. V.,** Aquatic Macrophytes and Their Relation to the Limnology of Florida Lakes, final report to the Bureau of Aquatic Plant Management, Florida Department of Natural Resources, Tallahassee, 1992, 599.
598. **Maceina, M. J. and Shireman, J. V.,** Influence of dense hydrilla infestation on black crappie growth, *Proc. Annu. Conf. Southeastern Assoc. Fish and Wildlife Agencies,* 36, 394, 1985.
599. **Wrenn, W. B.,** Grass carp stocking evaluation in a large embayment of a Tennessee River reservoir, in *Proc. 24th Annu. Meet. Aquatic Plant Control Research Program,* Huntsville, AL, November 13 to 16, 1989, 153.
600. **Shireman, J. V., Colle, D. E., and Rottmann, R. W.,** Size limits to predation of grass carp by largemouth bass, *Trans. Am. Fish. Soc.,* 107, 213, 1978.
601. **Ewel, K. C. and Fontaine, T. D.,** Proposed relationships between white amur and the aquatic ecosystem at Lake Conway, Florida, 159, 1977.
602. **Ewel, K. C. and Fontaine, T. D.,** A Model for the Evaluation of the Response of the Lake Conway, Florida, Ecosystem to Introduction of the White Amur. Baseline Studies. Large-Scale Operations Management Test of the Use Of the White Amur for Control of Problem Aquatic Plants. Technical report to Waterways Exp. Stn., Vicksburg, MS, 1981.
603. **Ewel, K. C. and Fontaine, T. D.,** III, Effects of white amur *(Ctenopharyngodon idella)* on a Florida lake: a model, *Ecol. Model.,* 16, 251, 1982.
604. **Wiley, M. J., Tazik, P. P., Sobaski, S. T., and Gorden, R. W.,** Biological Control of Aquatic Macrophytes by Herbivorous Carp. III. Stocking Recommendations for Herbivorous Carp and Description of the Illinois Herbivorous Fish Simulation System, Aquatic Biology Tech. Rep. 12, Illinois Natural History Survey, Champaign, 1984.
605. **Wiley, M. J. and Gorden, R. W.,** Biological Control of Aquatic Macrophytes by Herbivorous Carp. I. Executive Summary, Aquatic Biology Tech. Rep. 10, Illinois Natural History Survey, Champaign, 1984.
606. **Wiley, M. J., Gorden, R. W., Waite, S. W., and Powless, T.,** The relationship between aquatic macrophytes and sports fish production in Illinois ponds: a simple model, *N. Am. J. Fish. Manage.,* 4, 111, 1984.
607. **Wiley, M. J., Tazik, P. P., and Sobaski, S. T.,** Controlling Aquatic Vegetation with Triploid Grass Carp, Circ. No. 57, Illinois Natural History Survey, Champaign, 1987.
608. **Shireman, J. V.,** Cost analysis of aquatic weed control: fish versus chemicals in a Florida lake (grass carp, *Ctenopharyngodon idella,* hydrilla), *Prog. Fish-Cult.,* 44, 199, 1982.
609. **Dubbers, F. A. A., Ghoneim, S., Siemelink, M. E., El Gharably, Z., Pieterse, A. H., and Blom, E.,** Aquatic weed control in irrigation and drainage canals in Egypt by means of grass carp *(Ctenopharyngodon idella),* in *Proc. 5th Int. Symp. Biological Control of Weeds,* Melbourne, Australia, July 1980, Del Fosse, E. S., Ed., 1981, 261.

610. **Buckley, B. R. and Stott, B.,** Grass carp in a sport fishery, *Fish. Manage.,* 8, 8, 1977.
611. **Höne, U.,** Experiment in the use of grass carp for biological weed control in water-courses, *Weed Abstr.,* 24(1975), abstr. 3020, 1973.
612. **Bailey, W. M.,** Arkansas' Evaluation of the Desirability of Introducing the White Amur (*Ctenopharyngodon idella* Val.) for Control of Aquatic Weeds, paper presented at 102nd Annu. Meet. Am. Fish. Soc. Int. Assoc. Game and Fish Comm., Hot Springs, AR, 1972, 59.
613. **Terrell, J. W.,** Stocking Rates, Food Habits, and Catchability of Grass Carp (*Ctenopharyngodon idella* Val.), M.S. thesis, University of Georgia, Athens, 1975, 54.
614. **Wilson, J. L. and Cottrell, K. D.,** Catchability and organoleptic evaluation of grass carp in east Tennessee ponds, *Trans. Am. Fish. Soc.,* 108, 97, 1979.
615. **Henderson, S.,** Preliminary studies on the tolerance of the white amur, *Ctenopharyngodon idella,* to rotenone and other commonly used pond treatment chemicals, *Proc. Annu. Conf. Southeast. Assoc. Game Fish Comm.,* 27, 435, 1974.
616. **Colle, D. E., Shireman, J. V., Gasaway, R. D., Stettler, R. L., and Haller, W. T.,** Utilization of selective removal of grass carp *(Ctenopharyngodon idella)* used for biological control of *Hydrilla verticillata* from an 80-hectare Florida lake to obtain a population estimate, *Trans. Am. Fish. Soc.,* 107, 724, 1978.
617. **Hardin, S.,** Selective Removal and Population Estimation of Grass Carp in Lake Wales, Florida, Florida Game and Freshwater Fish Commission Report, Tallahassee, 1980, 12.
618. **Northcote, T. G.,** Fish in the structure and function of freshwater ecosystems: a "top-down" view, *Can. J. Fish. Aquat. Sci.,* 45, 361, 1988.
619. **Shapiro, J., Lammara, V., and Lynch, M.,** Biomanipulation: an ecosystem approach to lake restoration, in *Proc. Symp. Water Quality Management through Biological Control,* Brezowik, P. L. and Fox, J. L., Eds., University of Florida, Gainesville, 1975, 85.
620. **Gophen, M.,** Summary of the workshop on perspectives of biomanipulation in inland waters, *Hydrobiologia,* 191, 315, 1990.
621. **Carpenter, S. R., Kitchell, J. F., and Hodgson J. R.,** Cascading trophic interactions and lake productivity, *BioScience,* 35, 634, 1985.
622. **Wright, J. C.,** The limnology of Canyon Ferry Reservoir. I. Phytoplankton-zooplankton relationships in the euphotic zone during September and October 1956, *Limnol. Oceanogr.,* 3, 150, 1958.
623. **Hargrave, B. T. and Geen, G. H.,** Effects of copepod grazing on two natural phytoplankton populations, *J. Fish. Res. Bd. Can.,* 8, 1395, 1970.
624. **Lampert, W.,** The role of zooplankton: an attempt to quantify grazing, in *Proc. Int. Congr. Lake Pollution and Recovery,* European Water Pollution Control Association, Rome, 1985, 54.
625. **Hrbácek, J., Dvorakova, M., Korinek, V., and Prscházkóva, L.,** Demonstration of the effect of the fish stock on the species composition of zooplankton and the intensity of metabolism of the whole plankton association, *Verh. Int. Ver. Limnol.,* 14, 192, 1961.
626. **Hrbácek, J.,** Species composition and the amount of the zooplankton in relation to the fish stock, *Rozpr. Cesk. Akad. Ved Rada Mat. Prir. Ved,* 72, 1, 1962.
627. **Brooks, J. L. and Dodson, S. I.,** Predation, body size, and composition of plankton, *Science,* 150, 28, 1965.
628. **Straskraba, M.,** The effect of fish on the number of invertebrates in ponds and streams, *Mitt. Lut. Ver. Limnol.,* 3, 106, 1965.

629. **Lamarra, V. A.,** Digestive activities of carp as a major contributor to the nutrient loading of lakes, *Verh. Int. Ver. Limnol.,* 19, 2461, 1975.
630. **Persson, A. and Harin, S. F.,** Effects of cyprinids on the release of phosphorus from lake sediments, in Abstr. XXVSIL Int. Congr., Barcelona,, August 21 to 27, 1992.
631. **Verigin, B. V.,** The role of herbivorous fishes at reconstruction of icthyofauna under the conditions of antropogenic evolution of waterbodies, in *Proc. Grass Carp Conf.,* Shireman, J. V., Ed., University of Florida, Gainesville, 1979, 139.
632. **Vybornov, A. A.,** Effects of silver carp, *Hypophthalmichthys molitrix,* on production indices of phyto- and zooplankton under experimental conditions, *J. Ichthyol.,* 29, 136, 1989.
633. **Kajak, Z., Rybak, J. I., Spodinewska, I., and Godlewska-Lipowa, W. A.,** Influence of the planktonivorous fish *Hypopthalmichthys molitrix* (Val.) on the plankton and benthos of the eutrophic lake, *Pol. Arch. Hydrobiol.,* 22, 301, 1975.
634. **Barthelmes, D.,** Heavy Silver Carp (*Hypophthalmichthys molitrix* Val.) Stocking in Lakes and Its Influence on Indigenous Fish Stocks, EIFAC Tech. Pap./Doc. Tech. CECPI 42 Suppl. Vol. 2, 1984, 314.
635. **Barthelmes, D., Kozianowski, A., Haenstsch, Ch., Helms, C., Preez, H., and Kleibs, K.,** Ein Experiment zur Producktion von Silberkarpfen *(Hypophthalmichthys molitrix)* in seen bie einer Besatzdichte vo 1000 St./ha (Heiliger See bei Angermuende), *Z. Binnenfisch. DDR,* 33, 79, 1986.
636. **Grygierck, E.,** The influence of phytophagous fish on pond zooplankton, *Aquaculture,* 2, 197, 1973.
637. **Januszko, M.,** The effect of three species of phytophagous fish on algae development, *Pol. Arch. Hydrobiol.,* 21, 431, 1974.
638. **Opuszynski, K.,** Silver carp, *Hypopthalmichthys molitrix* (Val.), in carp ponds. III. Influence on ecosystem, *Ekol. Pol.,* 27, 117, 1979.
639. **Allen, H. L.,** Phytoplankton photosynthesis, micronutrient interaction, and inorganic carbon availability in a soft-water Vermont lake, in *Nutrients and Eutrophication,* Spec. Symp. I, American Society of Limnology and Oceanography, 1972, 63.
640. **Moss, B.,** The influence of environmental factors on distribution of freshwater algae: an experimental study, *J. Ecol.,* 61, 157, 1973.
641. **Colby, P. J., Spangler, G. R., Hurley, D. A., and McCombie, A. M.,** Effects of eutrophication on salmonid communities in oligotrophic lakes, *J. Fish. Res. Bd. Can.,* 29, 975, 1972.
642. **Hartmann, J.,** Fischereiliche Veränderungen in Kulturbedingt eutrophierenden Seen, *Schweiz. Hydrol.,* 39, 243, 1977.
643. **Opuszynski, K.,** The role of fishery management in the counteracting eutrophication processes, *Dev. Hydrobiol.,* 2, 263, 1980.
644. **Drenner, R. W., O'Brien, W. J., and Mummert, J. R.,** Filter-feeding rates of gizzard shad, *Trans. Am. Fish. Soc.,* 111, 210, 1982.
645. **Mummert, J. R. and Drenner, R. W.,** Effect of fish size on the filtering efficiency and selective particle ingestion of a filter-feeding clupeid, *Trans. Am. Fish. Soc.,* 115, 522, 1986.
646. **Drenner, R. W., deNoyelles, F., Jr., and Kettle, D.,** Selective impact of filter-feeding gizzard shad on zooplankton community structure, *Limnol. Oceanogr.,* 27, 965, 1982.
647. **Drenner, R. W., Taylor, S. B., Lazzaro, X., and Kettle, D.,** Particle-grazing and plankton community impact of an omnivorous cichlid, *Trans. Am. Fish. Soc.,* 113, 397, 1984.

648. **Drenner, R. W., Threkeld, S. T., and McCracken, M. D.,** Experimental analysis of the direct and indirect effects of an omnivorous filter-feeding clupeid on plankton community structure, *Can. J. Fish. Aquat. Sci.,* 43, 1935, 1986.
649. **Threlkeld, S. T. and Drenner, R. W.,** An experimental mesocosm study of residual and contemporary effects of an omnivorous, filter-feeding, clupeid fish on plankton community structure, *Limnol. Oceanogr.,* 32, 1331, 1987.
650. **Crisman, T. L.,** Algal control through trophic-level interactions: a subtropical perspective, in *Proc. Workshop on Algal Management and Control,* Tech. Rep. E-81–7, U.S. Environmental Protection Agency, Las Vegas, 1981, 131.
651. **Crisman, T. L. and Beaver, J. R.,** Applicability of planktonic biomanipulation for managing eutrophication in the subtropics, *Hydrobiologia,* 200/201, 177, 1990.
652. **Smith, D. W.,** Biological control of excessive phytoplankton growth and the enhancement of aquacultural production, *Can. J. Fish. Aquat. Sci.,* 42, 1940, 1985.
653. **Laws, E. A. and Weisburd, R. S. J.,** Use of silver carp to control algal biomass in aquaculture ponds, *Prog. Fish-Cult.,* 52, 1, 1990.
654. **Leventer, H. and Teltsch, B.,** The contribution of the silver carp *(Hypophthalmichthys molitrix)* to the biological control of the Netofa reservoirs, *Hydrobiologia,* 191, 47, 1990.
655. **Vinyard, G. L., Drenner, R. W., Gophen, M., Pollinger, U., Winkelman, D. L., and Hambright, K. D.,** An experimental study of the plankton community impacts of two omnivorous filter-feeding cichlids, *Tilapia galilaea* and *Tilapia aurea, Can. J. Fish. Aquat. Sci.,* 45, 685, 1988.
656. **Drenner, R. W., Hambright, K. D., Vinard, G. L., Gophen, M., and Pollingher, V.,** Experimental study of size-selected phytoplankton grazing by a filter-feeding cichlid and the cichlid's effect on plankton community structure, *Limnol. Oceanogr.,* 32, 1138, 1987.
657. **Milstein, A., Hepher, B., and Teltsch, B.,** Principal component analysis of interactions between fish species and the ecological conditions in fish ponds. I. Phytoplankton, *Aquacult. Fish. Manage.,* 16, 305, 1985.
658. **Milstein, A., Hepher, B., and Teltsch, B.,** Principal component analysis of interaction between fish species and the ecological conditions in fish ponds. II. Zooplankton, *Aquacult. Fish. Manage.,* 16, 319, 1985.
659. **Milstein, A., Hepher, B., and Teltsch, B.,** The effect of fish species combination in fish ponds on plankton composition, *Aquacult. Fish. Manage.,* 19, 127, 1988.
660. **Chen, S., Liu, X., and Hua L.,** The role of silver carp and bighead in the cycling of nitrogen and phosphorus in the East Lake ecosystem, *Acta Hydrobiol. Sin/Shuisheng Shengwu Xuebao,* 15, 8, 1991.
661. **Shi, W., Jin, W., Wang, D., Li, Y., Hong, J., and Zheng, Y.,** Effect of stocking silver carp and bighead carp on the eutrophication in waters, *J. Dalian Fish. Coll.,* 4, 11, 1989.
662. **Starling, F. L. R. M. and Rocha, A. J. A.,** Experimental study of the impacts of planktivorous fishes on plankton community and eutrophication of a tropical Brazilian reservoir, *Hydrobiologia,* 200/201, 581, 1990.
663. **Starling, F. L. R. M.,** Control of phytoplankton biomass by silver carp, *Hypophthalmichthys molitrix,* in the eutrophic Paranoá Reservoir (Brasília, Brazil): a mesocosm experiment, *Hydrobiologia,* 257, 143, 1993.
664. **Carruthers, A. D.,** Effects of silver carp on blue-green algal blooms in Lake Orakai, *N.Z. Minist. Agric. Fish. Res. Div. Fish. Environ. Rep.,* 68, 1, 1986.

665. **Scavia, D., Fahnenstiel, G. L., Evans, M. S., Jude, D. J., and Lehman, J. T.,** Influence of salmonine predation and weather on long-term water quality trends in Lake Michigan, *Can. J. Fish. Aquat. Sci.,* 43, 435, 1986.
666. **Benndorf, J., Schultz, H., Benndorf, A., Unger, R., Penz, E., Kneschke, H., Kossatz, K., Hornig, U., Kruspe, R., and Reichel, S.,** Food-web manipulation by enhancement of piscivorous fish stocks: long-term effects in the hypertrophic Bautzen Reservoir, *Limnologica (Berlin),* 19, 97, 1988.
667. **Golterman, H. L.,** Concluding remarks, *Hydrobiologia,* 191, 319, 1990.
668. **Mann, H.,** Organgenichte und Fettgehatt beim chinesischen grasfisch *(Ctenopharyngodon idella), Fischwirt,* 18, 85, 1968.
669. **Jaschew, L. and Bojadschew, A.,** Chranitelna stoinost i rokusowi katschestwa na bilija amur, *Ribuo Stopanstwo,* 16, 22, 1969.
670. **Klejmienov, J.,** Chemical and Weight Composition of Fishes in Waters of the USSR and Adjacent Countries, Isd. Rybn. Khoz. WNIRO, Moscow, 1962.
671. **Gomon, M. F. and Paxton, J. R.,** A revision of the Odacidae, a temperate Australian-New Zealand labroid fish family, *Indo-Pacific Fish.,* 8, 57, 1985.
672. **Choat, J. H. and Ayling, A. M.,** The relationship between habitat structure and fish faunas of New Zealand reefs, *J. Exp. Mar. Biol. Ecol.,* 110, 257, 1987.
673. **Horn, M. H., Murray, S. N., and Edwards, T. W.,** Dietary selectivity in the field and food preferences in the laboratory for two herbivorous fishes *(Cebidichthys violaceus* and *Xiphister mucosus)* from a temperate intertidal zone, *Mar. Biol.,* 67, 237, 1982.
674. **Lowe-McConnell, R. H.,** *Fish Communities in Tropical Freshwaters,* Longmans, London, 1975, 337.
675. **De Silva, S. S., Cumaranatunga, P. R. T., and De Silva, C. D.,** Food, feeding ecology and morphological features associated with feeding of four co-occurring cyprinids (Pisces: Cyprinidae), *Netherl. J. Zool.,* 30, 54, 1980.
676. **Moyle, P. B. and Cech, J. J.,** *Fishes: An Introduction to Ichthyology,* Prentice-Hall, Englewood Cliffs, NJ, 1982, chap. 20.
677. **Berra, T. M.,** *An Atlas of Distribution of the Freshwater Fish Families of the World,* University of Nebraska Press, Lincoln, 1981, 197.
678. **Welcomme, R. L.,** *Fisheries Ecology of Floodplain Rivers,* Longmans, London, 1979, 315.
679. **Gery, J.,** *Characoids of the World,* T. F. H. Publications, Hong Kong, 1977, 672.
680. **Greenwood, P. H.,** The cichlid fish of Lake Victoria, East Africa: the biology and evolution of a species flock, *Bull. Br. Mus. Nat. Hist.* (Zool.), Suppl. 6, 134, 1974.
681. **Montgomery, W. L., Leibfried, W., and Gooby, K.,** *Feeding by Rainbow Trout on Cladophora glomerata* at Lees Ferry, Colorado River, Arizona: The Roles of Cladophora and Epiphytic Diatoms in Trout Nutrition, a Final Report to the Bureau of Reclamation, Northern Arizona University, Flagstaff, 1986.
682. **Naiman, R. J.,** Productivity of a herbivorous pupfish population *(Cyprinodon neradensis)* in a warm desert stream, *J. Fish Biol.,* 9, 125, 1976.
683. **Merrick, J. R. and Schmida, G. E.,** *Australian Freshwater Fishes,* Griffin Press, Netley, South Australia, 1984, 273.
684. **Wootton, R. J.,** *Ecology of Teleost Fishes,* Chapman & Hall, London, 41, 55, 1990.
685. **De Silva, S. S. and Perera, M. K.,** Digestibility of an aquatic macrophyte by the cichlid (*Etroplus suratensis* Bloch) with observations on the relative merits of three indigenous components as markers and daily changes in protein digestibility, *J. Fish Biol.,* 23, 675, 1983.

686. **Yomaoka, K.,** Feeding relationships, in *Cichlid Fishes: Behaviour, Ecology and Evolution,* Keenleyside, M. H. A., Ed., Chapman & Hall, London, 1991, chap. 7.
687. **Eschmeyer, W. N.,** *Catalog of the Genera of Recent Fishes,* California Academy of Sciences, San Francisco, 1990, 697.
688. **Welcomme, R. L.,** Register of International Transfers of Inland Species, FAO Fish. Tech. Pap. No. 213, Food and Agricultural Organization, Rome, 1981, 120.
689. **Welcomme, R. L.,** International Introductions of Inland Aquatic Species, FAO Tech. Pap. No. 294, Food and Agriculture Organization, Rome, 1988, 318.
690. **Cassani, J. R.,** Feeding behaviour of underyearling hybrids of the grass carp, *Ctenopharyngodon idella* ♀, and the bighead, *Hypophthalmichthys nobilis* ♂, on selected species of aquatic plants, *J. Fish Biol.,* 8, 127, 1981.
691. **Edwards, D. J.,** Taking a bite at the waterweed problem, *N. Z. J. Agric.,* 130, 33, 1975.
692. **Johnson, M. and Laurence, J. M.,** Biological weed control with the white amur, in *Herbivorous Fish for Aquatic Plant Control,* U.S. Army Corps of Engineers Tech. Rep. 4, Gangstad, E. O., Ed., Aquatic Plant Control Program, Waterways Exp. Stn., Vicksburg, MS, 1973, E-1.
693. **Pentelow, F. T. K. and Stott, B.,** Grass carp for weed control, *Prog. Fish-Cult.,* 27, 210, 1965.
694. **Prabhavathy, G. and Sreenivasan, A.,** Cultural prospects of Chinese carps in Tamilnadu, *Proc. IPFC,* 17, 354, 1977.
695. **Stevenson, J. H.,** Observations on grass carp in Arkansas, *Prog. Fish-Cult.,* 27, 203, 1965.
696. **Sutton, D. L., Miley, W. W., II, and Stanley, J. G.,** Report to the Florida Department of Natural Resources on the Project: Onsight Inspection of the Grass Carp in the USSR and Other European Countries, University of Florida Agricultural Research Center, Fort Lauderdale, 1977, 48.
697. **Van Dyke, J. M.,** A Nutritional Study of the White Amur (*Ctenopharyngodon idella* Val.) Fed Duckweed, M.S. thesis, University of Florida, Gainesville, 1973, 35.
698. **Willey, R. G., Diskocil, M. J., and Lembi, C. A.,** Potential of the white amur (*Ctenopharyngodon idella* Val.) as a biological control for aquatic weeds in Indiana, *Proc. Indiana Acad. Sci.,* 83, 173, 1974.
699. **Rowe, D. K. and Schipper, C. M.,** An assessment of the Impact of the Grass Carp *(Ctenopharyngodon idella)* in New Zealand Waters, Rep. No. 58, Fisheries Research Division, Ministry of Agriculture and Fisheries, Rotorua, New Zealand, 1985, 177.
700. **Theriot, R. F. and Sanders, D. R.,** Food preferences of yearling hybrid carp, *Hyacinth Control J.,* 13, 51, 1975.
701. **Zolczynski, S. J. and Smith, B. W.,** Evaluation of white amur for control of lyngbya in a 32 hectare public fishing lake, *Proc. South. Weed Sci. Soc.,* 196, 1980.
702. **Cross, D. G.,** Aquatic weed control using grass carp, *J. Fish Biol.,* 1, 27, 1969.
703. **Duthu, G. S. and Kilgen, R. H.,** Aquarium studies on the selectivity of 16 aquatic plants as food by fingerling hybrids of the cross between *Ctenopharyngodon idella* ♂ *and Cyprinus carpio* ♀, *J. Fish Biol.,* 7, 203, 1975.
704. **Fischer, Z.,** Food selection in grass carp (*Ctenopharyngodon idella* Val.) under experimental conditions, *Pol. Arch. Hydrobiol.,* 15, 1, 1968.
705. **Venkatesh, B. and Shetty, H. P. C.,** Studies on the growth rate of the grass carp, *Ctenopharyngodon idella* (Valenciennes), fed on two aquatic weeds ad libitum, Mysore, *J. Agric. Sci.,* 12, 622, 1978.

706. **Urban, E. and Fisher, Z.,** Comparison of bioenergetic transformations in grass carp (*Ctenopharyngodon idella* Val.) in two age classes, under different feeding conditions, in *Proc. 2nd Int. Symp. Herbivorous Fish, Novi Sad, Yugoslavia,* European Weed Research Society, Wageningen, The Netherlands, 1982.
707. **Fischer, Z.,** The elements of energy balance in grass carp (*Ctenopharyngodon idella* Val.). III. Assimilability of proteins, carbohydrates, and lipids by fish fed with plant and animal food, *Pol. Arch. Hydrobiol.,* 19, 83, 1972.
708. **Gidumal, J. L.,** A survey of the biology of the grass carp, *Ctenopharyngodon idellus* (C. and V.), *Hong Kong Univ. Fish. J.,* 2, 1, 1958.
709. **Adams, A. E. and Titeko, V.,** A progress report on the introduction of grass carp *(Ctenopharyngodon idellus)* in Fiji, *Fiji Agric. J.,* 32, 43, 1970.
710. **Chaudhuri, H., Chakrabarty, R. D., Sen, P. R., Rao, N. G. S., and Jena, S.,** A new high in fish production in India with record yields by composite fish culture in freshwater ponds, *Aquaculture,* 6, 343, 1975.
711. **Sinha, V. R. P.,** From research institutions: polyculture of carps, *FAO Aquacult. Bull.,* 6, 6, 1973.
712. **Sinha, V. R. P. and Gupta, M. V.,** On the growth of grass carp, *Ctenopharyngodon idella* Val. in composite fish culture at Kalyani, West Bengal (India), *Aquaculture,* 5, 283, 1975.
713. **Yashouv, A.,** Acclimatization of new species in the fishponds of the station, *Bamidgeh,* 10, 75, 1958.
714. **Hickling, C. F.,** Observations on the growth rate of the Chinese grass carp, *Ctenopharyngodon idellus* (C. and V.), *Malay. Agric. J.,* 43, 49, 1960.
715. **Pike, T.,** Fish and shellfish introductions: grass carp in South Africa, *FAO Aquacult. Bull.,* 8, 20, 1977.
716. Annual Report, Alabama Department of Conservation, 1967.
717. **Crowder, J. P. and Snow, R. R.,** From research institutions: use of grass carp for weed control in ponds, *FAO Fish. Cult. Bull.,* 2, 6, 1969.
718. **Bizyaev, I. W. and Chesnokova, T. V.,** Experiments on rearing phytophagous fishes in rice fields, *Rybn. Khoz.,* 3, 19, 1966.
719. **Horoszewicz, L.,** Survival of starving common carp larvae at different water temperature, *Rocz. Nauk Roln.,* H-96(3), 45, 1974 (in Polish with English summary).
720. **Wolnicki, J. and Opuszynski, K.,** "Point of no return" in carp (*Cyprinus carpio* L.) and the phytophagous fish (*Ctenopharyngodon idella* Val., *Hypopthalmichthys molitrix* Val., *Aristichthys nobilis* Rch.) larvae, *Rocz. Nauk Roln.,* H-101(4), 61, 1988 (in Polish with English summary).
721. **Das, P., Kumar, D., and Guha Roy, M. K.,** National demonstration on composite fish culture in West Bengal, *J. Inland Fish. Soc. India,* 7, 112, 1975.
722. **Ebregt, E. M.,** Parasitic agents reported from grass carp, in *Recent Advances in Aquaculture,* Vol. 2, Muir, J. F. and Roberts, R. J., Eds., Croon Helm, London, 1985, 187.
723. **Li, L.-X.,** Studies on a new species of *Trichophrya* (Suctoria), *Trichophrya variformis* sp. nov. from the gills of *Ctenopharyngodon idella, Acta Hydrobiol. Sin.,* 9, 382, 1986.
724. **Shamsudin, M. N.,** Bacteriological examination of fingerlings of bighead carp *(Aristichthys nobilis)* and grass carp *(Ctenopharyngodon idella)* imported into Malaysia, in *The First Asian Fisheries Forum, Proc. 1st Asian Fisheries Forum, Manila, the Philippines, 26–31 May 1986,* Maclean, J. L., Dizon, L. B., and Hosillos, L. V., Eds., Asian Fisheries Society, Manila, the Philippines, 1986, 235.

725. **Feng, S. and Wang, J.,** Four new species of *Myxosporidia* from the freshwater fishes in Danjiangkou reservoir, *Acta Hydrobiol. Sin.,* 14, 68, 1990.
726. **Salih, N. E., Ali, N. M., and Abul-Ameer, K. N.,** Helminthic fauna of three species of carp raised in ponds in Iraq, *J. Biol. Sci. Res.,* 19, 369, 1988.
727. **Shah, K. L. and Tyagi, B. C.,** An eye disease in silver carp, *Hypophthalmichthys molitrix,* held in tropical ponds, associated with the bacterium *Staphylococcus aureus, Aquaculture,* 55, 1, 1986.
728. **El-Zarka, S.,** Aquaculture development: Afghanistan. Fish culture development in Afghanistan, *FAO Aquacult. Bull.,* 5, 12, 1974.
729. **Mastrarrigo, V.,** Fish introductions: introduction of grass carp in Argentina, *FAO Aquacult. Bull.,* 3, 14, 1971.
730. **Liepolt, R. and Weber, E.,** Studies with phytophagous fish *(Ctenopharyngodon idella), Rev. Roum. Biol. (Ser. Zool.),* 14, 127, 1969 (translated from German).
731. **Bari, A.,** Aquaculture development in Bangladesh, *Proc. Indo-Pacific Fisheries Council,* 17, 189, 1976.
732. **Anon.,** Recent introductions of fish, shrimps and oysters, *FAO Fish Cult. Bull.,* 2, 15, 1969.
733. **Anon.,** Introductions of fish, *FAO Aquacult. Bull.,* 3, 15, 1970.
734. **Blanc, M. et al.,** *European Inland Water Fish: A Multilingual Catalogue,* Fishing News Books, London, 1971 (in French, Spanish, German).
735. **Wurtz-Arlet, J.,** Biological Methods for the Control of Aquatic Plants, paper presented at the 3rd Int. Sym. Aquatic Weeds, Proc. Eur. Weed Res. Counc., 1971, 9 (translated by H. A. Lennon).
736. **Anon.,** Fish and crayfish introductions, *FAO Fish Cult. Bull.,* 1969.
737. **Jähnichen, H.,** The effectiveness of grass carp in the biological control of aquatic plants in the watercourses of the East German Republic, *Z. Binnenfisch. DDR,* 20, 14, 1973 (translated from German by L. Brownlow).
738. **Anon.,** From the research institutes: culture of Chinese carps, stocking rates, *FAO Fish Cult. Bull.,* 1, 3, 1968.
739. **Chaudhuri, H. et al.,** Role of Chinese carp, *Ctenopharyngodon idella* (Val.), in biological control of noxious aquatic weeds in India: a review, in *Aquatic weeds in South East Asia,* Varshney, C. K. and Rzoska, J., Eds., W. Junk, The Hague, 1976, 315.
740. **Ivanova, A.,** Introductions of fish: Chinese carp in Iran, *FAO Aquacult. Bull.,* 3, 15, 1970.
741. **Anon.,** Recent introductions: *Ctenopharyngodon idella* in Iraq, *FAO Fish Cult. Bull.,* 1, 12, 1969.
742. **Anon.,** Fish and shellfish introductions: recent introductions, *FAO Aquacult. Bull.,* 4, 16, 1972.
743. **Schuster, W. H.,** A provisional survey of the introduction and transplantation of fish throughout the Indo-Pacific region, *Proc. IPFC,* 3, 187, 1952.
744. **Anon.,** Aquaculture development: Kenya, Republic of Korea, and Thailand, *FAO Aquacult. Bull.,* 3, 11, 1970.
745. **Anon.,** Recent introductions and transplantations, *FAO Fish Cult. Bull.,* 1, 12, 1968.
746. **Chanthepha, S.,** Aquaculture development: Laos. Fish culture development in Laos, *FAO Aquacult. Bull.,* 5, 10, 1972.
747. **Gopinath, K.,** Freshwater fish farming in the Malay Archipelago, *J. Zool. Soc. India,* 2, 101, 1950.

748. **Gándara, J. A. M., Sánchez, P. M., and Herrera, F. R. V.,** Aquatic Weed Control in Central and Southern Mexico, 1975, 23 (in Spanish with English summary).
749. **Shrestha, S. B.,** Induced breeding of grass carp in Nepal, *Bamidgeh,* 25, 10, 1973.
750. **Anon.,** Fish and shellfish introductions: recent introductions, *FAO Aquacult. Bull.,* 5, 20, 1973.
751. **Anon.,** Giant perch exported, grass carp imported, *Fish. Newsl. Aust.,* 24, 27, 1965.
752. **Moses, B. S.,** Aquaculture development: Nigeria. Fish culture development in Nigeria, *FAO Aquacult. Bull.,* 5, 12, 1972.
753. Draft Environmental Impact Statement. The Introduction of White Amur into Canal Zone Waters to Control Aquatic Weeds, Panama Canal Company, Balboa Heights, Canal Zone, 1977, 74.
754. **Custer, P. E. et al.,** The white amur as a biological control agent of aquatic weeds in the Panama Canal, *Fisheries,* 3, 2, 1978.
755. **Datingling, B. Y.,** The potential of the freshwater fish culture resources of the Philippines, *Proc. IPFC,* 17, 120, 1976.
756. **Thorslund, A.,** Fish introductions: introduction of grass carp in Sweden, *FAO Aquacult. Bull.,* 3, 14, 1971.
757. **Lin, S. Y.,** Induced spawning of Chinese carps by pituitary injection in Taiwan (a survey of technique and application), *Fish. Ser. Chin.-Am. J. Comm. Rural Reconstr.,* 5, 28, 1965.
758. **Ovchynnyk, M. M.,** Fish culture in the U. S. S. R., *Fish. News Int.,* 2, 279, 1963.
759. **Jhingran, V. G. and Gopalakrishnan, V.,** A catalogue of cultivated aquatic organisms, *FAO Fish. Tech. Pap.,* 130, 83, 1974.
760. **Wiley, M. J. and Gorden, R. W.,** Biological Control of Aquatic Macrophytes by Herbivorous Carp. II. Biology and Ecology of Herbivorous Carp, Aquatic Biology Tech. Rep. 11, Illinois Natural History Survey, Champaign, 1984.
761. **Beach, M. L., Lazor, R. L., and Burkhalter, A. P.,** Some aspects of the environmental impact of the white amur (*Ctenopharyngodon idella* Val.) in Florida and its use for aquatic weed control, in *Proc. 4th Int. Symp. Biological Control of Weeds,* Freeman, T. E., Ed., Institute of Food and Agricultural Sciences, University of Florida, Gainesville, 1977, 269.
762. **Van Donk, E., Gulati, R. D., and Grimm, M. P.,** Restoration by biomanipulation in a small hypertrophic lake: first year results, *Hydrobiologia,* 191, 285, 1990.
763. **Elser, J. J. and Carpenter, S. R.,** Predation-driven dynamics of zooplankton and phytoplankton communities in a whole-lake experiment, *Oecologia,* 76, 148, 1988.
764. **Shapiro, J. and Wright, D. I.,** Lake restoration by biomanipulation: Round Lake, Minnesota, the first two years, *Freshw. Biol.,* 14, 371, 1984.
765. **Andersson, G., Berggren, H., Cranburg, G., and Gelin, C.,** Effects of planktivorous and benthivorous fish on organisms and water chemistry in eutrophic lakes, *Hydrobiologia,* 59, 9, 1978.
766. **Tang, Y. A.,** Evaluation of balance between fishes and available fish foods in multispecies fish culture ponds in Taiwan, *Trans Am. Fish. Soc.,* 4, 708, 1970.
767. **Kitchell, J. F., O'Neill, R. V., Webb, D., Gallepp, G. W., Bartell, S. M., Koonce, J. F., and Ausmus, B. S.,** Consumer regulation of nutrient cycling, *BioScience,* 29, 28, 1979.
768. **Hestand, R. S., III,** personal communication, 1993.
769. **Mizumoto, M.,** personal communication, 1993.
770. **Kingsford, M. J.,** personal communication.

INDEX

A

B

C

D

E

F

G

H

I

K

L

M

N

O

P

R

S

X

Y